ASTROPHYSICS
AND
COSMOLOGY

Proceedings of the 26th Solvay Conference on Physics

ASTROPHYSICS
AND
COSMOLOGY

Brussels, Belgium   9 – 11 October 2014

Editors

Roger Blandford
Kavli Institute for Particle Astrophysics and Cosmology
Stanford University, USA

David Gross
Kavli Institute for Theoretical Physics
University of California, Santa Barbara, USA

Alexander Sevrin
Vrije Universiteit Brussel
Also at Universiteit Antwerpen and KU Leuven
Deputy-director, International Solvay Institutes, Belgium

World Scientific

NEW JERSEY · LONDON · SINGAPORE · BEIJING · SHANGHAI · HONG KONG · TAIPEI · CHENNAI · TOKYO

Published by

World Scientific Publishing Co. Pte. Ltd.

5 Toh Tuck Link, Singapore 596224

USA office: 27 Warren Street, Suite 401-402, Hackensack, NJ 07601

UK office: 57 Shelton Street, Covent Garden, London WC2H 9HE

Library of Congress Cataloging-in-Publication Data

Names: Solvay Conference on Physics (26th : 2014 : Brussels, Belgium) | Blandford, Roger D., editor. |
    Gross, D. (David Jonathan), editor. | Sevrin, A. (Alexander), editor.
Title: Astrophysics and cosmology : proceedings of the 26th Solvay Conference on Physics,
    Brussels, Belgium, 9–11 October 2014 / editors: Roger Blandford (Stanford University),
    David Gross (University of California, Santa Barbara), Alexander Sevrin
    (Vrije Universiteit Brussel, Belgium & International Solvay Institutes, Belgium).
Description: Singapore ; Hackensack, NJ : World Scientific Publishing Co. Pte. Ltd., [2016] |
    2016 | Includes bibliographical references.
Identifiers: LCCN 2016009660| ISBN 9789814759175 (hardcover ; alk. paper) |
    ISBN 9814759171 (hardcover ; alk. paper)
Subjects: LCSH: Astrophysics--Congresses. | Cosmology--Congresses.
Classification: LCC QB460 .S64 2014 | DDC 523.01--dc23
LC record available at http://lccn.loc.gov/2016009660

British Library Cataloguing-in-Publication Data

A catalogue record for this book is available from the British Library.

Printed by FuIsland Offset Printing (S) Pte Ltd Singapore

The International Solvay Institutes

Honorary Members

Baron André Jaumotte
Honorary Director of the Solvay Institutes, Honorary Rector
Honorary President of the ULB

Mr. Jean-Marie Piret
Emeritus Attorney General of the Supreme Court of Appeal
and Honorary Principal Private Secretary to the King

Prof. Jean-Louis Vanherweghem
Former President of the Administrative Board of the ULB

Prof. Irina Veretennicoff
Emeritus Professor at the VUB

Guest Members

Prof. Marc Henneaux
Director and Professor at the ULB

Prof. Alexander Sevrin
Deputy Director for Physics, Professor at the VUB
Scientific Secretary of the Committee for Physics

Prof. Lode Wyns
Deputy Director for Chemistry
Former Vice-rector for Research at the VUB

Prof. Franklin Lambert
Emeritus Professor at the VUB

Prof. Anne De Wit
Professor ULB and Scientific Secretary of the Committee for Chemistry

Ms Marina Solvay

Prof. Hervé Hasquin
Permanent Secretary of the Royal Academy of Sciences, Letters and Fine Arts
of Belgium

Prof. Géry van Outryve d'Ydewalle
Permanent Secretary of the Royal Flemish Academy of Belgium for Sciences
and the Arts

Director

Prof. Marc Henneaux
Professor at the ULB

Solvay Scientific Committee for Physics

Prof. David Gross (chair)
Kavli Institute for Theoretical Physics (Santa Barbara, USA)

Prof. Roger Blandford
Stanford University (USA)

Prof. Steven Chu
Stanford University (USA)

Prof. Robbert Dijkgraaf
Director of the Institute For Advanced Study (Princeton, USA)
Universiteit van Amsterdam (The Netherlands)

Prof. Bert Halperin
Harvard University (Cambridge, USA)

Prof. Giorgio Parisi
Università la Sapienza (Roma, Italy)

Prof. Pierre Ramond
University of Florida (Gainesville, USA)

Prof. Gerard 't Hooft
Spinoza Instituut (Utrecht, The Netherlands)

Prof. Klaus von Klitzing
Max-Planck-Institut (Stuttgart, Germany)

Prof. Peter Zoller
Universität Innsbruck (Austria)

Prof. Alexander Sevrin (Scientific Secretary)
Vrije Universiteit Brussel (Belgium)

Hotel Métropole (Brussels), 9–11 October 2014

Astrophysics and Cosmology
Chair: Professor Roger Blandford

The 26th Solvay Conference on Physics took place in Brussels from October 9 through October 11, 2014. Its theme was "Astrophysics and Cosmology" and the conference was chaired by Roger Blandford. The conference was followed by a public event entitled *Astrophysics and Cosmology*. Conny Aerts, François Englert and Martin Rees each delivered a lecture and a panel of scientists – led by David Gross and comprising Conny Aerts, Roger Blandford, François Englert, James Peebles, Jean-Loup Puget and Martin Rees – answered questions from the audience.

The organization of the 26th Solvay Conference has been made possible thanks to the generous support of the *Solvay Family*, the *Solvay Group*, the *Université Libre de Bruxelles*, the *Vrije Universiteit Brussel*, the *Belgian National Lottery*, the *Foundation David and Alice Van Buuren*, the *Belgian Science Policy Office*, the *Brussels-Capital Region*, the *Communauté française de Belgique*, de *Actieplan Wetenschapscommunicatie* of the *Vlaamse Regering*, the *Belgian Science Policy Office*, the *City of Brussels*, and the *Hôtel Métropole*.

Participants

Tom	**Abel**	Stanford University
Conny	**Aerts**	KULeuven
Laura	**Baudis**	University Zürich
Mitchell	**Begelman**	University of Colorado
Roger	**Blandford**	Stanford University
Dick	**Bond**	University of Toronto
Lars	**Brink**	Chalmers University
John	**Carlstrom**	The University of Chicago
Thibault	**Damour**	Institut des Hautes Études Scientifiques
Paolo	**De Bernardis**	Università di Roma "La Sapienza"
Ger	**de Bruyn**	ASTRON
Robbert	**Dijkgraaf**	IAS, Princeton
Jo	**Dunkley**	Oxford University
George	**Efstathiou**	Cambridge
Daniel	**Eisenstein**	Harvard
Richard	**Ellis**	California Institute of Technology
François	**Englert**	Université Libre de Bruxelles
Andrew	**Fabian**	Cambridge
Carlos	**Frenk**	Durham
Steven	**Furlanetto**	UCLA
Reinhard	**Genzel**	MPI for Extraterrestrial Science
Peter	**Goldreich**	IAS, Princeton
David	**Gross**	UCSB
Alan	**Guth**	MIT
John	**Hawley**	University of Virginia
Marc	**Henneaux**	Université Libre de Bruxelles
Werner	**Hofmann**	MPI für Kernphysik
Marc	**Kamionkowski**	Johns Hopkins
Vicky	**Kaspi**	McGill University
Eiichiro	**Komatsu**	MPI for Astrophysics
Chryssa	**Kouveliotou**	NASA
Michael	**Kramer**	Manchester University
Shri	**Kulkarni**	California Institute of Technology
James	**Lattimer**	SUNY Stony Brook
Nazzareno	**Mandolesi**	INAF-IASF Bologna
Piero	**Madau**	UCSC
Viatcheslav	**Mukhanov**	Ludwig-Maximilians-Universität
Hitoshi	**Murayama**	Berkeley
Jeremiah	**Ostriker**	Princeton University

P. James	**Peebles**	Princeton University
Ue-Li	**Pen**	Canadian Institute for Theoretical Astrophysics
E. Sterl	**Phinney**	California Institute of Technology
Tsvi	**Piran**	Racah Institute of Physics
Philipp	**Podsiadlowski**	Oxford
Clem	**Pryke**	University of Minnesota
Jean-Loup	**Puget**	Institut d'astrophysique spatiale
Georg	**Raffelt**	MPI München
Pierre	**Ramond**	UF Gainesville
Martin	**Rees**	Cambridge
Roger	**Romani**	Stanford University
Uros	**Seljak**	Berkeley
Alexander	**Sevrin**	Vrije Universiteit Brussel
Eva	**Silverstein**	Stanford University
David	**Spergel**	Princeton University
Rashid	**Sunyaev**	MPI für Astrophysik Garching
Scott	**Tremaine**	IAS, Princeton
Ed	**van den Heuvel**	UvAmsterdam
Neil	**Weiner**	NYU
Simon	**White**	MPI für Astrophysik Garching
Ralph	**Wijers**	UvAmsterdam
Matias	**Zaldarriaga**	IAS, Princeton
Saleem	**Zaroubi**	Kapteyn Astronomical Institute

Auditors

Maarten	**Baes**	Universiteit Gent
Glenn	**Barnich**	Université Libre de Bruxelles
Nicolas	**Chamel**	Université Libre de Bruxelles
Geoffrey	**Compère**	Université Libre de Bruxelles
Ben	**Craps**	Vrije Universiteit Brussel
Leen	**Decin**	Katholieke Universiteit Leuven
Stephane	**Detournay**	Université Libre de Bruxelles
Frank	**Ferrari**	Université Libre de Bruxelles
Jean-Marie	**Frère**	Université Libre de Bruxelles
Gianfranco	**Gentile**	Universiteit Gent and Vrije Universiteit Brussel
Thomas	**Hambye**	Université Libre de Bruxelles
Thomas	**Hertog**	Katholieke Universiteit Leuven
Dominique	**Lambert**	Université de Namur
Philippe	**Spindel**	Université de Mons
Michel	**Tytgat**	Université Libre de Bruxelles
Danny	**Vanbeveren**	Vrije Universiteit Brussel
Hans	**Van Winckel**	Katholieke Universiteit Leuven

Contents

Opening Session

Astrophysics and Cosmology

The Opening Session was held in the Gracious Presence of His Royal Highness Philippe, King of the Belgians.

Opening Address by Marc Henneaux,
Director of the International Solvay Institutes

Your Majesty,
Mrs. Solvay,
Mr. Solvay,
Members of the Solvay Family,
Ladies and Gentlemen,
Dear Colleagues,
Dear Friends,

It is my great honour and pleasure to open the 26th Solvay Conference on Physics. Its theme is "Astrophysics and Cosmology".

A distinctive feature of the Solvay conferences, which make them unique, is that they benefit from the benevolent support of the Royal Family. This has been true right from the start, for the very first Solvay Conference that took place in 1911, for which King Albert I expressed very strong support and interest. The pictures of the Solvay Physics Committee meeting in Laeken with Queen Elisabeth for the preparation of the 1933 Solvay Conference, the last Solvay Conference to take place before the dramatic events that shook Germany and then Europe, are particularly moving. We are fortunate that this interest in the Solvay Conferences and fundamental science has been kept intact over the years.

Sire,

It is with a respectful gratitude that we acknowledge the continuation of this centenary tradition today. Your Presence with us this morning is an enormous encouragement for basic scientific research.

The Conference that begins today is the fourth Solvay Conference on astrophysics and cosmology, the last one having taken place more than 40 years ago. Here is the list:

- Solvay 11 (1958): La structure et l'évolution de l'univers
- Solvay 13 (1964): The Structure and Evolution of Galaxies
- Solvay 16 (1973): Astrophysics and Gravitation

Given the spectacular developments undergone by the discipline since 1973, on so many fronts, it was high time to organize again a Solvay conference on the challenging questions raised by the understanding of our universe. The International Solvay Institutes are grateful that this theme was selected by its International Scientific Committee and that Roger Blandford accepted the very demanding task of chairing it.

Let me recall the format of the Solvay Conferences. These are conferences by invitation-only, with a limited number of participants. There are few presentations but a lot of discussions. People come to the Solvay Conferences for the scientific interactions, which are indeed privileged, not for giving a talk.

For the discussions to be fruitful, an extremely careful preparation is needed. Here is how it goes. The subject and scientific chair of the Solvay Conferences are chosen by the Solvay International Scientific Committee for Physics, which has complete freedom in doing so. The scientific chair of the conference is then in charge of the invitations and of the program, and again has complete "carte blanche" for achieving this task.

I can tell you that this required indeed an enormous amount of work and was very much time-consuming. I have seen thousands of email exchanges concerning the organization. Thanks to this effort, I am confident that the Conference will be a great success, in the Solvay tradition.

I would therefore like to express my deepest thanks to the Solvay Scientific Committee for Physics, and in particular to its chair, David Gross, and to Roger Blandford, for accepting to organize the Conference.

Before giving the floor to the next speakers, let me make one announcement of a more practical nature: since the discussions are important, they are included in the proceedings. Again this is a distinctive feature of the Solvay Conferences. We have a scientific secretariat in charge of achieving this task, directed by our colleague Alexander Sevrin. To facilitate their work, please give your name each time you intervene in the discussions.

Thank you very much for your attention.

Opening Address by David Gross,
Chair of the Solvay Scientific Committee for Physics

Your Majesty, Mr. Solvay and members of the Solvay family and the Solvay Board of Directors, members of the Solvay Physics committee, colleagues and friends, I am pleased to welcome you all to the opening of the 26th Solvay Conference in Physics devoted to astrophysics and cosmology. There are many people to thank for making this conference possible, but especially I wish to thank Jean-Marie Solvay and the Solvay family for its unwavering support of the Solvay Institutes, and to commend Marc Henneaux for his remarkable efforts in reviving the Solvay Institutes and conferences.

Three years ago the Scientific Committee decided that, after 41 years, it was certainly the right time to hold the next Solvay conference on Astrophysics and Cosmology, a field that has witnessed enormous advances both in observation and in theory over the last few decades. The choice of chair, Roger Blandford, was also obvious. Roger has done a marvelous job in organizing the conference in a way that will produce, I trust, excellent talks and much spontaneous discussion. All of the discussions will be transcribed and will appear in the published proceedings. I urge you all to speak up and contribute to these discussions.

I look forward to the next few days in which we will summarize the remarkable developments, as well as the open challenges, in our understanding of the universe we live in.

Opening Address by Roger Blandford,
Chair of the 26th Solvay Conference of Physics

1. Introduction

Your Majesty, Mr Solvay, members of the Solvay Board of Directors, members of the Solvay Scientific Committee, fellow participants and auditors. My name is Roger Blandford and I have the honour to chair the 26th Solvay Conference in Physics which is devoted to astrophysics and cosmology. It is 41 years since the last Solvay conference on astrophysics, 50 years since galaxies were discussed and 56 years have elapsed since cosmology was on the program. It is an understatement to say that much has been learned since these meetings. In fact there has been so much progress in each of these areas that we had to restrict the meeting ruthlessly to five distinct, but inter-related, topics each of which is in a very exciting phase right now. Allow me to set the stage.

2. Neutron Stars

Our first topic is neutron stars. Neutron stars contain roughly as much mass as the sun, mostly in the form of particles called neutrons, compressed into a sphere with a diameter approximately that of Brussels (Figure 1). They were conjectured to exist soon after the neutron was discovered in 1932. However, they were only identified in 1968, as "pulsars", through their lighthouse-like radio emission created as they spin about their axes in some cases faster than 600 times in a second. Neutron stars have magnetic fields that can be over a million billion times stronger than the earth's field that pervades this room. We have now found three thousand neutron stars and it has been estimated that there are several hundred million of them in our Milky Way Galaxy alone.

Neutron stars are fascinating to physicists because so much physics finds serious application in our attempts to understand them. For example, they allow us to study the properties of nuclear matter in a manner that complements what we learn from particle accelerators and this matter appears to be much tougher stuff than many of us expected. Astrophysicists love neutron stars because they provide exquisitely precise clocks. They can be used to weigh stars accurately. They can also perform sophisticated tests of Albert Einstein's general theory of relativity which it has passed with impressive accuracy. Using our pulsar clocks, we have been able to show that binary neutron stars lose energy by emitting gravitational waves, just as predicted. The challenge now is to detect waves like these using giant lasers. Meanwhile, the pulsars may also be used as detectors of lower frequency waves emitted by binary black holes in the nuclei of distant galaxies. The race is on to be first to receive this new type of signal directly.

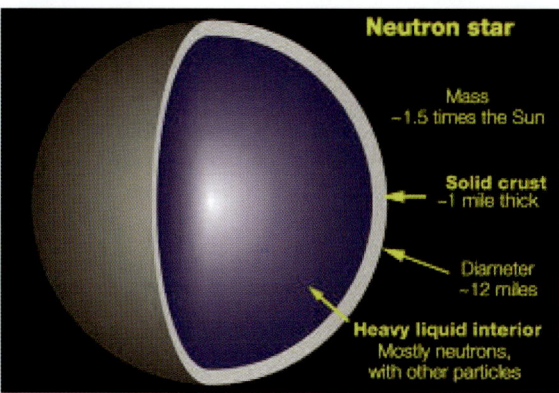

Fig. 1. Depiction of a neutron star. Much of the physics that goes into the detailed description of neutron stars is esoteric but well-understood. However, some of it is quite uncertain and can be probed using detailed observations. Credit: NASA/Marshall Space Flight Center.

3. Black Holes

Cosmic black holes, our next topic, are sources of gravitational fields that are so strong that not even light can escape them. Astronomical black holes are much simpler than neutron stars and can be characterised by just two numbers describing their mass and their rate of spin. They grow by devouring gas and stars which can heat and shine before crossing a surface of no-return, called the event horizon. Event horizons are commonly encountered in science fiction novels and movies. They are also of contemporary importance in the febrile imaginings of theoretical physicists as they raise important issues of principle in the interpretation of quantum mechanics which are proving to be very hard to settle. Despite these puzzles, we now know that black holes exist in abundance, like neutron stars, as the corpses of massive stars that live fast and die young.

Black holes are also found, in the nuclei of most normal galaxies including our own, with masses that can exceed ten billion times larger than the mass of our sun. We know there is a black hole in our Galaxy because we observe stars orbiting it just like planets orbit the sun. Now, when these big black holes are well-fed, they are bright and outshine galaxies. They are called quasars and can be seen across the universe. By contrast our black hole is being starved. Do not feel sorry for it; you would not want to live next door to a quasar!

Black holes can also be used to test general relativity. Again, everything checks out. We have also learned that black holes are often spinning very fast and can be orbited by disks of gas and threaded by magnetic field. They can create powerful outflowing jets of plasma moving with speed close to that of light. Here is a simulation of the formation of this jet in blue by the spinning black hole (Figure 2). The black lines represent magnetic field. Most exciting of all is the prospect that we can start to replace simulations like this with observations by combining telescopes

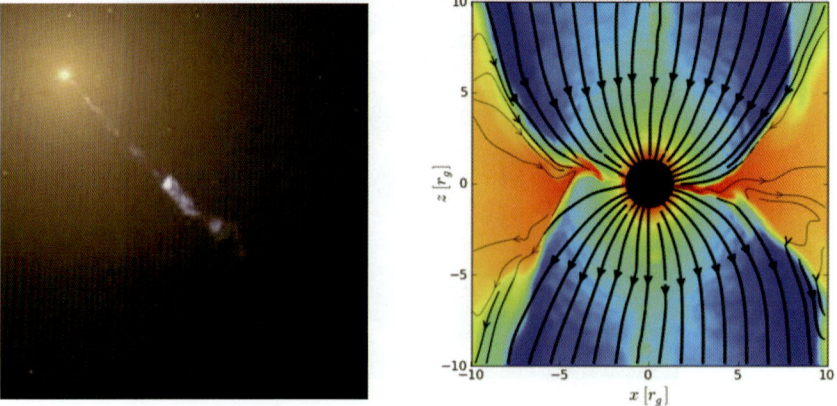

Fig. 2.   On the left hand side is a Hubble Space Telescope image of the galaxy M87 in the Virgo cluster of galaxies, exhibiting an outflowing, relativistic jet emanating from a spinning back hole with mass around five billion solar masses. (Credit Hubble Space Telescope.) On the right hand side is a frame from a general relativistic simulation of the electromagnetic processes that may be responsible for jets like those in M87.[1]

operating at millimeter wavelengths on different continents with the ALMA telescope in Chile.

4. Cosmic Dawn

One of many surprises about black holes is that a few of them were able to grow to enormous size inside quasars when the universe, was less than a billion years old. Like teenagers contemplating their parents, astronomers wonder if these quasars and their host galaxies were ever young! They must have grown very rapidly during the mysterious epoch of reionization, otherwise know as the cosmic dawn, when the universe was roughly 400 million years old. This era represents the transition from the ancient to the modern universe when the first galaxies, stars and planets were formed; a renaissance following the cosmic dark age. These newly formed objects illuminated the surrounding gas and caused its hydrogen atoms to break up into electrons and protons (Figure 3). We are now observing galaxies and at least one stellar object - a gamma ray burst - to even greater distances than the most distant and youngest quasars.

More tools are being developed and deployed to observe the cosmic dawn. They include the James Webb Space Telescope, which will be launched in 2018 and sensitive, low frequency radio telescopes in the Netherlands, Australia and North America designed to pick up faint signals from hydrogen atoms. Success in this endeavor will enable astronomers to complete the narrative history of the universe and to understand which came first, the galactic chickens or the black hole eggs. They will also satisfy our curiosity into the socially complex behavior of the adolescent galaxies that grew into the more staid counterparts we see in advanced middle age today.

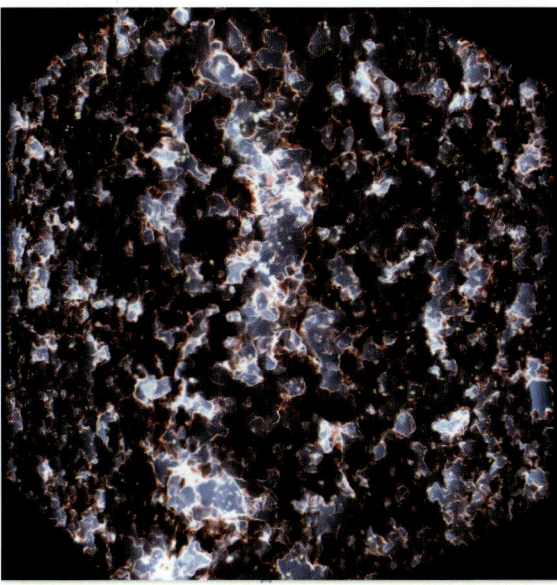

Fig. 3.   Single frame from a simulation of the epoch of reionization when the first stars and galaxies converted the surrounding hydrogen atoms to plasma. The ionized gas is depicted in blue.[2]

5. Dark Matter

This growth of form and structure in the universe from a remarkably smooth beginning was orchestrated by an entity that is now called 'dark matter'. The evidence that most of the stuff of the universe did not shine in the dark has also been around since the 1930s. We exhibit this on what we call a 'pie chart' (Figure 4). The orange slice, just five percent of the universe, is the regular matter that we see around us. The blue slice – over a quarter – is the dark matter. Over the past thirty years, we have demonstrated that dark matter has at most very weak, non-gravitational interactions. This extreme alienation suffices to describe the growth of structure using giant numerical simulations. The shadowy background depicts the dark matter. It can form long filaments. We call it the cosmic web. Dark matter provides the framework for the luminous galaxies stars and planets to grow. The bright features are young galaxies illuminated by their stars. It is most commonly presumed that dark matter comprises one or more elementary particles and much effort is being expended to identify their nature. Astronomers are eager to cover their embarrassment in going from believing that they knew what the universe was made of to understanding less than five percent of it. Physicists are desperate to wrestle with a second standard model of elementary particles, having just filled in the last puzzle piece of the first one, as Professor Englert will explain on Sunday.

The quest for dark matter is being conducted below ground, in deep mines where it is hoped to catch an occasional particle. It is also taking place on ground at the

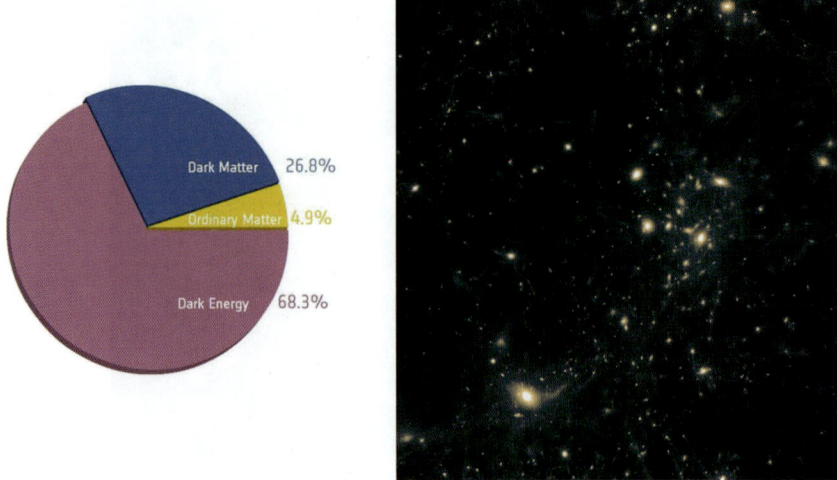

Fig. 4.   On the left hand side is a pie chart showing the three basic constituents of the modern universe, dark energy, dark matter and ordinary or baryonic matter. (Credit ESA/Planck.) On the right hand side is a frame from a simulation exhibiting the growth of structure orchestrated by dark matter in the expanding universe.[3]

Large Hadron Collider in Switzerland where we are trying to create these particles. Finally, orbiting satellites, like Fermi and AMS, tuned to the gamma rays and positrons that dark matter particles might occasionally emit, are searching above ground. All three approaches have been made much more sensitive than originally expected, which is a fine testament to the craft of experimental physics. None of them have yielded convincing detections of dark matter particles. It is hoped to make them a hundred times more sensitive. Even if they are unsuccessful, they will still tell us much about elementary particle physics.

6. Microwave Background

Finally, we will discuss the cosmic microwave background radiation, discovered half a century ago. Today, this radiation is very cold with a temperature about one percent that of the air in this room. We are effectively looking at the smooth inner surface of a very large sphere, some 40 billion light years in radius today, from its centre at a time when the universe was only about 400 thousand years old. When this radiation is examined in detail we find tiny – ten parts per million – fluctuations (Figure 5). What we actually observe are gravitational seeds that grew into the large scale structure in the distribution of galaxies we see around us today. The microwave background provides the single strongest piece of evidence that the universe began nearly fourteen billion years ago in a hot big bang. Superb telescopes such as the Planck satellite have observed it precisely and provide reliable measurements on the shape, size, age and structure of the universe. These have been supplemented by

Fig. 5.   On the left hand side is a representation of the microwave background over the whole sky showing the tiny fluctuations in its measured temperature.[4] On the right hand side is an image exhibiting the "B-mode" component of the polarization of the radiation from a small patch of the sky.[5]

other approaches to produce a standard model of the universe that parallels that devised for elementary particles. A small number of parameters suffices to fit a lot of data at the one to ten percent level. This is a remarkable advance, given the state of observational cosmology at the time of the last Solvay conference devoted to cosmology.

In this model, the majority – nearly 70 percent – of the contemporary universe has a form similar to that of a cosmological constant as introduced by Einstein in 1917. This is the purple slice of our pie. A cosmological constant or, more generally, dark energy, has the curious property that it causes the universe to accelerate as was shown explicitly by measurements of supernova explosions. In its simplest form, it condemns us to a fate of eternal, runaway expansion, dilution and decay – an agoraphobic's worst nightmare. In addition, we can use these observations to infer that the initial fluctuation spectrum was almost the same on all length scales and appears to be completely random. The universe began with a hum, not a fanfare. All of this discovery is consistent with a proposal that the universe underwent a much earlier epoch of runaway acceleration, called inflation, which terminated when the universe was, perhaps, 10^{-33} s old.

Recently, special patterns in the polarization of the microwave background, called B-modes, have been measured. There has been much debate over whether or not these patterns, which look like whorls in fingerprints, were made at the time of inflation or are imprinted more recently in the dust of our galaxy. Now Belgium is associated with great detectives such as Thomson and Thompson and although we may need their services one day, we are using physics from our own (J. J.) Thomson (who discovered the electron) and (William) Thomson (who developed techniques used by radio astronomers) to see if we have enough evidence to convict inflation.

In summary, and as a link to the next talk, I can do no better than quote the remarkably prescient Georges Lemaître writing as long ago as 1933. "The expansion thus took place in three phases, a first period of rapid expansion in which the atom universe was broken into atomic stars, a period of slowing down, followed by a third period of accelerated expansion."

7. Afterword

I would like to conclude by telling a short story. 71 years ago a young airman was flying over Belgium and his journey was interrupted. He ended up being accommodated and fed, under challenging conditions, in private homes in and around Liège. He remained there for almost six months before continuing on his way to Switzerland. Ordinary people forced by circumstance to perform extraordinary, brave acts. That airman was my father and for reasons that I trust you can now appreciate, my family has a longstanding, high regard for and deep gratitude to Belgians. My father passed away last year, on Armistice Day. Your Majesty, my father understood as well as anyone that the Europe of today is a very different and much better place than the Europe of 71 years ago. However, he also knew that I would visit Brussels and wanted me to take advantage of this singular opportunity to say "Merci, dank".

These are words I repeat on behalf of all of us to you, the Solvay family and our Belgian colleagues for all of your gracious hospitality here at the 26th Solvay Conference in Physics.

References

1. J. McKinney, A. Tchekhovskoy and R. Blandford. *Mon. Not. R. Astr. Soc.* **423** 3083 (2012).
2. R. Kaehler, M. Alvarez and T. Abel. *Numerical Modeling of Space Plasma Flows.* ed. N. Pogorelov, E. Audit and G. Zank. (San Francisco: ASP, 2010).
3. R. Kaehler, O. Hahn and T. Abel. *Instrumentations and Methods for Astrophysics. IEEE VGTC* **18** 12 (2012).
4. Planck Collaboration; P. Ade et al. *Astron. Astrophys.* **571** 1 (2014).
5. P. Ade et. al. *Phys. Rev. Lett.* **112** 241101 (2014).

Address by Thomas Hertog,
Georges Lemaître: A Visionary Belgian Cosmologist

Your Majesty, Mrs Solvay, dear members of the Solvay family, dear colleagues.

The 26th Solvay conference on Astrophysics and Cosmology is an opportunity to look back on a precious jewel in Belgian history, namely Georges Lemaître's discovery that our universe expands.

The history of modern cosmology can roughly be divided into six periods, which take us from the first explorations of Einstein's static universe starting in 1917 to the precision science we have today. Lemaître's contributions must be situated in the second, crucial period from 1927 to 1939 in which the basic framework of modern cosmology was developed.

Lemaître himself traces his interests in science and cosmology to his childhood years he spent in and around the city of Charleroi in the South of Belgium. Unfortunately, World War I intervened and, like so many of his contemporaries, Lemaître signed up to join the Belgian army to defend his country. After the War he entered the seminary for the priesthood and he was ordained as a priest in 1923.

In the seminary, Lemaître was granted special permission by Cardinal Mercier to study Relativity, Einstein's new theory of gravity. He wrote a dissertation on Einstein's new physics and his ideas on cosmology. On the basis of this work the Commission for Relief in Belgium, under the auspices of the American Educational Foundation, awarded Lemaître a fellowship to study abroad. That was the beginning of a unique scientific adventure.

He first went to the University of Cambridge where he deepened his knowledge of Relativity under the guidance of Sir Arthur Eddington, one of the foremost astronomers at the time. It is likely that the confluence of Eddington's interests both in the theory of Relativity and in astronomical observations has encouraged Lemaître to explore himself their intersection in his later work. Lemaître and Eddington had great admiration for each other. Later Eddington would write (in a letter to de Donder, Lemaître's mentor in Belgium) that he had found in Lemaître "a truly brilliant student, wonderfully quick and clear-sighted and of great mathematical ability". Coming from Eddington this really meant something!

In 1924 Lemaître went on to MIT and to the Harvard College Observatory to work with Shapley on Cepheids, variable stars. The timing of this visit was excellent because during that year the first observations which would challenge the age-old idea of an everlasting, static universe would be coming in. Lemaître was present e.g. at the celebrated 33rd meeting of the American Astronomical Society held in Washington in December 1924 where Russell announced Hubble's discovery that the great spiral nebulae are in fact other, distant galaxies. It is during this year that we also find Lemaître's first explorations of cosmology. He studied in particular the model of the universe proposed by the Dutch astronomer de Sitter - which incidentally was disguised as a static universe - and he showed, independently from

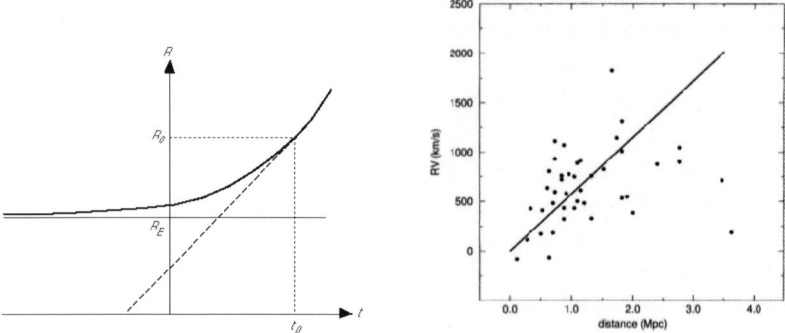

Fig. 1. Left: Lemaître's first model of 1927 of a dynamical universe, in which an expanding universe emerges from a nearly static Einstein universe in the far past. The radius R increases in time t from a finite value R_E in the infinite past. Right: the observations of extra-galactic nebulae that Lemaître used to verify the distance-velocity relation he had derived in this model.

Weyl, that in de Sitter's universe galaxies would recede from each other at a rate proportional to their separation. Starlight from distant galaxies would therefore be shifted to the red in de Sitter's universe, in line with the observational evidence at the time. But de Sitter's universe is empty, it contains neither galaxies nor observers! Therefore Lemaître abandoned this model.

In July 1925 Lemaître returned to Belgium to take up a faculty position at the Catholic University of Louvain. He continued to think about cosmology, and wondered in particular whether Relativity could accommodate a universe that retains the appealing features of both Einstein's static universe and de Sitter's empty universe. A universe, in other words, that contains matter in the form of galaxies but at the same time exhibits the reddening of distant galaxies. Lemaître's stroke of genius then was to abandon the idea of a static universe. He did so in 1927 in a seminal paper *Un univers homogène de masse constante et de rayon croissant, rendant compte de la vitesse radiale des nébuleuses extragalactiques*, which he chose to publish in the Annales de la Société Scientifique de Bruxelles.[1]

In this paper Lemaître first rediscovers Friedmann's equations that govern the evolution of a dynamical universe in Einstein's theory of Relativity. He then identifies a solution of those equations that describes an expanding universe interpolating between Einstein's static universe in the far past and de Sitter's empty universe in the distant future [cf Figure 1]. He shows further that if this were our universe then the expansion of space would cause starlight from distant galaxies to be shifted to the red, as if the light were Doppler shifted by the motion of galaxies away from us. Lemaître derives (in equation (24) of[1]) what would later become known as the Hubble law; a linear relation between the rate of separation of distant galaxies and their distance away from us. Moreover, seeking observational corroboration or falsification for his prediction of a redshift, Lemaître takes Slipher's redshifts and Hubble's distances for a sample of 42 extra-galactic nebulae to estimate the proportionality constant H in the distance-velocity relation. Because of the large uncertainties in the individual observations, particularly in the distances, Lemaître decides to divide the mean velocity by the mean distance in the sample, and in this way obtains the value $H = 575$ km/s per Megaparsec.

In short, Lemaître establishes in this paper *the* fundamental connection between the theory of Relativity and cosmology. He himself once recalled, in his characteristic light, humble style, that "I happened to be more mathematician than most astronomers, and more astronomer than most mathematicians".

However, most of the important figures in cosmology hardly took notice of Lemaître's groundbreaking work, and the few remarks that did reach Lemaître, were actually mostly negative. In the margin of the 1927 Solvay Conference, for instance, Lemaître had a brief discussion of his work with Einstein, who concluded this by saying "Your calculations, Monsieur Lemaître, are correct, but your physical insight is *tout à fait abominable*". Clearly the scientific community was not (yet) prepared to abandon the ancient, cherished idea of an eternal, static universe.

But in 1929, Hubble established observationally a linear distance-velocity relation for the spiral nebulae. Using more precise observations of 24 distant extragalactic nebulae obtained with the 100-inch telescope on Mt Wilson, the most powerful telescope at the time, Hubble obtained a proportionality constant of 513 km/s per Megaparec - not very different from the value found by Lemaître two years earlier. Hubble's work made no mention of the expansion of the universe. Instead he interpreted his observations in terms of a usual Doppler shift. But the scientific community recognised the potentially far-reaching implications of Hubble's observations and in particular the need to reconcile these with Relativity if the latter were to provide a viable theoretical framework for cosmology.

The problem of the reddening of distant nebulae was therefore high on the agenda at the London meeting of the Royal Astronomical Society on Friday, 10 January 1930, where Eddington famously said "We ought to put a little motion into Einstein's world of inert matter, or a little matter in de Sitter's primum mobile". Georges Lemaître was not present at this meeting, but when he read its proceedings in *The Observatory* a few weeks later he responded and reminded Eddington that two years before he had already found the intermediate, expanding solution that he and de Sitter were now looking for [cf Figure 2]. Lemaître also enclosed several copies of his original paper with his letter and asked Eddington to give one to de Sitter.

Eddington confessed that, although he had seen Lemaître's pioneering paper at the time, he had failed to realise its far-reaching consequences and he had forgotten about it until that moment. Around the same time Eddington himself independently showed that Einstein's static universe is unstable to either expansion or contraction. He was thus ready to adopt Lemaître's model of 1927, which became known as the Eddington - Lemaître universe.

Starting in May 1930 both Eddington and de Sitter generously recognised Lemaître's major discovery in their publications, and they enthusiastically sup-

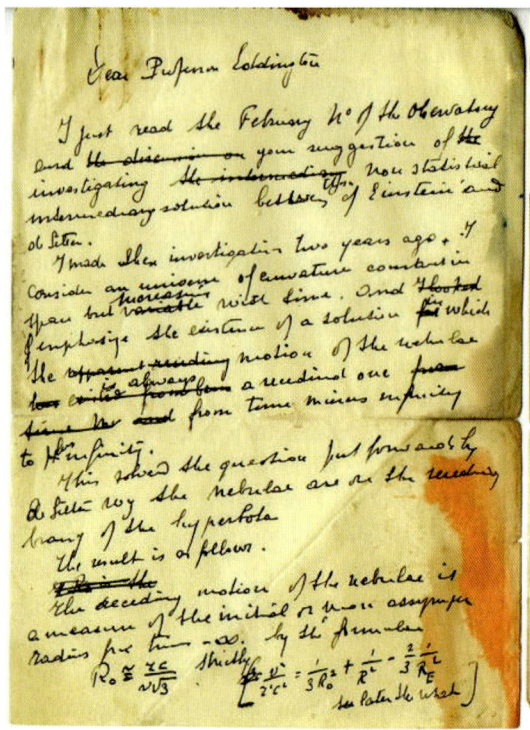

Fig. 2.   The first draft page of Lemaître's letter in which he tells Eddington that three years before he had found the expanding solution Eddington was looking for. (Source: Archives Georges Lemaître, Université catholique de Louvain, Louvain-la-Neuve.)

ported and disseminated the new concept of an expanding universe. In 1931, in an extraordinary sign of appreciation that shows the importance he attached to Lemaître's work, Eddington even ordered a translation of Lemaître's original paper to be published in the widely read Monthly Notices of the Royal Astronomical Society (MNRAS).[2]

But then something seemingly odd happened. The section in the original paper where Lemaître derives the 'Hubble constant' H was omitted in the translation, and replaced by a short note referring to 'available data'. This has led some historians to suggest Lemaître had been censored - perhaps even to advance Hubble's reputation?

However the case was settled in 2010 thanks to a careful investigation by Livio, who found in the archives of the Royal Astronomical Society a letter from Lemaître to Smart, the editor of the MNRAS, in which Lemaître writes that "he did not find advisable to reprint his provisional discussion of radial velocities". Lemaître's motivation to leave out this particular section was most likely that the uncertain observational material available in 1927, which nevertheless convinced him of the validity of his theoretically derived 'Hubble law', had by 1931 been superseded by

Fig. 3.   Lemaître's first sketch made around 1928 of a range of possible dynamical universes, includes the effect of a cosmological constant term on the universe's evolution. (Source: Archives Georges Lemaître, Université catholique de Louvain, Louvain-la-Neuve.)

better data from Hubble and Humason. And, of course, Lemaître was not interested in self-promotion anyway.

The translation of Lemaître's article in the MNRAS had a large impact, and his idea of an expanding, evolving universe rapidly became the central pillar of modern relativistic cosmology. Finally also Einstein came around. In the short article in which he accepted the expanding universe he also discarded the idea of a cosmological constant, which he had introduced in his equations in 1917 to make possible a static universe. In a letter to Tolman he wrote "Dies ist wirklich unvergleichlich befriedigender" (this is really incomparably more satisfactory), referring to his theory of Relativity without the cosmological constant term.

Interestingly Lemaître had a rather different view on the cosmological constant. He actually regarded this as a physical substance, which is nowadays known as dark energy. Consequently 'little lambda' (as the cosmological constant was referred to at the time) featured prominently in Lemaître's work on cosmology. The first known representations of an expanding universe, made by Lemaître around 1928 we believe, clearly illustrate this [cf Figure 3]. Around 1931 Lemaître settled on what he called a 'hesitating' universe. This is a universe which initially expands fast, then slows down so that large-scale structures such as stars and galaxies can form, and finally accelerates again, driven by the effect of a dark energy component. Being much more than Einstein guided in his work by observations, Lemaître was led to the idea of a 'hesitating' universe because the large value of the Hubble constant which he and Hubble had found, implied there had to be a preceding era of slower expansion in order for the universe to be old enough to harbour stars and galaxies at least as old as planet Earth. He maintained this vision of cosmological evolution - which is in excellent agreement with present-day precision observations - for the rest of his life.

Lemaître's hesitating universe also introduces a profoundly new feature in his cosmology: it replaces the nearly static phase in the far past of his original model of 1927 with a genuine origin. He referred to the state at the beginning as a primeval atom. (The term Big Bang was coined much later by Fred Hoyle.) By boldly proposing the world had a beginning Lemaître made it clear that a universe in expansion may well have been in a radically different physical state in the far past. He explained his view in what is perhaps his most visionary article *The Beginning of the World from the Point of View of Quantum Theory*, published in Nature in 1931.[3] In this short paper he argues, to my knowledge for the very first time in the history of modern cosmology, that our universe had an origin, which should be part of science, governed by physical laws we can discover. It is a beautiful, almost poetic paper in which Lemaître explores from a purely physical viewpoint how our universe could have come into existence - a question that would become one of the central research problems in quantum gravity and quantum cosmology more than half a century later.

Of course Lemaître did not put forward a theory or even a model for his primeval atom. In Relativity the origin of an expanding universe is a spacetime singularity where our usual notions of space and time cease to be meaningful, and Einstein's theory breaks down. Lemaître realised this, but suggested space and time emerged from a more fundamental, abstract quantum mechanical state which, he argued, stands 'before' space-time. In line with this view he regarded the beginning also as a closure of our universe - a horizon as it were beyond which lies a realm of reality that neither influences our observable universe nor will ever be accessible to our observations.

Incidentally Lemaître was led to consider a quantum origin of the world partly because he thought there should be a 'natural' beginning, and he reasoned that the indeterminacy of quantum theory could provide a potential mechanism to generate a complex universe from a natural, simple beginning. Today this idea is realised concretely in inflationary cosmology where the rapid expansion transforms the simplest initial state - the quantum vacuum - into the seeds of the complex configuration of large-scale structures we find in today's universe.

Lemaître realised however that a fuzzy quantum origin does not give rise to a unique world. Contemplating the implications of this, he wrote "Clearly the initial quantum could not conceal in itself the whole course of evolution. The story of the world need not have been written down in the first quantum like the song on a disc of a phonograph ... Instead from the same beginning widely different universes could have evolved"[3] - a worldview not unlike what we now call the multiverse.

In the light of Einstein's reluctance to accept cosmic evolution it will come as no surprise that he was not happy with Lemaître's primeval atom. At some point he even complained to the Belgian priest that this reminded him too much of Christian dogma!

Despite their differences, however, in the early 1930s Lemaître and Einstein interacted frequently with each other. Their discussions were friendly and stimulating. During these years Lemaître spent several terms in the United States, where he wrote a number of highly influential articles in which we find the seeds of many of the ideas that later became part of the standard model of relativistic cosmology. These include the construction, inspired by Tolman's work, of the first (spherical) models of the formation of galaxies in an expanding universe, and an interpretation of the cosmological constant as a vacuum energy. In response to a question from Einstein, Lemaître also demonstrated that under certain conditions a beginning of time is unavoidable in Relativity. This result would be proven in full generality by Hawking and Penrose only in the 1960s, and it emphasises the quantum mechanical nature of Lemaître's primeval atom. Finally in 1934, he suggested there should exist fossil relics of the hot dense state of the universe at early times, which might allow us to trace back our history and "reconstruct the vanished burst of formation of the worlds" as he put it.

Meanwhile Lemaître had become the darling of the American press. The public discovered to its amazement that the 'father of the big bang' was also a Catholic priest! Lemaître, however, patiently explained why he saw no conflict between what he called 'the two paths to truth' that he decided, at a very young age, to follow. "Once you realise", he argued, "the Bible does not purport to be a textbook on science, and once you realise Relativity is irrelevant for salvation, the old conflict between science and religion disappears." [cf Figure 4]

Lemaître carefully maintained a clear distinction between science and religion throughout all his life, respecting meticulously the differences in methodology and language between both. Far from the concordist interpretations that sought to derive the truths of faith from scientific results Lemaître insisted that science and religion have their own autonomy. He set out his position clearly and eloquently in his rapporteur talk at the Solvay Conference on Astrophysics in 1958 in which he argued the hot big bang model is nothing but a scientific hypothesis, to be verified or falsified by observations, which remains entirely outside the realm of metaphysics or religion.

Consequently, Lemaître was not amused with the Un'Ora address of Pope Pius XII to the Pontifical Academy of Sciences in 1951, in which the Pope suggested that modern cosmology gives credit to the doctrine of *ex nihilo* creation at the beginning of time (without, however, explicitly referring to Lemaître's work). In the early 1960s Monsignor Lemaître, as President of the Pontifical Academy, would strive to maintain the autonomy of the Academy to avoid any such mixing of science and theology.

Fig. 4.   The 'father of the big bang' maintained a clear distinction between science and religion. (Source: Archives Georges Lemaître, Université catholique de Louvain, Louvain-la-Neuve.)

Does this mean that in Lemaître's view cosmology has absolutely no meaning for philosophy or theology, or vice versa? Not exactly. Lemaître himself certainly experienced a deep unity between his spiritual and professional life, and I am tempted to think that the harmonious coexistence of his cosmology and his faith in his mind and in his actions may well have been an important source of inspiration and creativity that led him to conceive of a universe in evolution. We can find a hint of such a unity in the last paragraph of the manuscript of the article[3] in which he put forward his 'primeval atom' hypothesis where he writes "I think that everyone who believes in a supreme being supporting every being and every acting, believes also that God is essentially hidden and may be glad to see how present physics provides a veil hiding the creation." Lemaître crossed out this paragraph before he submitted his paper to Nature, most likely because he thought it inappropriate to state his personal philosophical position in a scientific article, especially one in which he made the case for a scientific inquiry of the universe's origin.

When the second World War engulfed the continent Lemaître stayed in Belgium where he focused on the needs of his students and tried to comfort his family and friends. During this period he was cut off from his international contacts and became scientifically isolated. He did not participate in the further development of the big bang model after the war, leading e.g. to Alpher's and Gamov's understanding of primordial nucleosynthesis and to the prediction by Alpher and Hermann of a thermal relic radiation of the hot big bang. Instead, Lemaître devoted most of his time to numerical computation, an old passion of him dating back at least to his time at MIT in the late 1920s. He famously called upon his students to carry a Burroughs E101 (one of the first electronic computing machines) which he had seen at the World Expo in Brussels in 1958, up to the attic of the physics building in Leuven, thereby establishing the university's first computing centre.

But observations that could vindicate Lemaître's hot big bang hypothesis remained elusive even in the 1950s. In those years his cosmology was actually not seldom regarded as old fashioned science that had been pursued in a spirit of concordism, his critics would say, and a rival theory, the steady state model of Bondi, Gold and Hoyle, entered the stage.

Lemaître's fortunes turned around in 1964 with the discovery of the cosmic microwave background by Penzias and Wilson and its cosmological interpretation by Dicke, Peebles, Roll and Wilkinson as remnant radiation of a hot Big Bang. Lemaître heard about this discovery on the 17th of June in 1966, a mere three days before his death, in the hospital, where a close friend brought him the news that the fossil relics that prove his theory right had finally been found.

Acknowledgments

I thank Liliane Moens of the Archives Georges Lemaître at the Université catholique de Louvain, Louvain-la-Neuve (Belgium) for her warm welcome and her assistance.

References

1. Lemaître, G. (1927). Un univers homogène de masse constante et de rayon croissant, rendant compte de la vitesse radiale des nébuleuses extragalactiques. *Annales de la Société scientifique de Bruxelles*. Série A, **47**, 49-59.
2. Lemaître, G. (1931). A homogeneous universe of constant mass and increasing radius accounting for the radial velocity of extra-galactic nebulae. *MNRAS*, **91**, 490-501.
3. Lemaître, G. (1931b). The beginning of the world from the point of view of quantum theory. *Nature*, **127**, 706.

Session 1

Neutron Stars

Chair: *Ed van den Heuvel*, Universiteit van Amsterdam, the Netherlands
Rapporteurs: *Vicky Kaspi*, MCGill University, Canada and *Michael Kramer*, Manchester University, UK
Scientific secretaries: *Nicolas Chamel*, Université Libre de Bruxelles, Belgium and *Hans Van Winckel*, KULeuven, Belgium

E. van den Heuvel We heard from Roger Blandford earlier today that we all love neutron stars and we do so because they are so extreme. To paraphrase what Melvin Ruderman once said: if you compress 400 000 times the mass of the Earth in a ball more or less the size of Brussels, you will have some idea of what the density in a neutron star is; to put it differently: the volume of a raindrop of neutron-star material contains as much matter as of all the seven billion people in the world together. So neutron stars are very extreme objects, with fantastic extreme physics involved. Also, as you will hear from the second rapporteur, they are extremely useful objects to test relativity and physical laws in general. They are also useful because we would not have been here if there had not been neutron stars. Their formation is so energetic that all the rest of the star is kicked out, including all the heavy elements produced inside the progenitor star. These elements are injected into the interstellar medium from which new stars form and around one of these stars, our planetary system formed as well. We therefore owe our existence to neutron stars. Now, I am not going to talk much more about that. We have two wonderful rapporteurs, and I first would like to hand the floor to Vicky Kaspi, who will talk about radio pulsars and the neutron star zoo, summarizing the latest knowledge about neutron stars. After that the second rapporteur, who is Michael Kramer. He will talk about the radio sky as a laboratory for fundamental physics: neutron stars, binary pulsars and also fast radio transients.

Rapporteur Talks by V. Kaspi and M. Kramer: Radio Pulsars: The Neutron Star Population & Fundamental Physics

Abstract

Radio pulsars are unique laboratories for a wide range of physics and astrophysics. Understanding how they are created, how they evolve and where we find them in the Galaxy, with or without binary companions, is highly constraining of theories of stellar and binary evolution. Pulsars' relationship with a recently discovered variety of apparently different classes of neutron stars is an interesting modern astrophysical puzzle which we consider in Part I of this review. Radio pulsars are also famous for allowing us to probe the laws of nature at a fundamental level. They act as precise cosmic clocks and, when in a binary system with a companion star, provide indispensable venues for precision tests of gravity. The different applications of radio pulsars for fundamental physics will be discussed in Part II. We finish by making mention of the newly discovered class of astrophysical objects, the Fast Radio Bursts, which may or may not be related to radio pulsars or neutron stars, but which were discovered in observations of the latter.

1. Introduction

The discovery of evidence for the neutron by Chadwick in 1932 was a major milestone in physics,[1] and was surely discussed with great excitement at the 1933 Solvay Conference titled "Structure et propriétés des noyaux atomiques." That same year, two now-famous astronomers, Walter Baade and Fritz Zwicky, suggested the existence of *neutron stars*, which they argued were formed when a massive star collapses in a "super-nova".[2] They argued that such a star "may possess a very small radius and an extremely high density." It took over 3 decades for this seemingly prophetic prediction to be confirmed: in 1967[a] then-graduate student Jocelyn Bell and her PhD supervisor Antony Hewish detected the first radio pulsar,[3] and Shklovksy suggested that the X-ray source Sco X-1 was an accreting neutron star.[4] The focus of this paper is on the former discovery, now a class of celestial objects of which over 2300 are known in our Galaxy (although the accreting variety will be mentioned in §3). Though radio pulsars were compellingly identified as neutron stars not long after their discovery,[5,6] the radio emission was unexpected, prompting the noted physicist and astronomer John Wheeler to remark his surprise that neutron stars come equipped with a handle and a bell.[b] Though the origin of the radio emission is not well understood today, it has nevertheless served as a valuable beacon with which we have learned vast amounts about the neutron star phenomenon. Using

---

[a]Coincidentally, the year both of these authors were born.
[b]This quote appears in Ref. 7 but its origin is unspecified.

this beacon as a tool also provides us with unique laboratories to study fundamental physics. In this first part of this contribution, we will review the diversity in the "neutron star zoo," before we discuss their applications for understanding the laws of nature, in particular gravity, in the second part.

Part I: The Different Manifestations of Neutron Stars

2. Radio Pulsars

Radio pulsars are rapidly rotating, highly magnetized neutron stars whose magnetic and rotation axes are significantly misaligned. It is believed that beams of radio waves emanate from the magnetic pole region and are observed as pulsations to fortuitously located observers, with one pulse per rotation. In some cases, two pulses per rotation may be visible if the source's magnetic and rotation axes are nearly

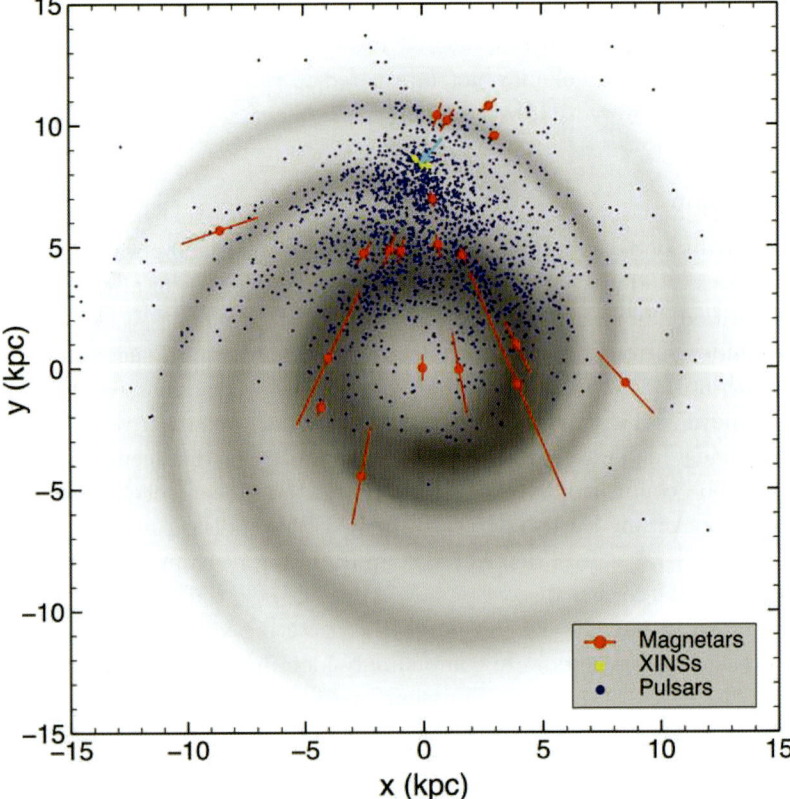

Fig. 1.   Spatial distribution of radio pulsars (in blue), magnetars (in red; see §5), XDINS (aka XINS, in yellow; see §6), projected on the Galactic disk. The location of the Earth is indicated by a cyan arrow. The underlying grey scale roughly traces the free electron distribution. Figure taken from Olausen & Kaspi (2014).

orthogonal. Pulsations are believed to be produced, and occasionally are observed, across the full electromagnetic spectrum (see §2.1), however the vast majority of known pulsars are observed exclusively in the radio band. The known pulsar population, currently consisting of over 2300 sources with numbers constantly increasing thanks to ongoing radio pulsar surveys,[8–10] is largely confined to the Galactic Plane, with a e^{-1} thickness of ~100 pc.[11] However, the pulsar scale height appears to increase with source age. This is presumably because pulsars are high velocity objects, with space velocities typically several hundred km/s.[11–14] Such high speeds are likely due to a birth kick imparted at the time of the supernova, due to a combination of binary disruption (for sources initially enjoying a binary companion) and asymmetry in the supernova explosion itself. It is important to note that the known radio pulsar population is very incomplete and subject to strong observational selection biases; this is clear in Figure 1 wherein the locations of the radio pulsars on the Galactic disk are seen to be strongly clustered near Earth. These biases include those imposed by dispersion and scattering of radio waves by free electrons in the interstellar medium, by preferential surveying in the Galactic Plane, as well as by practical limits on time and frequency resolution in radio pulsar surveys. See for example Refs. 11,15 for a discussion of selection effects in radio pulsar surveys.

Pulsars rotate rapidly, typically with rotation periods P of a few hundred ms. The presently known slowest radio pulsar has a period of ~8 s,[16] while the fastest has period 1.4 ms or 716 Hz.[17] All pulsars spin down steadily, a result of magnetic dipole braking, hence can be characterized by period P and its rate of change $\dot{P}$. The latter, though typically only tenths of microseconds per year, is eminently measurable for all known sources because of pulsars' famous rotational stability. Measurements of pulsar spin-down rate $\dot{P}$ are extremely useful, as they enable helpful estimates of key physical properties.

One such property is the surface dipolar magnetic field at the equator,

$$B = \left(\frac{3c^3 I}{8\pi^2 R^6} \right)^{1/2} \sqrt{P\dot{P}} = 3.2 \times 10^{19} \ \sqrt{P\dot{P}} \ \ G, \tag{1}$$

where I is the stellar moment of inertia, typically estimated to be 10^{45} g cm^2, and R is the neutron star radius, usually assumed to be 10 km (see Part II. for observational constraints.) This estimate assumes magnetic braking *in vacuo*, which was shown to be impossible early in the history of these objects[18] since rotation-induced electric fields dominate over the gravitational force, even for these compact stars, such that charges must surely be ripped from the stellar surface and form a dense magnetospheric plasma. Nevertheless, modern relativistic magnetohydrodynamic simulations of pulsar magnetospheres have shown that the simple estimate offered by Eq. 1 is generally only a factor of 2–3 off.[19] The observed distribution of radio pulsar magnetic fields is shown in Figure 2.

Measurement of P and $\dot{P}$ also enable an estimate of the source's age. The pulsar's

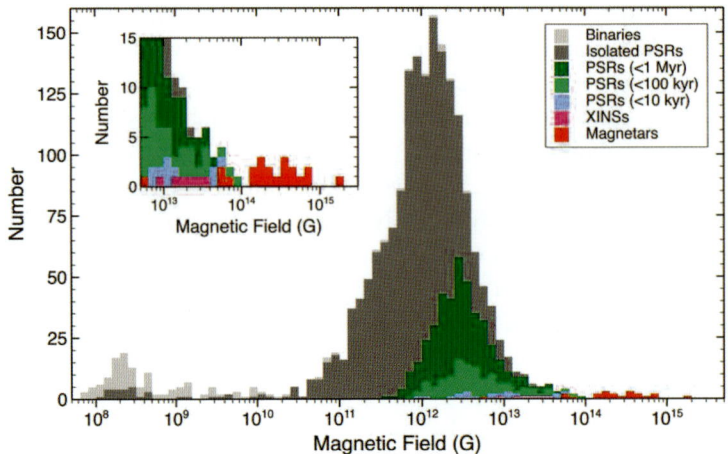

Fig. 2. Distribution of spin-inferred magnetic fields (using Eq. 1) of radio pulsars (in various shades of blue and grey), XDINS (aka XINS, in yellow; see §6), and magnetars (in red; see §5). From Olausen & Kaspi (2014).

characteristic age τ_c is defined as

$$\tau_c = \frac{P}{(n-1)\dot{P}}\left[1 - \left(\frac{P}{P_0}\right)^{(n-1)}\right] \simeq \frac{P}{2\dot{P}}, \qquad (2)$$

where n is referred to as the 'braking index' and is equal to 3 for simple magnetic dipole braking (see e.g. Ref. 7), though is observed to be less than 3 in the handful of sources for which a measurement of n has been possible.[20] P_0 is the spin period at birth and is generally assumed to be much smaller than the current spin period, although this is not always a valid assumption, particularly for young pulsars.[21]

Finally, a pulsar's spin-down luminosity L_{sd} (also known as $\dot{E}$, where $E \equiv \frac{1}{2}I\omega^2$ with $\omega \equiv 2\pi/P$ is the stellar rotational kinetic energy) can be estimated from P and $\dot{P}$ and is given by

$$\dot{L}_{sd} = \frac{d}{dt}\left(\frac{1}{2}I\omega^2\right) = I\omega\dot{\omega} = 4\pi^2 I \frac{\dot{P}}{P^3} = 4 \times 10^{31}\left(\frac{\dot{P}_{-15}}{P_1}\right) \quad \text{erg/s}, \qquad (3)$$

where $\dot{P}_{-15}$ is $\dot{P}$ in units of 10^{-15} and P_1 is the period in units of seconds. L_{sd} represents the power available for conversion into electromagnetic radiation, an upper limit on the (non-thermal; see §2.1) radiation a pulsar can produce. For this reason, radio pulsars are also known as 'rotation-powered pulsars.'

A traditional way of summarizing the pulsar population is via the P-$\dot{P}$ diagram (Figure 3). Here the spin periods of pulsars are plotted on the x-axis and $\dot{P}$ on the y-axis. The swarm of conventional radio pulsars clearly has its P peak near $\sim$500 ms, with typical $B \simeq 10^{11}$ G and characteristic age $\tau_c \simeq 10^7$ yr. The youngest radio pulsars have $\tau_c \simeq 1$ kyr and are generally found in supernova remnants;

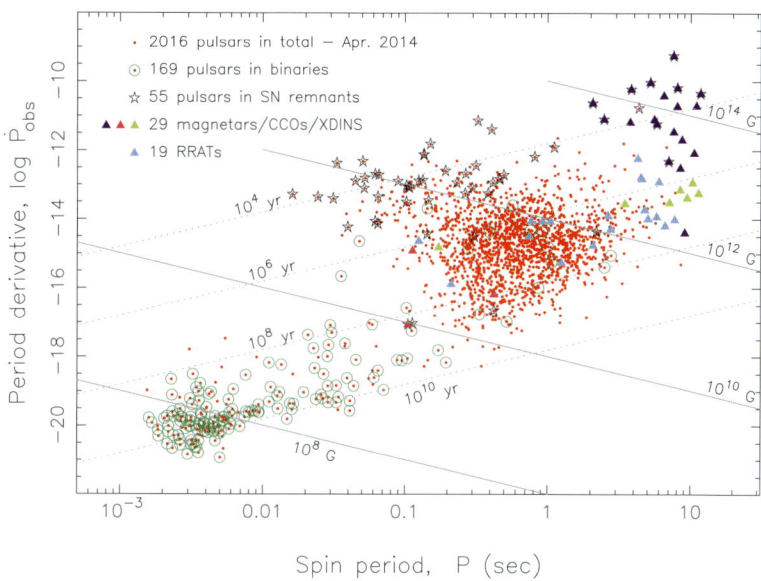

Fig. 3.   *P-Ṗ* diagram. Red dots indicate known radio pulsars as of April 2014. Blue circles represent binaries. Stars represent associated supernova remnants. Magnetars are represented by green triangles (see §5). The XDINS and CCOs (see §4,5) are pink and yellow triangles, respectively. RRATs (§9) are in cyan. Solid grey lines are of constant magnetic field (Eq. 1) and dotted lines are of constant characteristic age (Eq. 2). From Tauris et al. (2014).

the latter dissipate typically after 10–100 kyr, explaining why older pulsars are generally not so housed in spite of all having been born in core-collapse supernovae. The $P = 33$-ms Crab pulsar is arguably the most famous of the young pulsars, its birth having been recorded by Asian astrologers in 1054 A.D.[22] However, it is in fact not the youngest known pulsar; this honour presently goes to the 884-yr old PSR J1846−0258 in the supernova remnant Kes 75.[23] Also clear in the *P-Ṗ* diagram is the collection of binary pulsars, nearly all of which cluster in the lower left portion of the diagram, where the "millisecond pulsars" reside. This is no coincidence; although the rapid rotation of the young Crab-like pulsars is almost certainly a result of angular momentum conservation in the core collapse, that of the millisecond pulsars (MSPs) is intimately tied to their binarity. MSPs are believed to have been spun-up by an episode of mass accretion from their binary companion (see §3).

2.1. *Pulsar Emission*

Though rotation-powered pulsars are usually referred to as 'radio pulsars,' in reality these objects emit across the full electromagnetic spectrum. In fact, the radio emission (that in the ∼100 MHz to ∼100 GHz range), which must surely be of a non-thermal nature owing to the enormous brightness temperatures implied, usually represents a tiny fraction (typically $\sim 10^{-6}$) of L_{sd}, hence is energetically unimportant. The richness of radio observations and phenomenology has fuelled over the years significant theoretical effort into understanding its origin. However at present, there is no concensus and it remains an open question.[24,25] In spite of the lack of an understanding of the physics of the radio emission, pulsar astronomers are generally content to accept its existence as coming from a 'black box,' and use it as an incredibly useful beacon of the dynamical behaviour of the star as described in Part II.

The second most commonly observed emission from rotation-powered pulsars is in the X-ray band. See Ref. 26 or 27 reviews. The origin of pulsar X-rays is far better understood than is the radio emission and we describe it briefly here as it is instructive, particularly when considering other classes of neutron stars (see §4). X-rays originate from one of two possible mechanisms, which can sometimes both be operating. One is thermal emission from the surface, due either to the star being initially hot following its formation in a core-collapse event (in which case the thermal luminosity need not be constrained by L_{sd}), or from surface reheating by return currents in the magnetosphere. The latter is particularly common in millisecond pulsars, but may well be present in all pulsars and indeed can be an important complicating factor in efforts to constrain neutron-star core composition via studies of cooling. As the thermal emission is thought to arise from the surface, it is typically characterized by quasi-sinusoidal pulsations, likely broadened by general relativistic light bending. The second source of X-rays is purely magnetospheric. This emission has a strongly non-thermal spectrum and appears highly beamed, as observed via very short duty-cycle pulsations. The non-thermal emission is ultimately powered by spin-down (as is the thermal return-current emission) so its luminosity must be limited by L_{sd}. Note that additional X-ray emission can be present in pulsars' immediate vicinity due to *pulsar wind nebulae* – sometimes spectacular synchrotron nebulae that result from pulsars' relativistic winds being confined by the ambient medium. See Ref. 28 for a review of these objects.

Space limitations preclude discussion of the third-most commonly observed emission band for rotation-powered pulsars – the gamma-ray regime. For a recent review of this interesting and highly relevant area of radio pulsar astrophysics, see Ref. 29.

3. Binary Radio Pulsars

As seen in Figure 3, pulsars with a binary companion generally, but not exclusively, inhabit the lower left of the P-$\dot{P}$ diagram, where spin periods are short and spin-down rates low. Indeed the vast majority of millisecond pulsars are in binary systems and have among the lowest magnetic field strengths of the pulsar population (see the small peak at the very low end in the B-field distribution in Figure 2). These facts are not coincidental. According to the standard model,[30–32] although the vast majority of radio pulsars originated from progenitors that were in binaries, most of these systems were disruped by the supernova. Of the few that survived, subsequent evolution of the pulsar binary companion, under the right circumstances, resulted in Roche-lobe overflow and the transferring of matter and angular momentum onto the neutron star, in the process spinning it up. Such spun-up pulsars are often called "recycled" as they are effectively given a new life by their companion; without the latter they would have spun down slowly, alone, until ultimately the radio emission mechanism ceased as it evenutally must. The mass transfer phase, observed as a bright accreting X-ray source powered by the release of gravitational energy as the transferred matter falls onto the neutron star,[32] has a final result that depends strongly on the nature of the companion and its proximity to the neutron star. For low-mass companions, this mass transfer phase can last long enough to spin the pulsar up to millisecond periods. For higher-mass companions, only tens of millisecond periods can be achieved as these companions have shorter lifetimes. Simultaneous with the spin-up is an apparent quenching of the magnetic field, a process whose physics are poorly understood, but for which there is strong observational evidence. The above is a very broad-brush description of a very rich field of quantitative research that blends orbital dynamics with stellar evolution and neutron-star physics. One pictorial example of evolutionary scenarios that can lead to the formation of recycled pulsars is shown in Figure 4.[33]

One outstanding mystery in the standard evolutionary model is the existence of isolated MSPs. These can be seen scattered in the lower left-hand part of Figure 3. Indeed, the first discovered MSP, PSR B1937+21, is isolated.[34] If binarity is key to recycling and spin-up, where are the isolated MSPs companions? One plausible answer may lie in the apparent companion 'ablation' that appears to be in progress in some close (orbital periods of a few hours) MSP binaries, notably those in which the radio pulsar is regularly eclipsed by material that extends well beyond the surface of the companion.[35–37] The companion's mass loss is believed to be fueled by the impingement of the intercepted relativistic pulsar particle wind which is ultimately powered by L_{sd}.

Another newly identified mystery is the discovery of eccentric MSP binaries. Key to the recycling process is rapid and efficient circularization of orbits and indeed some MSP binaries[38,39] have eccentricities well under 10^{-6}. The discovery of a eccentricity 0.44 MSP in a 95-day orbit in the Galactic disk[40] thus was difficult to understand; one possibility is that it formed as part of a hierarchical triple system[41] in which the inner companion was eventually ablated. The recent unambiguous detection of an MSP in a hierarchical triple system[42] supports the existence of such systems, and suggests that binary evolution may be an incomplete picture of the paths to making MSPs.[43] Today there are 3 more MSP binary systems having eccentricities $\gtrsim 0.1$ (Refs. 8,44,45) and though origins in triple systems are still on the table, other scenarios for their production, including accretion-induced collapse of a super-Chandrasekhar mass oxygen-neon-magnesium white dwarf in a close binary[46] and dynamical interaction with a circumbinary disk,[47] have been proposed.

Very recently, there has been a series of spectacular confirmations of key aspects of binary evolution theory. One is in the form of the discovery of a binary radio

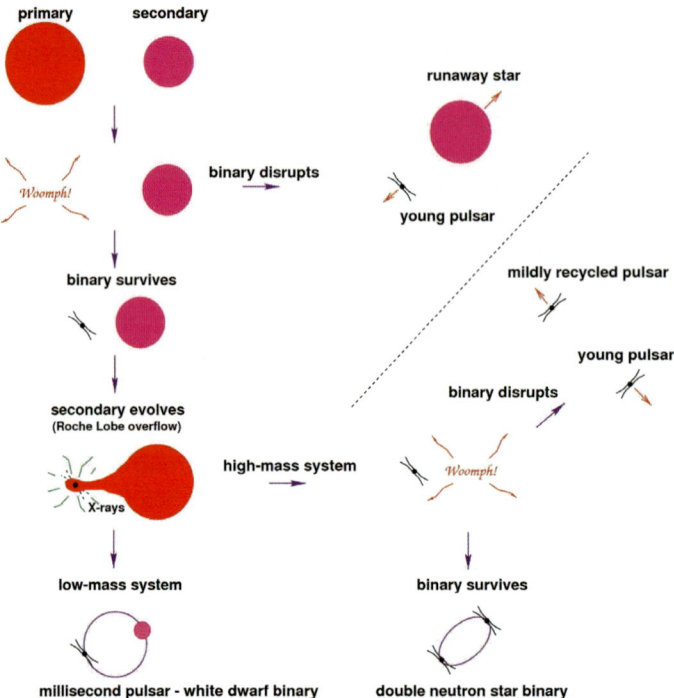

Fig. 4.   Two neutron star binary evolution scenarios, one forming a millisecond pulsar – white dwarf binary, and the other a double neutron star binary. The primary deciding factor in the end state of the neutron star is the mass of its companion, with the white dwarf binary forming from a low-mass companion and the double neutron star from a high-mass companion. From Lorimer (2008).

MSP, PSR J1023+0038, which had been observed to have an accretion disk in the previous decade.[37] Then there came the discovery of repeated swings between radio pulsations and bright accretion-powered X-ray pulsations in a different source.[48] Interestingly, the radio pulsations from PSR J1023+0038 have subsequently vanished[49] and a far brighter X-ray source has turned on,[50] suggesting some form of accretion, possibly in the propeller regime, is ongoing. Yet a third similar X-ray binary/radio MSP transitioning source has also recently been identified.[51] This flurry of discoveries has brought us into a new era for making progress on the physics of accretion and accretion flows, the nature of the end of the recycling process and the formation of radio MSPs.

Finally, it is important to note the handful of radio pulsar binaries that sit among the regular population in the P-$\dot{P}$ diagram, i.e. young binaries in which the pulsar has not yet been recycled, and in which the companion is a massive main-sequence star. Only a few such objects are known[52–54] likely owing to their short lifetimes. Unsurprisingly, these binaries are highly eccentric, resulting from a kick likely imparted at the time of the supernova explosion that formed the pulsar, but which (barely) did not unbind the orbit. These systems are interesting for a variety of reasons, including unusual dynamics present due to spin-induced quadrupole moments in the massive star, such as coupling between the stellar and orbital angular momenta.[55] This can cause precession in the system which can be used to detect misalignment between the stellar and orbital angular momenta, which provides strong evidence for a kick at the time of the neutron-star formation.[56–58] Also, these systems provide a unique way to constrain the nature of massive star winds.[59,60] One is also a γ-ray emitter,[61,62] and serves as a possible 'Rosetta Stone' for a different class of γ-ray-emitting binaries in which the nature of the compact object is unknown.[63]

4. Diversity in Neutron Stars

The last decade has shown us that the observational properties of neutron stars are remarkably diverse: Wheeler's 'handle and bell,' invoked to describe emission from radio pulsars, now appears to be occasionally accompanied or sometimes substituted by a horn, a basket, a flashing light and/or a flag. It turns out, radio pulsars are just one observational manifestation of neutron stars. Today we have identified multiple other classes (or possibly sub-classes): magnetars (which have been sub-classified into 'anomalous X-ray pulsars (AXPs)' and 'soft gamma repeaters (SGRs)'), X-ray dim isolated neutron stars (XDINS), Central Compact Objects (CCOs), and Rotating Radio Transients (RRATs). In addition to an explosion of acronyms, we have an explosion of phenomenology. See Ref.[64] for a review. An important current challenge in neutron-star astrophysics is to establish an overarching physical theory of neutron stars and their birth properties that can explain this great diversity. Next we discuss each of these classes in turn.

5. Magnetars

Magnetars are without doubt the most dramatic of the neutron star population, with their hallmark observational trademark the emission of brief but intense – often greatly hyper-Eddington – X-ray and soft γ-ray bursts. This class of neutron stars was first noted in 1979 with the detection of repeated soft γ-ray bursts from two different sources by space-based detectors[65,66] – hence the name 'soft gamma repeater' (SGR). Today there are 23 confirmed magnetars; the first magnetar catalog has been published[67] and is available online.[c] See Ref. 68 for a very recent review. Three magnetars have shown particularly powerful "giant flares;"[69,70] in the first 0.2 s of one such event, from SGR 1806−20, more energy was released than the Sun produces in a quarter of a million years[69] and in the first 0.125 s, the source outshone by a factor of 1000 all the stars in our Galaxy, with peak luminosity upwards of 2×10^{47} erg s^{-1} and total energy released approximately 4×10^{46} erg.

---

[c]http://www.physics.mcgill.ca/~pulsar/magnetar/main.html

Apart from their signature X-ray and soft γ-ray bursts, magnetars have the following basic properties. They are persistent X-ray pulsars, with periods for known objects in the range 2–12 s and are all spinning down, such that application of the standard magnetic braking formula (Eq. 1) yields field strengths typically in the range 10^{14}-10^{15} G. In the past, two sub-classes have been referred to in the literature: the SGRs, and the 'anomalous X-ray pulsars' (AXPs) which, prior to 2002, had similar properties to the SGRs except did not seem to burst (but see below). Roughly 1/3 of all these sources are in supernova remnants, which clearly indicates youth; in very strong support of this is the tight confinement of Galactic magnetars (two are known in the Magellanic Clouds) to the Galactic Plane, with a scale height of just 20–30 pc.[67] This, along with some magnetar associations with massive star clusters,[71] strongly suggests that magnetars are preferentially produced by very massive ($\gtrsim 30 M_\odot$) stars that might otherwise have naively have been though to produce black holes. Note that the magnetar spatial distribution in the Galaxy is subject to far fewer selection effects than is that of radio pulsars (see Figure 1), because magnetars are typically found via their hard X-ray bursts (on which the interstellar medium has no effect) using all-sky monitors that have little to no preference for direction.

Importantly, and at the origin of their name, is that in many cases their X-ray luminosities and/or their burst energy outputs (and certainly the giant flare energy outputs!) are orders of magnitude larger than what is available from their rotational kinetic energy loss, in stark contrast with conventional radio pulsars. Thus the main puzzle regarding these sources initially was their energy source. Accretion from a binary companion was ruled out early on given the absence of any evidence for binarity.[72] Thompson and Duncan[73–75] first developed the magnetar theory by arguing that an enormous internal magnetic field would be unstable to decay and could heat the stellar interior,[76] thereby stressing the crust from within, resulting in occasional sudden surface and magnetospheric magnetic restructuring that could explain the bursts. That same high field, they proposed, could explain magnetars' relatively long spin periods in spite of their great youth, as well as confine the energy seen in relatively long-lived tails of giant flares. The direct measurement of the expected spin-down rates[77] (and the implied spin-inferred magnetic fields mentioned above) came, crucially, after this key prediction. This provided the most powerful confirmation of the magnetar model; additional strong evidence came from the detection of magnetar-like bursts from the AXP source class[78,79] which had previously been explicitly called out in Ref. 75 as being likely magnetars.

Although the magnetar model is broadly accepted by the astrophysical community, as for radio pulsars, a detailed understanding of their observational phenomena is still under development. Following the seminal theoretical work in Refs. 73–75, later studies have shown that magnetar magnetospheres likely suffer various degrees of 'twisting,' either on a global scale[80] or, more likely, in localised regions that have come to be called 'j-bundles'.[81] The origin of sudden X-ray flux enhancements at the times of outburst may be in the development of these twists, with subsequent radiative relaxation coupled with field untwisting. On the other hand, interior heat depositions can also account for the observed flux relaxations post outburst, and, in this interpretation, can potentially yield information on crustal composition.[82–84] Interesting open questions surround magnetar spectra, which are very soft below 10 keV, consisting of a thermal component that is rather hot ($kT \simeq 0.4$ keV) compared with those of radio pulsars (§2.1), and a non-thermal component that may arise from resonant Compton scattering of thermal photons by charges in magnetospheric currents. A sharp upturn in the spectra of magnetars above ~15 keV[85,86] was unexpected but may be explainable of coronal outflow of $e^{\pm}$ pairs which undergo resonant scattering with soft X-ray photons and lose their kinetic energy at high altitude.[87] Another magnetar mystery is that they are prolific glitchers,[88] in spite of apparently high interior temperatures that previously were invoked in the young and presumably hot Crab pulsar to explain its paucity of glitches.[89,90] Also, some magnetar glitch properties are qualitatively different from those of radio pulsars, starting with their frequent (but not exclusive) association with bright X-ray outbursts.[79,91,92]

5.1. *High-B Radio Pulsars and Magnetars*

One particularly interesting issue is how especially high-B radio pulsars relate to magnetars. Figure 2 shows histograms of the spin-inferred magnetic field strengths of radio pulsars (coloured by age) and magnetars. Although, generally speaking, magnetar field strengths are far higher than those of radio pulsars, there is a small overlap region in which there exist otherwise ordinary radio pulsars having magnetar-strength fields, and magnetars having rather low B fields.[84,93] This is also easy to see in Figure 3. A partial answer to this comes from an event in 2006 in which the otherwise ordinary (though curiously radio quiet) rotation-powered pulsar PSR J1846−0258, albeit one with a moderately high B of 5×10^{13} G, suddenly underwent an apparent 'magnetar metamorphosis,' brightening by a factor of > 20 in the X-ray band and emitting several magnetar-like bursts.[94] This outburst lasted ~6 weeks, and then the pulsar returned to (nearly) its pre-outburst state. (See Ref. 95 for the post-outburst status.) This suggests that in high-B rotation-powered pulsars, there is the capacity for magnetar-type instabilities. Recent theoretical work on magnetothermal evolution in neutron stars supports this.[96,97] Conversely, radio emission has now been detected from 4 magnetars,[98–101] although it has interestingly different properties from that typical of radio pulsars. Notably it is often more variable, has an extremely flat radio spectrum, is essentially 100% linearly polarized and appears to be present only after outbursts, fading away slowly on time scales of months to years. One particularly interesting radio magnetar is SGR J1745−2900, found in the Galactic Centre, within 3″ of Sgr A*.[101–104] Though plausibly gravitationally bound to the black hole, its rotational instabilities (typical for magnetars) will likely preclude dynamical experiments.[101,105] Nevertheless it is of considerable interest as its radio emission suffers far less interstellar scattering than expected given its location, suggesting future searches of the Galactic Centre region for more rotationally stable radio pulsars may bear fruit and allow sensitive dynamical experiments as described in Part II.

6. XDINS

The 'X-ray Dim Isolated Neutron Stars' (XDINS; also sometimes known more simply as Isolated Neutron Stars, INSs) are sub-optimally named neutron stars because (i) the term 'dim' is highly detector specific, and (ii) most radio pulsars are both neutron stars and 'isolated.' Nevertheless this name has stuck and refers to a small class that has the following defining properties: quasi-thermal X-ray emission with relatively low X-ray luminosity, great proximity, lack of radio counterpart, and relatively long periodicities (P =3–11 s). For past reviews of XDINSs, see Refs. 27,106,107. XDINSs may represent an interestingly large fraction of all Galactic neutron stars;[108] we are presently only sensitive to the very nearest such objects (see Figure 1). Timing observations of several objects have revealed that they are spinning down regularly, with inferred dipolar surface magnetic fields of typically a $\sim 1 - 3 \times 10^{13}$ G,[107,109] and characteristic ages of ~1–4 Myr (see Figure 3). Such fields are somewhat higher than the typical radio pulsar field. This raises the interesting question of why the closest neutron stars should have preferentially higher B-fields. The favoured explanation for XDINS properties is that they are actually radio pulsars viewed well off from the radio beam. Their X-ray luminosities are thought to be from initial cooling and they are much less luminous than younger thermally cooling radio pulsars because of their much larger ages. However, their luminosities are too large for conventional cooling, which suggests an additional source of heating, such as magnetic field decay, which is consistent with their relatively high magnetic fields.

7. 'Grand Unification' of Radio Pulsars, Magnetars and XDINS: Magnetothermal Evolution

Recent theoretical work suggests that radio pulsars, magnetars and XDINS can be understood under a single physical umbrella as having such disparate properties simply because of their different birth magnetic fields and their present ages. Motivated largely by mild correlations between spin-inferred B and surface temperature in a wide range of neutron stars, including radio pulsars, XDINSs and magnetars[110] (but see Ref. 111), a model of 'magneto-thermal evolution' in neutron stars has been developed in which thermal evolution and magnetic field decay are inseparable.[97,110,112–114] Temperature affects crustal electrical resistivity, which in turn affects magnetic field evolution, while the decay of the field can produce heat that then affects the temperature evolution. In this model, neutron stars born with large magnetic fields ($> 5 \times 10^{13}$ G) show significant field decay, which keeps them hotter longer. The magnetars are the highest B sources in this picture, consistent with observationally inferred fields; the puzzling fact that XDINSs, in spite of their great

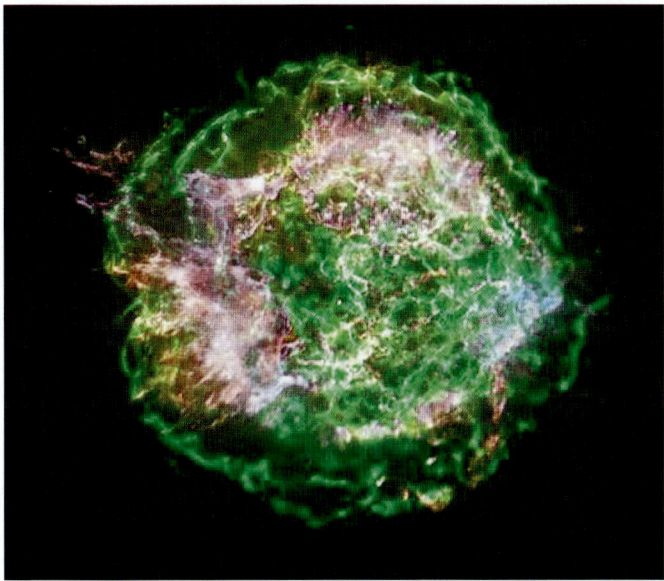

Fig. 5. X-ray image of the Cas A supernova remnant, obtained with the *Chandra* X-ray Observatory, showing the mysterious compact object at the centre. Image from http://chandra.harvard.edu/photo/2013/casa/.

proximity, appear to have high inferred Bs relative to radio pulsars is explained as the highest B sources remain hottest, hence most easily detected, longest.

8. CCOs

A census of neutron-star classes should mention the so-called Central Compact Objects (CCO).[d] CCOs are a small, heterogeneous collection of X-ray emitting neutron-star-like objects at the centres of supernova remnants, but having puzzling properties which defy a clean classification. Properties common among CCOs are absence both of associated nebulae and of counterparts at other wavelengths. The poster-child CCO, discovered in the first-light observation of the *Chandra* observatory (Figure 5), is the mysterious central object in the young oxygen-rich supernova remnant Cas A. Particularly puzzling is its lack of X-ray periodicity, lack of associated nebulosity, and unusual X-ray spectrum.[115–118]

Other objects that have been previously designated CCOs have been revealed to have low-level X-ray pulsations and surprisingly small spin-down rates. PSR J1852+0040 is at the centre of the SNR Kes 79.[119,120] This undoubtedly young pulsar, observed only in X-rays, has $P = 105$ ms yet a magnetic field strength of

---

[d] Again, a rather poor name that has stuck: the Crab pulsar is certainly 'central' to its nebula and compact, nevertheless is *not* considered a CCO!

only $B = 3.1 \times 10^{10}$ G. Its characteristic age, $\tau_c = 192$ Myr,[121] is many orders of magnitude larger than the SNR age, and much older than would be expected for an object of its X-ray luminosity (which greatly exceeds the spin-down luminosity). Interestingly, the object sits in a sparsely populated region of the P-$\dot{P}$ diagram (Figure 3), among mostly recycled binary pulsars. A similar case is the CCO in the SNR PKS 1209−52, 1E 1207.4−5209. This 0.4-s X-ray pulsar[122,123] has a spin-down rate that implies $B = 9.8 \times 10^{10}$ G and age again of orders of magnitude greater than the SNR age and inconsistent with a so large X-ray luminosity.[124] Yet another such low-B CCO is RX J0822−4300 in Puppis A,[125] with $P = 112$ ms and $B = 2.9 \times 10^{10}$ G.[124] Ref. 121 presents a synopsis of other sources classified as CCOs and argues that they are X-ray bright thanks to residual thermal cooling following formation, with the neutron star having been born spinning slowly. If so, the origin of the non-uniformity of the surface thermal emission is hard to understand. Even more puzzling however is the very high implied birthrate of these low-B neutron stars coupled with their extremely slow spin-downs: although none of these objects has yet shown radio emission, if one did, it should 'live' a very long time compared to higher-B radio pulsars, yet the region of the P-$\dot{P}$ diagram where CCOs should evolve is greatly underpopulated in spite of an absence of selection effects against finding them (see also Ref. 64). This argues that for some reason, CCO-type objects must never become radio pulsars, which is puzzling, as there exist otherwise ordinary radio pulsars with CCO-like spin properties.

9. Rotating Radio Transients

No neutron-star census today is complete without a discussion of the so-called Rotating Radio Transients, or RRATs. RRATs are a curious class of Galactic radio sources[126] in which only occasional pulses are detectable, with conventional periodicity searches showing no obvious signal. Nevertheless, the observed pulses are inferred to occur at multiples of an underlying periodicity that is very radio-pulsar-like. Indeed, patient RRAT monitoring has shown that they also spin down at rates similar to radio pulsars. The number of known RRATs is now approximately 90[e] although just under 20 have spin-down rates measured. At first thought to be possibly a truly new class of neutron star, it now appears most reasonable that RRATs are just an extreme form of radio pulsar, which have long been recognized as exhibiting sometimes very strong modulation of their radio pulses.[127,128] Indeed several RRATs sit in unremarkable regions of the P-$\dot{P}$ diagram (Figure 3). Interesting though is the mild evidence for longer-than-average periods and higher-than-typical B fields among the RRATs than in the general population. Regardless of whether RRATs are substantially physically different from radio pulsars, their discovery is important as it suggests a large population of neutron stars that was previously missed by radio surveys which looked only for periodicities. This may have important im-

---

[e]See the online "RRatalog" at http://astro.phys.wvu.edu/rratalog/rratalog.txt

plications for the neutron-star birth rate and its consistency with the core-collapse supernova rate.[108,126]

10. Fast Radio Bursts: A New Mystery

Finally, a newly discovered class of radio sources – or rather, radio events – merits mention, even though they may or may not be related to neutron stars. Fast Radio Bursts (FRBs) are single, short (few ms), bright (several Jy), highly dispersed radio pulses whose dispersion measures suggest an origin far outside our Galaxy and indeed at cosmological distances.[f] The first FRB reported[130] consisted of a single broadband radio burst lasting no longer than 5 ms from a direction well away from the Galactic Plane. The burst was extremely bright, with peak flux of 30 Jy, appearing for that moment as one of the brightest radio sources in the sky. The burst dispersion measure was a factor of 15 times the expected contribution from our Galaxy. Thornton et al. (2013) reported[131] 4 more FRBs (see Figure 6), demonstrating the existence of a new class of astrophysical events. Concerns that FRBs could be an instrumental phenomenon (since the Lorimer FRB and those reported by Thornton were all found using the Parkes Observatory in Australia) have recently been laid to rest by the discovery[132] of an FRB using the Arecibo telescope. Another FRB discovered in real-time[133] was found to be $14 - 20\%$ circularly polarised on the leading edge. No linear polarisation was detected, although depolarisation due to Faraday rotation caused by passing through strong magnetic fields and/or high-density environments cannot be ruled out. The apparent avoidance of the Galactic Plane by FRBs is consistent with a cosmological origin[134] and an event rate of $\sim 10^4$ per sky per day,[131] a surprisingly large number, albeit still based on small number statistics. Recent further data analysis and discoveries may suggest that this number may be a little smaller but still consistent with the previously estimated uncertainties (Champion, priv. comm.).

One may wonder, why it took six years since the first "Lorimer Burst" to discover further FRBs. This is due to the requirement to cover large areas of the sky with sufficient time and frequency resolution, combined with a need for sufficient computing power – areas, where recent modern surveys that are all based on digital hardware, are superior to their predecessors. Thus pulsar and RRAT hunters today are in unique positions to find FRBs, in particular with new instruments coming online that allow much larger fields-of-view.

The inferred large event rate and other FRB properties (DMs, widths, the presence of a scattering tail in some cases; see Figure 6) demand an explanation. The locations on the sky of the known FRBs are determined only to several arcminutes, a region that typically contains many galaxies. Hence identification of a host galaxy – key for understanding the nature of the burster and its environment – has been impossible. Nevertheless, some models have been proposed; papers in the

---

[f]Note that FRBs are different in their properties from so-called "perytons", which turned out to be caused by local radio interference at the radio telescope site.[129]

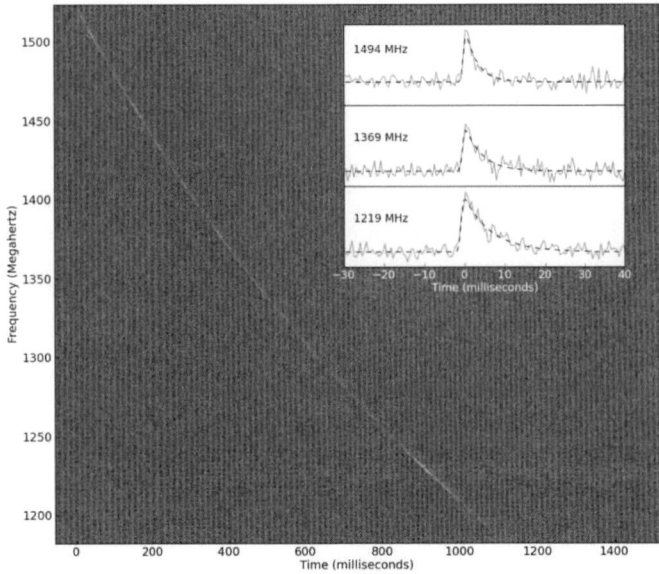

Fig. 6.   One of the Thornton et al. (2013) FRBs. The telltale dispersion sweep of this single pulse is obvious across the radio band, and the burst profile at different radio frequencies is shown in the inset, where the radio-frequency dependence of the observed pulse broadening due to scattering is clear.

refereed literature have appeared faster than FRB detections! We discuss some of those models in Part II with reference of their importance to fundamental physics.

FRBs are thus highly reminiscent of the now-famous 'Gamma Ray Burst' problem of the 1970s and 1980s – sudden, unpredictable burst events on the sky and difficult to localize – though with FRBs having the added difficulty of dispersion and the attendant great delay in detection presently due to computational demands. We cannot presently rule out that FRBs may represent a hitherto unrecognized type of astrophysical object, although as described below, neutron stars are also a plausible possibility.

Part II. - Neutron Stars as Laboratories for Fundamental Physics

As described above, the vast majority of neutron stars have been discovered in the radio regime in the form of radio pulsars. Putting aside astrophysical population and pulsar emission issues, radio observations of pulsars are important for totally independent reasons: they add to other techniques and methods employed to study fundamental physics with astronomical means. The latter include the study of a possible variation of fundamental constants across cosmic time using molecular spectroscopy of emission that originates from distant quasars. One can study the radio photons of the Cosmic Microwave Background (CMB) in great detail, as is being done as part of this conference. One can also use the coherent emission of water maser sources to obtain an accurate distance ladder to measure the local expansion of the Universe. Table 1 gives an overview of such experiments with references for further reading.

In the following, we will concentrate mostly on the study of gravitational physics where neutron star observations provide us with the best tests and constraints existing todate. Most of these tests are possible due to the rotational stability of neutron stars; the very large amount of stored rotational energy ($\sim 10^{44}$ W), in particular that of the fast rotating millisecond pulsars, makes them effective flywheels, delivering a radio "tick" per rotation with a precision that rivals the best atomic clocks on Earth. At the same time they are strongly self-gravitating bodies, enabling us to test not only the validity of general relativity, but also to probe effects predicted by alternative theories of gravity. They act as sources of gravitational wave (GW) emission, if they are in a compact orbit with a binary companion, but they may also act as detectors of low-frequency GWs in a so-called "pulsar timing array" (PTA) experiment, as we discuss next.

11. Tests of Theories of Gravity

The idea behind the usage of pulsars for testing general relativity (GR) and alternative theories of gravity is straightforward: if the pulsar is in orbit with a binary companion, we use the measured variation in the arrival times of the received signal to determine and trace the orbit of the pulsar about the common centre of mass as the former moves in the local curved space time and in the presence of spin effects. In alternative theories, self-gravity effects are often expected, modifying also the orbital motion to be observed.

This "pulsar timing" experiment is simultaneously clean, conceptually simple and very precise. The latter is true since when measuring the exact arrival time of pulses at our telescope on Earth, we do a ranging experiment that is vastly superior in precision than a simple measurement of Doppler-shifts in the pulse period. This is possible since the pulsed nature of our signal links tightly and directly to the rotation of the neutron star, allowing us to count every single rotation. Furthermore, in this experiment we can consider the pulsar as a test mass that has a precision clock attached to it.

While, strictly speaking, binary pulsars move in the weak gravitational field of a companion, they do provide precision tests of the (quasi-stationary) strong-field regime. This becomes clear when considering that the majority of alternative theories predicts strong self-field effects which would clearly affect the pulsars' orbital motion. Hence, tracing their fall in a gravitational potential, we can search for tiny deviations from GR, which can provide us with unique precision strong-field tests of gravity.

As a result, a wide range of relativistic effects can be observed, identified and studied. These are summarised in Table 2 in the form of limits on the parameters in the "Parameterised Post-Newtonian" (PPN) formalism (Ref. 135) and include

Table 1.   Selected aspects of fundamental physics studied with radio astronomical techniques compared to other methods. Note that some solar system tests have better numerical precision but are derived in weak gravitational field of the Solar System. In contrast, binary pulsar limits may sometimes be less constraining in precision, but they are derived for strongly self-gravitating bodies where deviations are expected to be larger. References are given for more information or further reading. For a general review see Will (2014), and for pulsar-related limits see Wex (2014).

Tested phenomena	Method	Radio astronomy	Ref.
Variation of fundamental constants:			
Fine structure constant $(e^2/\hbar c)$	Clock comparison, radio active decays, limit depending on redshift, $< \sim 10^{-16}$ yr^{-1}	Quasar spectra, $< 10^{-16}$ yr^{-1}	135
e-p mass ratio	Clock comparison, $< 3.3 \times 10^{-15}$ yr^{-1}	Quasar spectra, $< 3 \times 10^{-15}$ yr^{-1}	136,137
Gravitational constant, $\dot{G}/G$	Lunar Laser Ranging (LLR), $(-0.7 \pm 3.8) \times 10^{-13}$ yr^{-1}	Binary pulsars, $(-0.6 \pm 3.2) \times 10^{-12}$ yr^{-1}	138–140
Universality of free fall:	LLR, Nordvedt parameter, $|\eta_N| = (4.4 \pm 4.5) \times 10^{-4}$	Binary Pulsars, $\Delta < 5.6 \times 10^{-3}$	135,140,141
Universal preferred frame for gravity:		see Table 2	
PPN parameters and related phenomena:		see Table 2	
Gravitational wave properties:		Binary pulsars	140
Verification of GR's quadrupole formula		Double Pulsar, $< 3 \times 10^{-4}$	142
Constraints on dipolar radiation		PSR-WD systems, $(\alpha_A - \alpha_B)^2 < 4 \times 10^{-6}$	139,143
Geodetic precession	Gravity Probe B, 0.3%	PSR B1913+16; Double Pulsar, 13%; PSR B1534+12, 17%	144–146
Equation-of-State	e.g. thermal emission from X-ray binaries	fast spinning pulsars; massive neutron stars	17,143,147–149
Cosmology	e.g. Supernova distances	CMB	this conference

Table 2. Best limits for the parameters in the PPN formalism. Note that 6 of the 9 independent PPN parameters are best constrained by radio astronomical techniques. Five of them are derived from pulsar observations. Adapted from Will (2014) but see also Wex (2014) for details.

Par.	Meaning	Method	Limit	Remark/Ref.
$\gamma - 1$	How much space-curvature produced by unit rest mass?	time delay	2.3×10^{-5}	Cassini tracking/135
		light deflection	2×10^{-4}	VLBI/135
$\beta - 1$	How much "non-linearity" in the superposition law for gravity?	perihelion shift	8×10^{-5}	using $J_{2\odot} = (2.2 \pm 0.1) \times 10^{-7}$/135
		Nordtvedt effect	2.3×10^{-4}	$\eta_N = 4\beta - \gamma - 3$ assumed/135
ξ	Preferred-location effects?	spin precession	4×10^{-9}	Isolated MSPs/150
α_1	Preferred-frame effects?	orbital polarisation	4×10^{-5}	PSR-WD, PSR J1738+0333/151
α_2		spin precession	2×10^{-9}	Using isolated MSPs/152
α_3		orbital polarisation	4×10^{-20}	Using ensemble of MSPs/141
ζ_1	Violation of conservation of total momentum?	Combining PPN bounds	2×10^{-2}	135
ζ_2		binary acceleration	4×10^{-5}	Using $\ddot{P}$ for PSR B1913+16/135
ζ_3		Newton's 3rd law	10^{-8}	lunar acceleration/135
ζ_4		not independent parameter		$6\zeta_4 = 3\alpha_3 + 2\zeta_1 - 3\zeta_3$

concepts and principles deeply embedded in theoretical frameworks. If a specific alternative theory is developed sufficiently well, one can also use radio pulsars to test the consistency of this theory. Table 3 lists a number of theories where this has been possible. Sometimes, however, gravitational theories are put forward to explain certain observational phenomena without having studied the consequences of these theories in other areas of parameter space. In particular, alternative theories of gravity are sometimes proposed without having worked out their radiative properties, while in fact, tests for gravitational radiation provide a very powerful and sensitive probe for the consistency of the theory with observational data. In other words, every successful theory has to pass the binary pulsar experiments.

Table 3. Constraining specific (classes of) gravity theories using radio pulsars. See text and also Wex (2014) for more details.

Theory (class)	Method	Ref.
Scalar-tensor gravity:		
Jordan-Fierz-Brans-Dicke	limits by PSR J1738+0333 and PSR J0348+0432, comparable to best Solar system test (Cassini)	139 Freire priv. comm.
Quadratic scalar-tensor gravity	for $\beta_0 < -3$ and $\beta_0 > 0$ best limits from PSR-WD systems, in particular PSR J1738+0333 and PSR J0348+0432	139 Krieger et al. in prep., Freire priv. comm.
Massive Brans-Dicke	for $m_\varphi \sim 10^{-16}$ eV: PSR J1141−6545	153
Vector-tensor gravity:		
Einstein-Æther	combination of pulsars (PSR J1141−6545, PSR J0348+0432, PSR J0737−3039, PSR J1738+0333)	154
Hořava gravity	combination of pulsars (see above)	154
TeVeS and TeVeS-like theories:		
Bekenstein's TeVeS	excluded using Double Pulsar	142
TeVeS-like theories	excluded using PSR 1738+0333	139

The various effects or concepts to be tested require sometimes rather different types of laboratory. For instance, in order to test the important radiative properties of a theory, we need compact systems, usually consisting of a pair of neutron stars. As we have seen, double neutron star systems (DNSs) are rare but they usually produce the largest observable relativistic effects in their orbital motion and, as we will see, produce the best tests of GR for strongly self-gravitating bodies. On the other hand, to test the violation of the Strong Equivalence Principle, one would like to use a binary system that consists of different types of masses (i.e. with different gravitational self-energy), rather than a system made of very similar bodies, so that we can observe how the different masses fall in the gravitational potential of the companion and of the Milky Way. For this application, a pulsar-black hole system would be ideal. Unfortunately, despite past and ongoing efforts, we have not yet found a pulsar orbiting a stellar black hole companion or orbiting the supermassive black hole in the centre of our Galaxy.[155] Fortunately, we can use pulsar-white dwarf (PSR-WD) systems, as white dwarfs and neutron stars differ very significantly in their structure and, consequently, self-energies. Furthermore, some PSR-WD systems can also be found in relativistic orbits.[9,156]

12. The First Binary Pulsar – A Novel Gravity Laboratory

The first binary pulsar to ever be discovered happened to be a rare double neutron star system. It was discovered by Russel Hulse and Joe Taylor in 1974 (Ref. 157). The pulsar, B1913+16, has a period of 59 ms and is in an eccentric ($e = 0.62$) orbit around an unseen companion with an orbital period of less than 8 hours. Soon after the discovery, Taylor and Hulse noticed that the pulsar does not follow the movement expected from a simple Keplerian description of the binary orbit, but that it shows the impact of relativistic effects. In order to describe the relativistic effects in a theory-independent fashion, one introduces so-called "Post-Keplerian" (PK) parameters that are included in a timing model to describe accurately the measured pulse times-of-arrival (see e.g. Ref. 158 for more details).

For the Hulse-Taylor pulsar, a relativistic advance of its periastron was soon measured analogous to what is seen in the solar system for Mercury, albeit with a much larger amplitude. The value measured today, $\dot{\omega} = 4.226598 \pm 0.000005$ deg/yr,[159] is much more precise than was originally measured, but even early on the precision was sufficient to permit meaningful comparisons with GR's prediction. The value depends on the Keplerian parameters and the masses of the pulsar and its companion:

$$\dot{\omega} = 3T_\odot^{2/3} \left(\frac{P_b}{2\pi} \right)^{-5/3} \frac{1}{1 - e^2} (m_p + m_c)^{2/3}. \tag{4}$$

Here, $T_\odot = GM_\odot/c^3 = 4.925490947\mu s$ is a constant, P_b the orbital period, e the eccentricity, and m_p and m_c the masses of the pulsar and its companion, respectively. See Ref. 158 for further details.

The Hulse-Taylor pulsar also shows the effects of gravitational redshift (including a contribution from a second-order Doppler effect) as the pulsar moves in its elliptical orbit at varying distances from the companion and with varying speeds. The result is a variation in the clock rate with an amplitude of $\gamma = 4.2992 \pm 0.0008$ ms (Ref. 159). In GR, the observed value is related to the Keplerian parameters and the masses as

$$\gamma = T_\odot^{2/3} \left(\frac{P_b}{2\pi} \right)^{1/3} e \frac{m_c(m_p + 2m_c)}{(m_p + m_c)^{4/3}}. \tag{5}$$

We can now combine these measurements. We have two equations with a measured left-hand side. On the right-hand side, we measured everything apart from two unknown masses. We solve for those and obtain, $m_p = 1.4398 \pm 0.0002 \, M_\odot$ and $m_c = 1.3886 \pm 0.0002 \, M_\odot$.[159] These masses are correct if GR is the right theory of gravity. If that is indeed the case, we can make use of the fact that (for point masses with negligible spin contributions), the PK parameters in each theory should only be functions of the *a priori* unknown masses of pulsar and companion, m_p and m_c, and the easily measurable Keplerian parameters (Ref. 160).[g] With the two masses now being determined using GR, we can compare any observed value of a third PK parameter with the predicted value. A third such parameter is the observed decay of the orbit which can be explained fully by the emission of gravitational waves. And indeed, using the derived masses, along with the prediction of GR, i.e.

$$\dot{P_{\rm b}} = -\frac{192\pi}{5}T_\odot^{5/3}\left(\frac{P_{\rm b}}{2\pi}\right)^{-5/3}\frac{\left(1+\frac{73}{24}e^2+\frac{37}{96}e^4\right)}{(1-e^2)^{7/2}}\frac{m_{\rm p}m_{\rm c}}{(m_{\rm p}+m_{\rm c})^{1/3}}, \qquad (6)$$

one finds an agreement with the observed value of $\dot{P}_{\rm b}^{\rm obs} = (2.423 \pm 0.001) \times 10^{-12}$ (Ref. 159) – however, only if a correction for a relative acceleration between the pulsar and the solar system barycentre is taken into account. As the pulsar is located about 7 kpc away from Earth, it experiences a different acceleration in the Galactic gravitational potential than does the solar system (see e.g. Ref. 158). The precision of our knowledge to correct for this effect eventually limits our ability to compare the GR prediction to the observed value. Nevertheless, the agreement of observations and prediction, today within a 0.2% (systematic) uncertainty,[159] represented the first evidence for the existence of gravitational waves. Today we know many more binary pulsars in which we can detect the effects of gravitational wave emission. In one particular case, the measurement uncertainties are not only more precise, but also the systematic uncertainties are much smaller, as the system is much more nearby. This system is the Double Pulsar.

13. The Double Pulsar

The Double Pulsar was discovered in 2003.[161,162] It not only shows larger relativistic effects and is much closer to Earth (about 1 kpc) than the Hulse-Taylor pulsar, allowing us to largely neglect the relative acceleration effects, but the defining unique property of the system is that it does not consist of one active pulsar and its *unseen* companion, but that it harbours two *active* radio pulsars.

One pulsar is mildly recycled with a period of 23 ms (named "A"), while the other pulsar is young with a period of 2.8 s (named "B"). Both orbit the common centre of mass in only 147-min with orbital velocities of 1 Million km per hour. Being also mildly eccentric ($e = 0.09$), the system is an ideal laboratory to study gravitational physics and fundamental physics in general. A detailed account of the exploitation for gravitational physics has been given, for instance, by Refs. 163–165. An update on those results is in preparation,[142] with the largest improvement undoubtedly given by a large increase in precision when measuring the orbital decay. Not even ten years after the discovery of the system, the Double Pulsar provides the best test for the accuracy of the gravitational quadrupole emission prediction by GR far below the 0.1% level.

---

[g]For alternative theories of gravity this statement may only be true for a given equation-of-state.

In order to perform this test, we first determine the mass ratio of pulsar A and B from their relative sizes of the orbit, i.e. $R = x_B/x_A = m_A/m_B = 1.0714 \pm 0.0011$.[163] Note that this value is theory-independent to the 1PN level.[166] The most precise PK parameter that can be measured is a large orbital precession, i.e. $\dot{\omega} = 16.8991 \pm 0.0001$ deg/yr. Using Eq. 4, this measured value and the mass ratio, we can determine the masses of the pulsars, assuming GR is correct, to be $m_A = (1.3381 \pm 0.0007) \, M_{\odot}$ and $m_B = (1.2489 \pm 0.0007) \, M_{\odot}$. The masses are shown, together with others determined by this and other methods, in Figure 7.

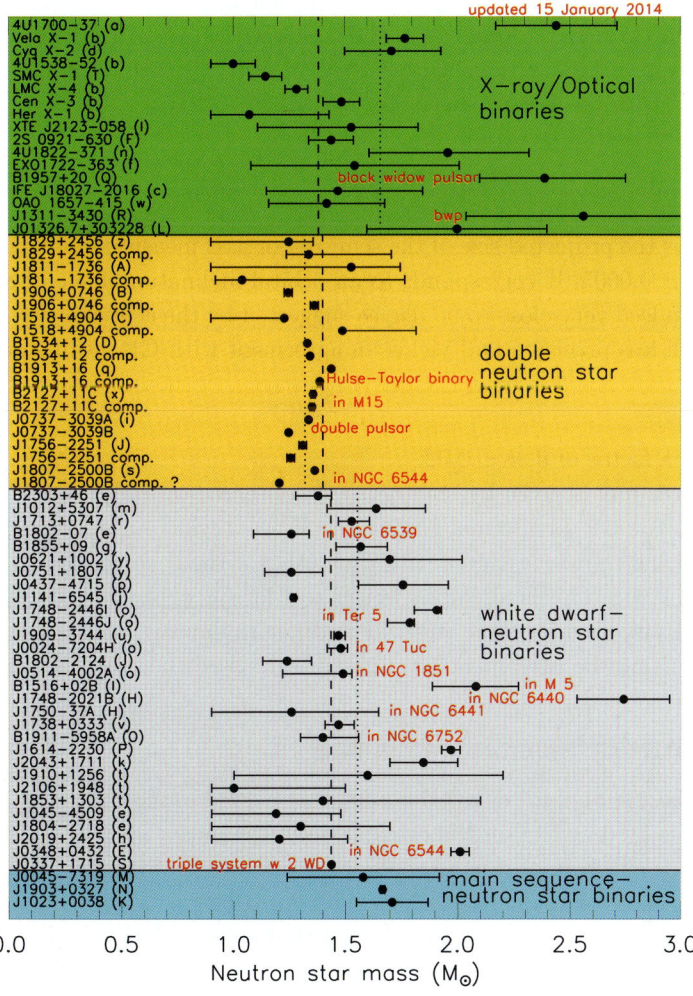

Fig. 7. Neutron star mass measurements compiled by J. Lattimer. *Annu. Rev. Nucl. Part. Sci.* 62, 485 (2012), and available at www.stellarcollapse.org.

We can use these masses to compute the expected amplitude for the gravitational redshift, γ, if GR is correct. Comparing the result with the observed value of $\gamma = 383.9 \pm 0.6$ μs, we find that theory (GR) agrees with the observed value to a ratio of 1.000 ± 0.002, as a first of five tests of GR in the Double Pulsar.

The Double Pulsar also has the interesting feature that the orbit is seen nearly exactly edge-on. This leads to a 30-s long eclipse of pulsar A due to the blocking magnetosphere of B that we discuss further below, but it also leads to a "Shapiro delay": whenever the pulse needs to propagate through curved space-time, it takes a little longer than travelling through flat space-time. At superior conjunction, when the signal of pulsar A passes the surface of B in only 20,000 km distance, the extra path length due to the curvature of space-time around B leads to an extra time delay of about 100 μs. The shape and amplitude of the corresponding Shapiro delay curve yield two PK parameters, s and r, known as *shape* and *range*, allowing two further tests of GR. s is measured to $s = \sin(i) = 0.99975 \pm 0.00009$ and is in agreement with the GR prediction of

$$s = T_\odot^{-1/3} \left(\frac{P_\mathrm{b}}{2\pi} \right)^{-2/3} x \frac{(m_A + m_B)^{2/3}}{m_B}, \tag{7}$$

(where x is the projected size of the semi-major axis measured in lt-s) within a ratio of 1.0000 ± 0.0005. It corresponds to an orbital inclination angle of 88.7 ± 0.2 deg, which is indeed very close to 90 deg as suggested by the eclipses. r can be measured with much less precision and yields an agreement with GR's value given by

$$r = T_\odot m_B, \tag{8}$$

to within a factor of 0.98 ± 0.02.

A fourth test is given by comparing an observed orbital decay of 107.79 ± 0.11 ns/day to the GR prediction. Unlike the Hulse-Taylor pulsar, extrinsic effects are negligible and the values agree with each other without correction to within a ratio of 1.000 ± 0.001. This is already a better test for the existence of GW than possible with the Hulse-Taylor pulsar and will continue to improve with time. Indeed, at the time of writing the agreement has already surpassed the 0.03% level.[142]

14. Relativistic Spin-orbit Coupling

Apart from the Shapiro-delay, the impact of curved space-time is also immediately measurable by its effect on the orientation of the pulsar spin in a gyroscope experiment. This effect, known as geodetic precession or de Sitter precession represents the effect on a vector carried along with an orbiting body such that the vector points in a different direction from its starting point (relative to a distant observer) after a full orbit goes around the central object. Experimental verification has been achieved by precision tests in the solar system, e.g. by Lunar Laser Ranging (LLR) measurements, or recently by measurements with the Gravity Probe-B satellite mission (see Table 1). However, these tests are done in the weak field conditions of the solar system. Thus Pulsars currently provide the only access beyond weak-field, i.e. the quasi-stationary strong-field regime.

In binary systems one can interpret the observations, depending on the reference frame, as a mixture of different contributions to relativistic spin-orbit interaction. One contribution comes from the motion of the first body around the centre of mass of the system (de Sitter-Fokker precession), while the other comes from the dragging of the internal frame at the first body due to the translational motion of the companion.[167] Hence, even though we loosely talk about geodetic precession, the result of the spin-orbit coupling for binary pulsars is more general, and hence we will call it *relativistic spin-precession*. The consequence of relativistic spin-precession is a precession of the pulsar spin about the total angular moment vector, changing the orientation of the pulsar relative to Earth.

Since the orbital angular momentum is much larger than the spin of the pulsar, the orbital angular momentum practically represents a fixed direction in space, defined by the orbital plane of the binary system. Therefore, if the spin vector of the pulsar is misaligned with the orbital spin, relativistic spin-precession leads to a change in viewing geometry, as the pulsar spin precesses about the total angular momentum vector. Consequently, as many of the observed pulsar properties are determined by the relative orientation of the pulsar axes towards the distant observer on Earth, we should expect a modulation in the measured pulse profile properties, namely its shape and polarisation characteristics.[168] The precession rate is another PK parameter and given in GR by (e.g. Ref. 158)

$$\Omega_{\mathrm{p}} = T_{\odot}^{2/3} \left(\frac{2\pi}{P_{\mathrm{b}}} \right)^{5/3} \frac{m_{\mathrm{c}}(4m_{\mathrm{p}} + 3m_{\mathrm{c}})}{2(m_{\mathrm{p}} + m_{\mathrm{c}})^{4/3}} \frac{1}{1 - e^2}. \tag{9}$$

In order to see a measurable effect in any binary pulsar, (a) the spin axis of the pulsar needs to be misaligned with the total angular momentum vector and (b) the precession rate must be sufficiently large compared to the available observing time to detect a change in the emission properties. Considering these conditions, relativistic spin precession has now been detected in *all* systems where we can realistically expect this.

As the most relativistic binary system known to date, we expect a large amount of spin precession in the Double Pulsar system. Despite careful studies, profile changes for A have not been detected, suggesting that A's misalignment angle is less than a few degrees.[169] In contrast, changes in the light curve and pulse shape on secular timescales[170] reveal that this is not the case for B. In fact, B had been becoming progressively weaker and disappeared from our view in 2009.[171] Making the assumption that this disappearance is solely caused by relativistic spin precession, it will only be out of sight temporarily until it reappears later. Modelling suggests that, depending on the beam shape, this will occur in about 2035 but an earlier time cannot be excluded.[171] The geometry that is derived from this modelling is consistent with the results from complementary observations of spin precession, visible via a rather unexpected effect described in the following.

The change on the orientation of B also changes the observed eclipse pattern in the Double Pulsar, where we can see periodic bursts of emission of A during the dark eclipse phases, with the period being the full- or half-period of B. As this pattern is caused by the rotation of B's blocking magnetospheric torus that allows light to pass B when the torus rotates to be seen from the side, the resulting pattern is determined by the three-dimensional orientation of the torus, which is centred on the precessing pulsar spin. Eclipse monitoring over the course of several years shows exactly the expected changes, allowing a determination of the precession rate to $\Omega_{p,B} = 4.77^{+0.66}_{-0.65}$ deg/yr. This value is fully consistent with the value expected in GR, providing a fifth test.[146] This measurement also allows us to test alternative theories of gravity and their prediction for relativistic spin-precession in strongly self-gravitating bodies for the first time (see Ref. 165 for details).

15. Alternative Theories of Gravity

Despite the successes of GR, a range of observational data has fuelled the continuous development of alternative theories of gravity. Such data include the apparent observation of "dark matter" or the cosmological results interpreted in the form of "inflation" and "dark energy," as also discussed at this conference. Confronting alternative theories with data also in other areas of the parameter space (away from the CMB or Galactic scales), requires that these theories are developed sufficiently in order to make predictions. As mentioned, a particularly sensitive criterion is if the theory is able to make a statement about the existence and type of gravitational waves emitted by binary pulsars. Most theories cannot do this (yet), but a class of theories where this has been achieved is the class of tensor-scalar theories as discussed and demonstrated by Damour and Esposito-Farèse in a series of works (e.g. Ref. 172). For corresponding tests, the choice of a double neutron star system is not ideal, as the difference in scalar coupling, (that would be relevant, for instance, for the emission of gravitational *dipole* radiation) is small. The ideal laboratory would be a pulsar orbiting a black hole, as the black hole would have zero scalar charge. The next best laboratory is a pulsar-white dwarf system. Indeed, such binary systems are able to provide constraints for alternative theories of gravity that are equally good or even better than solar system limits.[139]

The previously best example for such a system was presented by Ref. 139, who reported the results of a 10-year timing campaign on PSR J1738+0333, a 5.85-ms pulsar in a practically circular 8.5-h orbit with a low-mass white dwarf companion. A large number of precision pulse time-of-arrival measurements allowed the determination of the intrinsic orbital decay due to gravitational wave emission. The agreement of the observed value with the prediction of GR introduces a tight upper limit on dipolar gravitational wave emission, which can be used to derive the most stringent constraints ever on general scalar-tensor theories of gravity. The new bounds are more stringent than the best current Solar system limits over most of the parameter space, and constrain the matter-scalar coupling constant α_0^2 to be below the 10^{-5} level. For the special case of the Jordan-Fierz-Brans-Dicke theory, the authors obtain a one-sigma bound of $\alpha_0^2 < 2 \times 10^{-5}$, which is within a factor of two of the Solar-System Cassini limit.[139,173] Moreover, their limit on dipolar gravitational wave emission can also be used to constrain a wide class of theories of gravity which are based on a generalisation of Bekenstein's Tensor-Vector-Scalar (TeVeS) gravity, a relativistic formulation of Modified Newtonian Dynamics (MOND).[174] They find that in order to be consistent with the results for PSR J1738+0333, these TeVeS-like theories have to be fine-tuned significantly (see Table 3). We expect the latest Double Pulsar results to close a final gap of parameter spaced left open by the PSR-WD systems.[139,142]

A recently studied pulsar-white dwarf system[9,143] turned out to be a very exciting laboratory for various aspects of fundamental physics: PSR J0348+0432 harbours a white dwarf whose composition and orbital motion can be precisely derived from optical observations. The results allow us to measure the mass of the neutron star, showing that it has a record-breaking value of $2.01 \pm 0.04 M_\odot$![143] This is not only the most massive neutron star known (at least with reliable precision), providing important constraints on the "equation-of-state" (see below) but the 39-ms pulsar and the white dwarf orbit each other in only 2.46 hours, i.e. the orbit is only 15 seconds longer than that of the Double Pulsar. Even though the orbital motion is nearly circular, the effect of gravitational wave damping is clearly measured. Hence, the high pulsar mass and the compact orbit make this system a sensitive laboratory of a previously untested strong-field gravity regime. Thus far, the observed orbital decay agrees with GR, supporting its validity even for the extreme conditions present in the system.[143] The precision of the observed agreement is already sufficient to add significant confidence to the usage of GR templates in the data analysis for gravitational wave (GW) detectors.

16. Pulsars as Gravitational Wave Detectors

The observed orbital decay in binary pulsars detected via precision timing experiments so far offers the best evidence for the existence of gravitational wave (GW) emission. Intensive efforts are therefore ongoing world-wide to make a direct detection of gravitational waves that pass over the Earth. Ground-based detectors like GEO600, VIRGO or LIGO use massive mirrors, the relative separations of which are measured by a laser interferometer set-up, while the envisioned space-based LISA detector uses formation flying of three test-masses that are housed in satellites. For a summary of these efforts, see, e.g. Ref. 135.

The change of the space-time metric around the Earth also influences the arrival times of pulsar signals measured at the telescope. Therefore, pulsars do not only act as sources of GWs, but they may eventually also lead to their direct detection. Fundamentally, the GW frequency range that pulsar timing is sensitive to, is bound by the cadence of the timing observations on the high frequency side, and by the length of the data set on low-frequency part. Hence, typically GWs with periods of the order of one year or more could be detected. Since GWs are expected to produce a characteristic quadrupole signature on the sky, the timing residuals from various pulsars should be correlated correspondingly,[175] so that the comparative timing of several pulsars can be used to make a detection. The sensitivity of such a "Pulsar Timing Array" (PTA) increases with the number of pulsars and should be able to detect gravitational waves in the nHz regime, hence below the frequencies to which LIGO (~kHz and higher) and LISA (~mHz) are sensitive. Sources in the nHz range (see, e.g. Ref. 176) include astrophysical objects (i.e. super-massive black hole binaries resulting from galaxy mergers in the early Universe), cosmological sources (e.g. the vibration of cosmic strings), and transient phenomena (e.g. phase transitions).

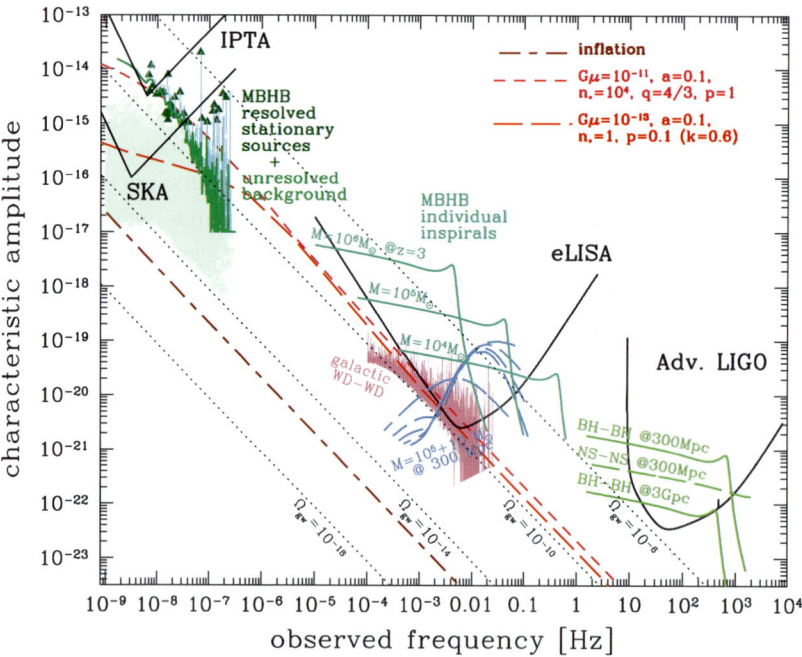

Fig. 8.   The gravitational wave spectrum with expected sources. Shown is the characteristic amplitude vs. frequency as presented by Janssen et al. (2015): In the nHz regime, individually resolvable systems and the level of the unresolved background are indicated. Nominal sensitivity levels for the IPTA and SKA are also shown. In the mHz frequency range, the eLISA sensitivity curve is shown together with typical circular SMBHB inspirals at z=3 (pale blue), the overall signal from Galactic WD-WD binaries (yellow) and an example of extreme mass ratio inspiral (aquamarine). In the kHz range an advanced LIGO curve is shown together with selected compact object inspirals (purple). The brown, red and orange lines running through the whole frequency range are expected cosmological backgrounds from standard inflation and selected string models, as labeled in figure.

A number of PTA experiments are ongoing, namely in Australia, Europe and North America (see Ref. 177 for a summary). The currently derived upper limits on a stochastic GW background (e.g. Refs. 178) are very close to the theoretical expectation for a signal that originates from binary supermassive black holes expected from the hierarchical galaxy evolution model.[179,180]

But the science that can eventually be done with the PTAs goes far beyond simple GW detection – a whole realm of astronomy and fundamental physics studies will become possible. The dominant signal in the nHz regime is expected to be a stochastic background due to merging supermassive black holes and many constraints can be placed on this source population, including their frequency in cosmic history, the relation between the black holes and their hosts, and their coupling with the stellar and gaseous environments.[181,182] Detection of gas disks surrounding merging supermassive black holes and related eccentricities in such systems is possible;[183,184] PTAs should be able to constrain the solution to the famous 'last parsec problem'.[185] In addition to detecting a *background* of GW emission, PTAs can detect *single* GW sources. We can, for instance, expect to detect anisotropies in a GW background, due to the signals of single nearby supermassive black hole binaries.[186,187] Considering the case when the orbit is effectively not evolving over the observing span, we can show that, by using information provided by the "pulsar term" (i.e. the retarded effect of the GW acting on the pulsar's surrounding spacetime), we may be able to achieve interesting (~ 1 arcmin) source localisation.[186] Even astrophysical measurements of more local relevance can be done with PTAs; for example an independent determination of the masses of the Jovian planetary system has already been made (Ref. 188) and additional future, improved measurements for Jupiter and other planets should be possible. On the fundamental physics side, departures from GR during supermassive black hole mergers should be measurable via different angular dependences of pulsar timing residuals on the sky such as for example from gravitational wave polarization properties that differ from those predicted by GR.[189–191] It may even be possible to constrain the mass of the graviton from the angular correlation of pulsar timing residuals.[192] If the ongoing PTA experiments do not detect GWs in the next few years, a first detection is virtually guaranteed with the more sensitive Phase I of the Square Kilometer Array.[176] With even further increased sensitivity of SKA Phase II, it should also be possible to study the fundamental properties of gravitational waves.

17. Black Holes or the Centre of the Galaxy as a Gravity Lab

What makes a binary pulsar with a black-hole companion so interesting is that it has the potential of providing a superb new probe of relativistic gravity. As pointed out by Ref. 193, the discriminating power of this probe might supersede all its present and foreseeable competitors. The reason lies in the fact that such a system would clearly expose the self-field effects of the body orbiting the black hole, hence making it an excellent probe for alternative theories of gravity.

But also for testing the black hole properties predicted by GR, a pulsar-BH system will be a superb laboratory. Ref. 194 was the first to provide a detailed recipe for how to exploit a pulsar-black hole system. They showed that the measurement of spin-orbit coupling in a pulsar-BH binary in principle allows us to determine the spin and the quadrupole moment of the black hole. This could test the "cosmic censorship conjecture" and the "no-hair theorem". While Ref. 194 showed that with current telescopes such an experiment would be almost impossible to perform (with the possible exception of pulsars about the Galactic centre black hole), Ref. 195 pointed out that the SKA sensitivity should be sufficient. Indeed, this experiment benefits from the SKA sensitivity in multiple ways. It provides the required timing precision while also enabling deep searches, enabling a Galactic Census which should eventually deliver the desired sample of pulsars with a BH companion. As shown recently,[196] with the SKA or the *Five-hundred-meter Aperture Spherical radio Telescope* (FAST) project[197] one could test the cosmic censorship conjecture by measuring the spin of a stellar black hole, though it is still unlikely to find a system that can enable the measurement of the quadrupole moment.

As the effects become easier to measure with more massive black holes, the best laboratory would be a pulsar orbiting the central black hole in the Milky Way, Sgr A*.[194,195,198] Indeed, Ref. 155 continued the work of Ref. 194 and studied this possibility in detail. They showed that it should be "fairly easy" to measure the spin of the GC black hole with a precision of $10^{-4} - 10^{-3}$. Even for a pulsar with a timing precision of only 100 μs, characteristic periodic residuals would enable tests of the no-hair theorem with a precision of one percent or better!

17.1. *Pulsars in the Galactic Centre*

Unfortunately, searches for pulsars near Sgr A* have been unsuccessful for the last 30 years - until April 2013. As described in Section 5.1, a radio signal of the 3.7-s magnetar J1745−2900 was detected.[101,104] The source has the highest dispersion measure of any known pulsar, is highly polarised and has a rotation measure that is larger than that of any other source in the Galaxy, apart from Sgr A*. This, and the fact that VLBI images of the magnetar show scattering identical to that in the radio image of Sgr A* itself,[199] support the idea that the source is indeed only ~ 0.1 pc away from the central black hole. Initially, measurements of the scatter-broadening of the single radio pulses[200] suggested that scattering due to the inner interstellar medium is too small to explain the lack of pulsar detection in previous survey. Recent preliminary results, enabled by the puzzling fact that the radio emission remains unabated in spite of significant source fading in the X-ray band,[201] show an increase of scattering, indicating that the conditions are instead highly changeable (Spitler et al. in prep.). The fact that a rare object like a radio-emitting magnetar is found in such proximity to Sgr A* suggests that estimates like that of Ref. 202, predicting as many as 1000 pulsars in the inner central parsec, may indeed be true. Further searches are ongoing but may require observations at very high (i.e. ALMA) frequencies, i.e. > 40 GHz to beat the extreme scattering, which decreases as $\sim \nu^{-4}$.

17.2. *The Event Horizon Telescope & BlackHoleCam*

Telescopes operating at high radio frequencies may not only allow us to find a pulsar in the Galactic Centre, but combined with other radio telescopes, they can also form an interferometer to take an image of Sgr A* that can resolve the "shadow" of the supermassive black hole in the centre of our Milky Way. With a mass of about $4.3 \times 10^6 M_\odot$[203,204] it is not very large in size compared to those in the centres of other galaxies, but it is the closest. The image to be taken by the so-called "Event Horizon Telescope" and "BlackHoleCam" experiments (see e.g. Ref. 205 for a recent review) will depend on the magnitude and direction of Sgr A*'s spin, i.e. information available by the discovery of pulsar around the central black hole, as described above. Combined measurements probe simultaneously the near- and far-field of Sgr A*, promising a unique probe of gravity.

18. Physics at Extreme Densities

The density of pulsars and neutron stars is so large that their matter cannot be reproduced in terrestrial observatories. Therefore, in order to understand how matter behaves under very extreme condition, observations of pulsars provide unique insight. On one hand, mass measurements constrain the Equation-of-State (EOS) at the highest densities, which also affects the maximum possible spin frequency of pulsars[17] and sets bounds to the highest possible density of cold matter (see contribution by J. Lattimer). Because a given EOS describes a specific mass-radius relationship (see, e.g. Ref. 206), measurements of the radii of neutron stars also set constrains on the EOS near nuclear saturation density and yield information about the density dependence of the nuclear symmetry energy.[149,207] In practice, mass measurements are easier to achieve than radius measurements – or the discovery of sub-millisecond pulsars with significantly faster spin-periods than currently known.[17] Specifically, while there are about 40 neutron star masses known with varying accuracy (see Figure 7), there are no precise simultaneous measurements of mass and radius for any neutron star.

For now, some of the best constraints for the EOS come simply from the maximum observed neutron star mass. Unlike in Newtonian physics, in GR a maximum mass exists as for any causal EOS as the isothermal speed of sound must never exceed the speed of light. Currently, the largest masses are measured for PSR J1614+2230 with $M = 1.94 \pm 0.04 M_\odot$[148] and PSR J0348+0432 with $M = 2.01 \pm 0.04 M_\odot$.[143] These independent measurements confirm the existence of high-mass neutron stars, ruling out a number of soft EOS already (see Figure 9). However, as explained, for instance, in Ref. 207, this lower limit on the maximum mass also provides constraints on the EOS at lower densities and on the radii of intermediate mass neutron stars. In general, however, most radii estimates come from estimates inferred from photospheric radius expansion bursts and thermal X-ray emission from neutron star surfaces. A Bayesian analysis of the existing data suggests a radius range of 11.3–12.1 km for a $1.4 M_\odot$ neutron star.[149,207]

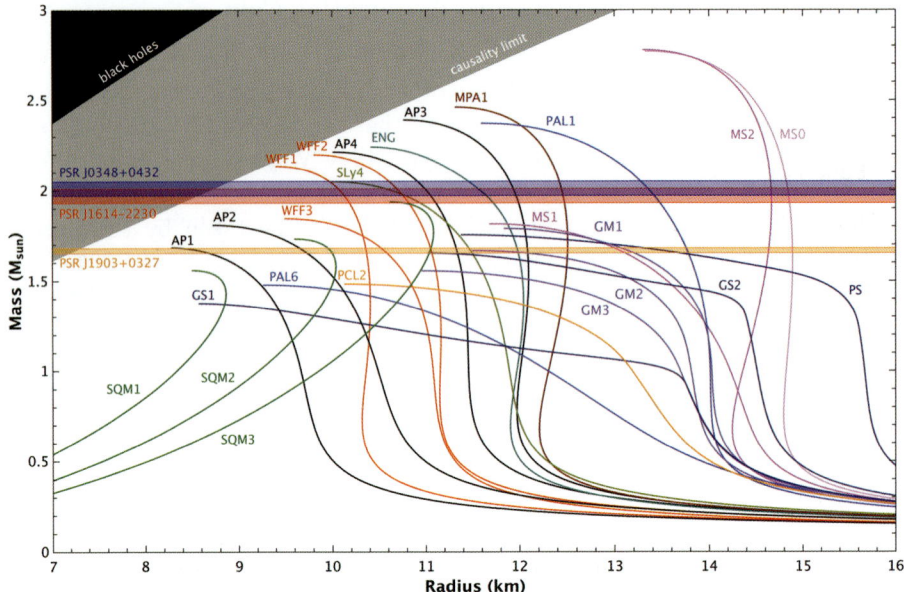

Fig. 9.   Constraints on the equation-of-state provided by mass measurements of the most massive neutron stars. Figure provided by N. Wex. For details see e.g. Demorest et al. (2010).

In terms of information about fundamental properties of super-dense matter, the *maximum* mass of neutron stars is clearly important. Small mass measurements, in particular those below $1.20 M_\odot$, are nevertheless extremely interesting from a neutron-star formation point-of-view as they would call into question the gravitational-collapse formation scenario.[207] One way to form such low-mass neutron stars is through electron-capture supernovae. Here, a white dwarf with an oxygen-neon-magnesium (O-Ne-Mg) core collapses to a low-mass neutron star due to electron captures on Ne and/or Mg, as was proposed for the formation of the light companion in the Double Pulsar system, PSR J0737−3039B.[208] It was suggested that electron capture could be triggered in particular in close binaries. Assuming minimal mass loss, the final mass should be determined by the mass of the progenitor star minus the binding energy. As for any given EOS one can calculate the relation between the gravitational mass and the baryonic mass, one can in principle use the observed mass and the small mass range expected for an e-capture progenitor ($M_0 \sim 1.366 - 1.375 M_\odot$) to constrain the EOS.[208] However, alternative ways of producing such light neutron stars, e.g. via ultra-stripped Type Ic Supernovae from close binary evolution,[209] have been proposed also.

The Double Pulsar may also allow us to actually measure the moment-of-inertia of a neutron star. As this combines the mass and the radius of a neutron star in one observable directly, such a measurement would be very significant in determining the correct EOS.[162,210] Indeed, a measurement of the moment-of-inertia of pulsar A in the Double Pulsar, even with moderate accuracy ($\sim 10\%$), would provide important constraints.[210–212] Recent timing results revealing 2PN-effects at the required level give hope that this goal can be reached eventually.[142] See Ref. 165 for a detailed review on the prospects for making such measurement.

19. Fast Radio Bursts, Revisited

In Part I we introduced a new type of transient radio sources now known as Fast Radio Bursts (FRBs). In the context of fundamental physics, we are interested in exploring their nature on the one hand, and their usage probes on the other. As their origin is still unclear, we will only attempt to give an overview of the existing, fast growing literature. We start with looking at the origin of FRBs.

All FRBs detected follow a perfect ν^{-2}-dispersion law, as it is expected from signal propagation in a cold ionized medium. In the discussion, whether the signals are Galactic or extra-galactic, Ref. 213 proposed FRBs may actually be Galactic flare stars wherein the large dispersion measure is due to dense plasma in low-mass star atmospheres, rather than a demonstration of a large distance traversed. However Refs. 214–216 reject this Galactic model using radiation transfer arguments; e.g. such high plasma densities should produce enormous intrinsic absorption that should render them undetectable, or produce free-free emission that is not seen, or result in a break-down of the cold plasma dispersion law, which contradicts observations. Moreover, a number of FRBs also show signs for interstellar scattering. Where it has been possible to measure (e.g. Ref. 131; see Figure 6), the frequency dependence of the scattering time follows a ν^{-4}-law, as expected for propagation in interstellar and intergalactic space. With the dispersion measure (typically vastly) exceeding the contribution expected from the Milky Way, an extra-galactic origin is the most likely explanation, with distances corresponding to redshifts of the order of $z \sim 1$ as inferred from an estimate of the intergalactic free electron content.[131]

From the combination of temporal brevity and great luminosity (inferred from their large distances), we then immediately infer that the sources must embody a physically extreme environment, likely involving very high gravitational or magnetic fields. Possibilities being discussed include interacting magnetospheres of coalescing neutron stars, coalescing white dwarfs, evaporating black holes, supernovae, and super-giant pulses (see Refs. 131,217 and references therein). More exotic models propose signals from (bare) strange stars,[218] white holes,[219] or super-conducting cosmic strings.[220] FRB emission must almost certainly be from a coherent process as the implied brightness temperature for a thermal process is impossibly large given the small size implied by the short durations; considering less exotic models, one would there expect that FRBs originate from some sort of compact object – white dwarf, neutron star or black hole. One possibility that appears particularly appealing based on expected event rates is giant magnetar flares.[216,221]

Whatever FRBs turn out to be, as extragalactic transient signals, they promise to be very useful cosmological probes. For example, their dispersion measure enables us to account for the ionized baryons between us and the FRB sources and to measure the curvature of spacetime through which the radiation propagates. A number of recent publications discuss these possibilities, many of which are very well summarized in Ref. 222. Generally, they fall in three categories, i.e. FRBs as locators of the "missing" baryons in the low ($z \leq 2$) redshift universe, high-redshift cosmic rulers which have the potential to determine the equation-of-state parameter w over a large fraction of cosmic history, or potential probes of primordial (intergalactic) magnetic fields and turbulence. See Ref. 222 for more details.

20. Summary and Conclusions

As we hope we have shown in this review, the field of neutron star research, and in particular radio pulsars, is extremely active, and addresses a very broad diversity of physical and astrophysical questions. These range from the structure and physics of dense supra-nuclear matter, to the fate and evolution of massive stars, to the nature of gravity and the origins of the Universe and the structure therein. We challenge our Solvay conference colleagues to identify an astrophysical area more replete with results and impact! The future for this domain of astrophysical research appears to be growing only brighter, buoyed in particular by the development and proliferation of multiple major new radio telescopes, including LOFAR, MWA, ALMA, Meerkat, ASKAP, CHIME, FAST, and in the next decade, SKA. Moreover, this science goes hand-in-hand with the blossoming field of astrophysical transients, whether considering magnetar bursts as possible FRB progenitors, or considering NS-NS mergers as aLIGO/VIRGO sources. We look forward to either participating in or hearing the results reported at the next Solvay astrophysics meeting (which will hopefully take place in fewer years than have passed since the last!) by which time we predict there will have been major discoveries in gravitational wave physics, in gravity in general, and in neutron-star astrophysics.

References

1. J. Chadwick, *Nature* **192** (1932) 312.
2. W. Baade and F. Zwicky, *Phys. Rev.* **45** (1934) 138.
3. A. Hewish, S. J. Bell, J. D. H. Pilkington, P. F. Scott and R. A. Collins, *Nature* **217** (1968) 709.
4. I. S. Shklovsky, *ApJL* **148** (1967) L1.
5. T. Gold, *Nature* **221** (1969) 25.
6. F. Pacini, *Nature* **219** (1968) 145.
7. R. N. Manchester and J. H. Taylor, *Pulsars* (Freeman, San Francisco, 1977).
8. E. D. Barr, D. J. Champion, M. Kramer, R. P. Eatough, P. C. C. Freire, R. Karup-pusamy, K. J. Lee, J. P. W. Verbiest, C. G. Bassa, A. G. Lyne, B. Stappers, D. R. Lorimer and B. Klein, *MNRAS* **435** (2013) 2234.
9. R. S. Lynch, J. Boyles, S. M. Ransom, I. H. Stairs, D. R. Lorimer, M. A. McLaughlin, J. W. T. Hessels, V. M. Kaspi, V. I. Kondratiev, A. M. Archibald, A. Berndsen, R. F. Cardoso, A. Cherry, C. R. Epstein, C. Karako-Argaman, C. A. McPhee, T. Pennucci, M. S. E. Roberts, K. Stovall and J. van Leeuwen, *ApJ* **763** (2013) 81.
10. P. Lazarus, A. Brazier, J. W. T. Hessels, C. Karako-Argaman, V. M. Kaspi, R. Lynch, E. Madsen, C. Patel, S. M. Ransom, P. Scholz, J. Swiggum, W. W. Zhu, B. Allen,

S. Bogdanov, F. Camilo, F. Cardoso, S. Chatterjee, J. M. Cordes, F. Crawford, J. S. Deneva, R. Ferdman, P. C. C. Freire, F. A. Jenet, B. Knispel, K. J. Lee, J. van Leeuwen, D. R. Lorimer, A. G. Lyne, M. A. McLaughlin, X. Siemens, L. G. Spitler, I. H. Stairs, K. Stovall and A. Venkataraman, *ArXiv/1504.02294* (2015).

11. C.-A. Faucher-Giguère and V. M. Kaspi, *ApJ* **643** (2006) 332.
12. A. G. Lyne and D. R. Lorimer, *Nature* **369** (1994) 127.
13. B. Hansen and E. S. Phinney, *MNRAS* **291** (1997) 569.
14. G. Hobbs, D. R. Lorimer, A. G. Lyne and M. Kramer, *MNRAS* **360** (2005) 974.
15. S. D. Bates, D. R. Lorimer, A. Rane and J. Swiggum, *MNRAS* **439** (2014) 2893.
16. M. D. Young, R. N. Manchester and S. Johnston, *Nature* **400** (1999) 848.
17. J. W. T. Hessels, S. M. Ransom, I. H. Stairs, P. C. C. Freire, V. M. Kaspi and F. Camilo, *Science* **311** (2006) 1901.
18. P. Goldreich and W. H. Julian, *ApJ* **157** (1969) 869.
19. A. Spitkovsky, *ApJL* **648** (2006) L51.
20. M. A. Livingstone, V. M. Kaspi, F. P. Gavriil, R. N. Manchester, E. V. G. Gotthelf and L. Kuiper, *ApSS* **308** (2007) 317.
21. V. M. Kaspi, M. S. E. Roberts, G. Vasisht, E. V. Gotthelf, M. Pivovaroff and N. Kawai, *ApJ* **560** (2001) 371.
22. D. H. Clark and F. R. Stephenson, *The Historical Supernovae* (Pergamon, Oxford, 1977).
23. M. A. Livingstone, V. M. Kaspi, E. V. Gotthelf and L. Kuiper, *ApJ* **647** (2006) 1286.
24. D. B. Melrose, Coherent radio emission from pulsars, in *Pulsars as Physics Laboratories*, eds. J. M. Shull and H. A. Thronson (1993), pp. 105–115.
25. D. B. Melrose, *J. Astrophys. Astr.* **16** (1995) 137.
26. R. Beck, *Philos. Trans. Roy. Soc. London A* **358** (2000) 777.
27. V. M. Kaspi, M. S. E. Roberts and A. K. Harding, Isolated neutron stars, pp. 279–339.
28. B. M. Gaensler and P. O. Slane, *Ann. Rev. Astr. Ap.* **44** (2006) 17.
29. P. A. Caraveo, *araa* **52** (2014) 211.
30. D. Bhattacharya and E. P. J. van den Heuvel, *Phys. Rep.* **203** (1991) 1.
31. E. S. Phinney and S. R. Kulkarni, *Ann. Rev. Astr. Ap.* **32** (1994) 591.
32. T. M. Tauris and E. P. J. van den Heuvel, Formation and evolution of compact stellar X-ray sources (Compact stellar X-ray sources, 2006), pp. 623–665.
33. D. R. Lorimer, *Living Reviews in Relativity* (2008) http://www.livingreviews.org/lrr-2008-8.
34. D. C. Backer, S. R. Kulkarni, C. Heiles, M. M. Davis and W. M. Goss, *Nature* **300** (1982) 615.
35. A. S. Fruchter, D. R. Stinebring and J. H. Taylor, *Nature* **333** (1988) 237.
36. B. W. Stappers, M. Bailes, A. G. Lyne, R. N. Manchester, N. D'Amico, T. M. Tauris, D. R. Lorimer, S. Johnston and J. S. Sandhu, *ApJ* **465** (1996) L119.
37. A. M. Archibald, I. H. Stairs, S. M. Ransom, V. M. Kaspi, V. I. Kondratiev, D. R. Lorimer, M. A. McLaughlin, J. Boyles, J. W. T. Hessels, R. Lynch, J. van Leeuwen, M. S. E. Roberts, F. Jenet, D. J. Champion, R. Rosen, B. N. Barlow, B. H. Dunlap and R. A. Remillard, *Science* **324** (2009) 1411.
38. D. R. Lorimer, A. J. Faulkner, A. G. Lyne, R. N. Manchester, M. Kramer, M. A. McLaughlin, G. Hobbs, A. Possenti, I. H. Stairs, F. Camilo, M. Burgay, N. D'Amico, A. Corongiu and F. Crawford, *MNRAS* **372** (2006) 777.
39. J. P. W. Verbiest, M. Bailes, W. A. Coles, G. B. Hobbs, W. van Straten, D. J. Champion, F. A. Jenet, R. N. Manchester, N. D. R. Bhat, J. M. Sarkissian, D. Yardley, S. Burke-Spolaor, A. W. Hotan and X. P. You, *MNRAS* **400** (2009) 951.

40. D. J. Champion, S. M. Ransom, P. Lazarus, F. Camilo, C. Bassa, V. M. Kaspi, D. J. Nice, P. C. C. Freire, I. H. Stairs, J. van Leeuwen, B. W. Stappers, J. M. Cordes, J. W. T. Hessels, D. R. Lorimer, Z. Arzoumanian, D. C. Backer, N. D. R. Bhat, S. Chatterjee, I. Cognard, J. S. Deneva, C.-A. Faucher-Giguère, B. M. Gaensler, J. Han, F. A. Jenet, L. Kasian, V. I. Kondratiev, M. Kramer, J. Lazio, M. A. McLaughlin, A. Venkataraman and W. Vlemmings, *Science* **320** (2008) 1309.
41. P. C. C. Freire, C. G. Bassa, N. Wex, I. H. Stairs, D. J. Champion, S. M. Ransom, P. Lazarus, V. M. Kaspi, J. W. T. Hessels, M. Kramer, J. M. Cordes, J. P. W. Verbiest, P. Podsiadlowski, D. J. Nice, J. S. Deneva, D. R. Lorimer, B. W. Stappers, M. A. McLaughlin and F. Camilo, *MNRAS* **412** (2011) 2763.
42. S. M. Ransom, I. H. Stairs, A. M. Archibald, J. W. T. Hessels, D. L. Kaplan, M. H. van Kerkwijk, J. Boyles, A. T. Deller, S. Chatterjee, A. Schechtman-Rook, A. Berndsen, R. S. Lynch, D. R. Lorimer, C. Karako-Argaman, V. M. Kaspi, V. I. Kondratiev, M. A. McLaughlin, J. van Leeuwen, R. Rosen, M. S. E. Roberts and K. Stovall, *Nature* **505** (2014) 520.
43. T. M. Tauris and E. P. J. van den Heuvel, *ApJL* **781** (2014) L13.
44. J. S. Deneva, K. Stovall, M. A. McLaughlin, S. D. Bates, P. C. C. Freire, J. G. Martinez, F. Jenet and M. Bagchi, *ApJ* **775** (2013) 51.
45. B. Knispel, A. G. Lyne, B. W. Stappers, P. C. C. Freire, P. Lazarus, B. Allen, C. Aulbert, O. Bock, S. Bogdanov, A. Brazier, F. Camilo, F. Cardoso, S. Chatterjee, J. M. Cordes, F. Crawford, J. S. Deneva, H.-B. Eggenstein, H. Fehrmann, R. Ferdman, J. W. T. Hessels, F. A. Jenet, C. Karako-Argaman, V. M. Kaspi, J. van Leeuwen, D. R. Lorimer, R. Lynch, B. Machenschalk, E. Madsen, M. A. McLaughlin, C. Patel, S. M. Ransom, P. Scholz, X. Siemens, L. G. Spitler, I. H. Stairs, K. Stovall, J. K. Swiggum, A. Venkataraman, R. S. Wharton and W. W. Zhu, *ArXiv/1504.03684* (2015).
46. P. C. C. Freire and T. M. Tauris, *MNRAS* **438** (2014) L86.
47. J. Antoniadis, *ApJ* **797** (2014) L24.
48. A. Papitto, C. Ferrigno, E. Bozzo, N. Rea, L. Pavan, L. Burderi, M. Burgay, S. Campana, T. di Salvo, M. Falanga, M. D. Filipović, P. C. C. Freire, J. W. T. Hessels, A. Possenti, S. M. Ransom, A. Riggio, P. Romano, J. M. Sarkissian, I. H. Stairs, L. Stella, D. F. Torres, M. H. Wieringa and G. F. Wong, *Nature* **501** (2013) 517.
49. B. W. Stappers, A. M. Archibald, J. W. T. Hessels, C. G. Bassa, S. Bogdanov, G. H. Janssen, V. M. Kaspi, A. G. Lyne, A. Patruno, S. Tendulkar, A. B. Hill and T. Glanzman, *ApJ* **790** (2014) 39.
50. S. P. Tendulkar, C. Yang, H. An, V. M. Kaspi, A. M. Archibald, C. Bassa, E. Bellm, S. Bogdanov, F. A. Harrison, J. W. T. Hessels, G. H. Janssen, A. G. Lyne, A. Patruno, B. Stappers, D. Stern, J. A. Tomsick, S. E. Boggs, D. Chakrabarty, F. E. Christensen, W. W. Craig, C. A. Hailey and W. Zhang, *ApJ* **791** (2014) 77.
51. S. Bogdanov, A. Patruno, A. M. Archibald, C. Bassa, J. W. T. Hessels, G. H. Janssen and B. W. Stappers, *ApJ* **789** (2014) 40.
52. S. Johnston, R. N. Manchester, A. G. Lyne, M. Bailes, V. M. Kaspi, G. Qiao and N. D'Amico, *ApJ* **387** (1992) L37.
53. V. M. Kaspi, S. Johnston, J. F. Bell, R. N. Manchester, M. Bailes, M. Bessell, A. G. Lyne and N. D'Amico, *ApJ* **423** (1994) L43.
54. I. H. Stairs, R. N. Manchester, A. G. Lyne, V. M. Kaspi, F. Camilo, J. F. Bell, N. D'Amico, M. Kramer, F. Crawford, D. J. Morris, N. P. F. McKay, S. L. Lumsden, L. E. Tacconi-Garman, R. D. Cannon, N. C. Hambly and P. R. Wood, *MNRAS* **325** (2001) 979.
55. D. Lai, L. Bildsten and V. M. Kaspi, *ApJ* **452** (1995) 819.

56. V. M. Kaspi, M. Bailes, R. N. Manchester, B. W. Stappers and J. F. Bell, *Nature* **381** (1996) 584.
57. N. Wex, *MNRAS* **298** (1998) 67.
58. E. C. Madsen, I. H. Stairs, M. Kramer, F. Camilo, G. B. Hobbs, G. H. Janssen, A. G. Lyne, R. N. Manchester, A. Possenti and B. W. Stappers, *MNRAS* **425** (2012) 2378.
59. S. Johnston, L. Ball, N. Wang and R. N. Manchester, *MNRAS* **358** (2005) 1069.
60. V. M. Kaspi, T. Tauris and R. N. Manchester, *ApJ* **459** (1996) 717.
61. A. A. Abdo, et al., *ApJL* **736** (2011) L11.
62. HESS run/Collaboration, *A&A* **551** (2013) A94.
63. I. F. Mirabel, *Science* **335** (2012) 175.
64. V. M. Kaspi, *Proceedings of the National Academy of Science* **107** (2010) 7147.
65. E. P. Mazets, S. V. Golenetskii and Y. A. Gur'yan, *Sov. Astron. Lett.* **5** (1979) 343.
66. E. P. Mazets, S. V. Golenetskii, V. N. Ilinskii, R. L. Apetkar and Y. A. Guryan, *Nature* **282** (1979) 587.
67. S. A. Olausen and V. M. Kaspi, *ApJS* **212** (2014) 6.
68. S. Mereghetti, J. A. Pons and A. Melatos, *Space Sciences Review, arXiv/1503.06313* (2015).
69. K. Hurley, S. E. Boggs, D. M. Smith, R. C. Duncan, R. Lin, A. Zoglauer, S. Krucker, G. Hurford, H. Hudson, C. Wigger, W. Hajdas, C. Thompson, I. Mitrofanov, A. Sanin, W. Boynton, C. Fellows, A. von Kienlin, G. Lichti, A. Rau and T. Cline, *Nature* **434** (2005) 1098.
70. D. M. Palmer, S. Barthelmy, N. Gehrels, R. M. Kippen, T. Cayton, C. Kouveliotou, D. Eichler, R. A. M. J. Wijers, P. M. Woods, J. Granot, Y. E. Lyubarsky, E. Ramirez-Ruiz, L. Barbier, M. Chester, J. Cummings, E. E. Fenimore, M. H. Finger, B. M. Gaensler, D. Hullinger, H. Krimm, C. B. Markwardt, J. A. Nousek, A. Parsons, S. Patel, T. Sakamoto, G. Sato, M. Suzuki and J. Tueller, *Nature* **434** (2005) 1107.
71. M. P. Muno, J. S. Clark, P. A. Crowther, S. M. Dougherty, R. de Grijs, C. Law, S. L. W. McMillan, M. R. Morris, I. Negueruela, D. Pooley, S. Portegies Zwart and F. Yusef-Zadeh, *ApJ* **636** (2006) L41.
72. S. Mereghetti, G. L. Israel and L. Stella, *MNRAS* **296** (1998) 689.
73. T. Damour and J. H. Taylor, *Phys. Rev. D* **45** (1992) 1840.
74. C. Thompson and R. C. Duncan, *MNRAS* **275** (1995) 255.
75. C. Thompson and R. C. Duncan, *ApJ* **473** (1996) 322.
76. P. Goldreich and A. Reisenegger, *ApJ* **395** (1992) 250.
77. C. Kouveliotou, S. Dieters, T. Strohmayer, J. van Paradijs, G. J. Fishman, C. A. Meegan, K. Hurley, J. Kommers, I. Smith, D. Frail and T. Murakami, *Nature* **393** (1998) 235.
78. F. P. Gavriil, V. M. Kaspi and P. M. Woods, *Nature* **419** (2002) 142.
79. V. M. Kaspi, F. P. Gavriil, P. M. Woods, J. B. Jensen, M. S. E. Roberts and D. Chakrabarty, *ApJ* **588** (2003) L93.
80. C. Thompson, M. Lyutikov and S. R. Kulkarni, *ApJ* **574** (2002) 332.
81. A. M. Beloborodov and C. Thompson, *ApJ* **657** (2007) 967.
82. Y. Lyubarsky, D. Eichler and C. Thompson, *ApJ* **580** (2002) L69.
83. C. Kouveliotou, D. Eichler, P. M. Woods, Y. Lyubarsky, S. K. Patel, E. Göğüş, M. van der Klis, A. Tennant, S. Wachter and K. Hurley, *ApJ* **596** (2003) L79.
84. P. Scholz, V. M. Kaspi and A. Cumming, *ApJ* **786** (2014) 62.
85. L. Kuiper, W. Hermsen and M. Mendez, *ApJ* **613** (2004) 1173.
86. L. Kuiper, W. Hermsen, P. den Hartog and W. Collmar, *ApJ* **645** (2006) 556.
87. A. M. Beloborodov, *ApJ* **777** (2013) 114.
88. R. Dib, V. M. Kaspi and F. P. Gavriil, *ApJ* **673** (2008) 1044.

89. P. W. Anderson and N. Itoh, *Nature* **256** (1975) 25.

90. M. A. Alpar, K. S. Cheng and D. Pines, *ApJ* **346** (1989) 823.

91. R. Dib and V. M. Kaspi, *ApJ* **784** (2014) 37.

92. R. F. Archibald, V. M. Kaspi, C.-Y. Ng, K. N. Gourgouliatos, D. Tsang, P. Scholz, A. P. Beardmore, N. Gehrels and J. A. Kennea, *Nature* **497** (2013) 591.

93. N. Rea, P. Esposito, R. Turolla, G. L. Israel, S. Zane, L. Stella, S. Mereghetti, A. Tiengo, D. Götz, E. Göğüş and C. Kouveliotou, *Science* **330** (2010) 944.

94. F. P. Gavriil, M. E. Gonzalez, E. V. Gotthelf, V. M. Kaspi, M. A. Livingstone and P. M. Woods, *Science* **319** (2008) 1802.

95. M. A. Livingstone, C.-Y. Ng, V. M. Kaspi, F. P. Gavriil and E. V. Gotthelf, *ApJ* **730** (2011) 66.

96. R. Perna and J. A. Pons, *ApJ* **727** (February 2011) L51.

97. D. Viganò, N. Rea, J. A. Pons, R. Perna, D. N. Aguilera and J. A. Miralles, *MNRAS* **434** (September 2013) 123.

98. F. Camilo, S. M. Ransom, J. P. Halpern, J. Reynolds, D. J. Helfand, N. Zimmerman and J. Sarkissian, *Nature* **442** (2006) 892.

99. F. Camilo, S. M. Ransom, J. P. Halpern and J. Reynolds, *ApJ* **666** (2007) L93.

100. L. Levin, M. Bailes, S. Bates, N. D. R. Bhat, M. Burgay, S. Burke-Spolaor, N. D'Amico, S. Johnston, M. Keith, M. Kramer, S. Milia, A. Possenti, N. Rea, B. Stappers and W. van Straten, *ApJ* (2010).

101. R. P. Eatough, H. Falcke, R. Karuppusamy, K. J. Lee, D. J. Champion, E. F. Keane, G. Desvignes, D. H. F. M. Schnitzeler, L. G. Spitler, M. Kramer, B. Klein, C. Bassa, G. C. Bower, A. Brunthaler, I. Cognard, A. T. Deller, P. B. Demorest, P. C. C. Freire, A. Kraus, A. G. Lyne, A. Noutsos, B. Stappers and N. Wex, *Nature* **501** (2013) 391.

102. K. Mori, E. V. Gotthelf, S. Zhang, H. An, F. K. Baganoff, N. M. Barrière, A. M. Beloborodov, S. E. Boggs, F. E. Christensen, W. W. Craig, F. Dufour, B. W. Grefenstette, C. J. Hailey, F. A. Harrison, J. Hong, V. M. Kaspi, J. A. Kennea, K. K. Madsen, C. B. Markwardt, M. Nynka, D. Stern, J. A. Tomsick and W. W. Zhang, *ApJ* **770** (2013) L23.

103. N. Rea, P. Esposito, J. A. Pons, R. Turolla, D. F. Torres, G. L. Israel, A. Possenti, M. Burgay, D. Vigano', R. Perna, L. Stella, G. Ponti, F. Baganoff, D. Haggard, A. Papitto, A. Camero-Arranz, S. Zane, A. Minter, S. Mereghetti, A. Tiengo, R. Schoedel, M. Feroci, R. Mignani and D. Gotz, *ApJL* **775** (2013) L34.

104. R. M. Shannon and S. Johnston, *MNRAS* **435** (2013) L29.

105. V. M. Kaspi, R. F. Archibald, V. Bhalerao, F. Dufour, E. V. Gotthelf, H. An, M. Bachetti, A. M. Beloborodov, S. E. Boggs, F. E. Christensen, W. W. Craig, B. W. Grefenstette, C. J. Hailey, F. A. Harrison, J. A. Kennea, C. Kouveliotou, K. K. Madsen, K. Mori, C. B. Markwardt, D. Stern, J. K. Vogel and W. W. Zhang, *ApJ* **786** (2014) 84.

106. F. Haberl, *ApJS* (2007) 73.

107. D. L. Kaplan and M. H. van Kerkwijk, *ApJ* **705** (2009) 798.

108. E. F. Keane and M. Kramer, *MNRAS* **391** (2008) 2009.

109. D. L. Kaplan and M. H. van Kerkwijk, *ApJ* **635** (2005) L65.

110. J. A. Pons, B. Link, J. A. Miralles and U. Geppert, *Phys. Rev. Lett.* **98** (2007) 071101.

111. W. Zhu, V. M. Kaspi, M. E. Gonzalez and A. G. Lyne, *ApJ* **704** (2009) 1321.

112. D. N. Aguilera, J. A. Pons and J. A. Miralles, *ApJ* **673** (2008) L167.

113. J. A. Pons, J. A. Miralles and U. Geppert, **496** (2009) 207.

114. K. N. Gourgouliatos and A. Cumming, *MNRAS* **438** (February 2014) 1618.

115. G. G. Pavlov, V. E. Zavlin, B. Aschenbach, J. Truemper and D. Sanwal, *ApJ* **531** (1999) L53.

116. D. Chakrabarty, M. J. Pivovaroff, L. E. Hernquist, J. S. Heyl and R. Narayan, *ApJ* **548** (2001) 800.

117. S. Mereghetti, A. Tiengo and G. L. Israel, *ApJ* **569** (2002) 275.

118. G. G. Pavlov and G. J. M. Luna, *ApJ* **703** (2009) 910.

119. E. V. Gotthelf, J. P. Halpern and F. D. Seward, *ApJ* **627** (2005) 390.

120. J. P. Halpern, E. V. Gotthelf, F. Camilo and F. D. Seward, *ApJ* **665** (2007) 1304.

121. J. P. Halpern and E. V. Gotthelf, *ApJ* **709** (2010) 436.

122. V. E. Zavlin, G. G. Pavlov, D. Sanwal and J. Trümper, *ApJ* **540** (2000) L25.

123. G. G. Pavlov, V. E. Zavlin, D. Sanwal and J. Trümper, *ApJ* **569** (2002) L95.

124. E. V. Gotthelf, J. P. Halpern and J. Alford, *ApJ* **765** (2013) 58.

125. E. V. Gotthelf and J. P. Halpern, *ApJ* **695** (2009) L35.

126. M. A. McLaughlin, A. G. Lyne, D. R. Lorimer, M. Kramer, A. J. Faulkner, R. N. Manchester, J. M. Cordes, F. Camilo, A. Possenti, I. H. Stairs, G. Hobbs, N. D'Amico, M. Burgay and J. T. O'Brien, *Nature* **439** (2006) 817.

127. P. Weltevrede, B. W. Stappers, J. M. Rankin and G. A. E. Wright, *ApJ* **645** (2006) L149.

128. E. F. Keane, M. Kramer, A. G. Lyne, B. W. Stappers and M. A. McLaughlin, *MNRAS* **415** (2011) 3065.

129. E. Petroff, E. F. Keane, E. D. Barr, J. E. Reynolds, J. Sarkissian, P. G. Edwards, J. Stevens, C. Brem, A. Jameson, S. Burke-Spolaor, S. Johnston, N. D. R. Bhat, P. Chandra, S. Kudale and S. Bhandari, *ArXiv e-prints/1504.02165* (2015).

130. D. R. Lorimer, M. Bailes, M. A. McLaughlin, D. J. Narkevic and F. Crawford, *Science* **318** (2007) 777.

131. D. Thornton, B. Stappers, M. Bailes, B. Barsdell, S. Bates, N. D. R. Bhat, M. Burgay, S. Burke-Spolaor, D. J. Champion, P. Coster, N. D'Amico, A. Jameson, S. Johnston, M. Keith, M. Kramer, L. Levin, S. Milia, C. Ng, A. Possenti and W. van Straten, *Science* **341** (2013) 53.

132. L. G. Spitler, J. M. Cordes, J. W. T. Hessels, D. R. Lorimer, M. A. McLaughlin, S. Chatterjee, F. Crawford, J. S. Deneva, V. M. Kaspi, R. S. Wharton, B. Allen, S. Bogdanov, A. Brazier, F. Camilo, P. C. C. Freire, F. A. Jenet, C. Karako-Argaman, B. Knispel, P. Lazarus, K. J. Lee, J. van Leeuwen, R. Lynch, S. M. Ransom, P. Scholz, X. Siemens, I. H. Stairs, K. Stovall, J. K. Swiggum, A. Venkataraman, W. W. Zhu, C. Aulbert and H. Fehrmann, *ApJ* **790** (2014) 101.

133. E. Petroff, M. Bailes, E. D. Barr, B. R. Barsdell, N. D. R. Bhat, F. Bian, S. Burke-Spolaor, M. Caleb, D. Champion, P. Chandra, G. Da Costa, C. Delvaux, C. Flynn, N. Gehrels, J. Greiner, A. Jameson, S. Johnston, M. M. Kasliwal, E. F. Keane, S. Keller, J. Kocz, M. Kramer, G. Leloudas, D. Malesani, J. S. Mulchaey, C. Ng, E. O. Ofek, D. A. Perley, A. Possenti, B. P. Schmidt, Y. Shen, B. Stappers, P. Tisserand, W. van Straten and C. Wolf, *MNRAS* **447** (2015) 246.

134. E. Petroff, W. van Straten, S. Johnston, M. Bailes, E. D. Barr, S. D. Bates, N. D. R. Bhat, M. Burgay, S. Burke-Spolaor, D. Champion, P. Coster, C. Flynn, E. F. Keane, M. J. Keith, M. Kramer, L. Levin, C. Ng, A. Possenti, B. W. Stappers, C. Tiburzi and D. Thornton, *ApJ* **789** (2014) L26.

135. C. M. Will, *Living Reviews in Relativity* **17** (2014) 4.

136. S. Blatt, A. D. Ludlow, G. K. Campbell, J. W. Thomsen, T. Zelevinsky, M. M. Boyd, J. Ye, X. Baillard, M. Fouché, R. Le Targat, A. Brusch, P. Lemonde, M. Takamoto, F.-L. Hong, H. Katori and V. V. Flambaum, *Phys. Rev. Lett.* **100** (2008) 140801.

137. A. Ivanchik, P. Petitjean, D. Varshalovich, B. Aracil, R. Srianand, H. Chand, C. Ledoux and P. Boissé, *A&A* **440** (2005) 45.

138. F. Hofmann, J. Müller and L. Biskupek, *A&A* **522** (2010) L5.

139. P. C. C. Freire, N. Wex, G. Esposito-Farèse, J. P. W. Verbiest, M. Bailes, B. A. Jacoby, M. Kramer, I. H. Stairs, J. Antoniadis and G. H. Janssen, *MNRAS* **423** (2012) 3328.

140. N. Wex, in the Brumberg Festschrift, edited by S. M. Kopeikein, Gruyter, Berlin, *ArXiv/1402.5594* (2014).

141. I. H. Stairs, A. J. Faulkner, A. G. Lyne, M. Kramer, D. R. Lorimer, M. A. McLaughlin, R. N. Manchester, G. B. Hobbs, F. Camilo, A. Possenti, M. Burgay, N. D'Amico, P. C. C. Freire, P. C. Gregory and N. Wex, *ApJ* **632** (2005) 1060.

142. M. Kramer and et al., *Phys. Rev. D* (2015) to be submitted.

143. J. Antoniadis, P. C. C. Freire, N. Wex, T. M. Tauris, R. S. Lynch, M. H. van Kerkwijk, M. Kramer, C. Bassa, V. S. Dhillon, T. Driebe, J. W. T. Hessels, V. M. Kaspi, V. I. Kondratiev, N. Langer, T. R. Marsh, M. A. McLaughlin, T. T. Pennucci, S. M. Ransom, I. H. Stairs, J. van Leeuwen, J. P. W. Verbiest and D. G. Whelan, *Science* **340** (2013) 448.

144. C. W. F. Everitt, D. B. Debra, B. W. Parkinson, J. P. Turneaure, J. W. Conklin, M. I. Heifetz, G. M. Keiser, A. S. Silbergleit, T. Holmes, J. Kolodziejczak, M. Al-Meshari, J. C. Mester, B. Muhlfelder, V. G. Solomonik, K. Stahl, P. W. Worden, Jr., W. Bencze, S. Buchman, B. Clarke, A. Al-Jadaan, H. Al-Jibreen, J. Li, J. A. Lipa, J. M. Lockhart, B. Al-Suwaidan, M. Taber and S. Wang, *Phys. Rev. Lett.* **106** (2011) 221101.

145. M. Kramer, *ApJ* **509** (1998) 856.

146. R. P. Breton, V. M. Kaspi, M. Kramer, M. A. McLaughlin, M. Lyutikov, S. M. Ransom, I. H. Stairs, R. D. Ferdman, F. Camilo and A. Possenti, *Science* **321** (2008) 104.

147. J. M. Lattimer and A. W. Steiner, *ApJ* **784** (2014) 123.

148. P. B. Demorest, T. Pennucci, S. M. Ransom, M. S. E. Roberts and J. W. T. Hessels, *Nature* **467** (2010) 1081.

149. S. Guillot, M. Servillat, N. A. Webb and R. E. Rutledge, *ApJ* **772** (July 2013) 7.

150. L. Shao and N. Wex, *Class. Quant Grav.* **30** (2013) 165020.

151. L. Shao and N. Wex, *Class. Quant Grav.* **29** (2012) 215018.

152. L. Shao, R. N. Caballero, M. Kramer, N. Wex, D. J. Champion and A. Jessner, *Class. Quant Grav.* **30** (2013) 165019.

153. J. Alsing, E. Berti, C. M. Will and H. Zaglauer, *Phys. Rev. D* **85** (2012) 064041.

154. K. Yagi, D. Blas, E. Barausse and N. Yunes, *Phys. Rev. D* **89** (2014) 084067.

155. K. Liu, N. Wex, M. Kramer, J. M. Cordes and T. J. W. Lazio, *ApJ* **747** (2012) 1.

156. V. M. Kaspi, A. G. Lyne, R. N. Manchester, F. Crawford, F. Camilo, J. F. Bell, N. D'Amico, I. H. Stairs, N. P. F. McKay, D. J. Morris and A. Possenti, *ApJ* **543** (2000) 321.

157. R. A. Hulse and J. H. Taylor, *ApJ* **195** (1975) L51.

158. D. R. Lorimer and M. Kramer, *Handbook of Pulsar Astronomy* (Cambridge University Press, 2005).

159. J. M. Weisberg, D. J. Nice and J. H. Taylor, *ApJ* **722** (2010) 1030.

160. T. Damour and J. H. Taylor, *ApJ* **366** (1991) 501.

161. M. Burgay, N. D'Amico, A. Possenti, R. N. Manchester, A. G. Lyne, B. C. Joshi, M. A. McLaughlin, M. Kramer, J. M. Sarkissian, F. Camilo, V. Kalogera, C. Kim and D. R. Lorimer, *Nature* **426** (2003) 531.

162. A. G. Lyne, M. Burgay, M. Kramer, A. Possenti, R. N. Manchester, F. Camilo, M. A. McLaughlin, D. R. Lorimer, N. D'Amico, B. C. Joshi, J. Reynolds and P. C. C. Freire, *Science* **303** (2004) 1153.

163. M. Kramer, I. H. Stairs, R. N. Manchester, M. A. McLaughlin, A. G. Lyne, R. D.

Ferdman, M. Burgay, D. R. Lorimer, A. Possenti, N. D'Amico, J. M. Sarkissian, G. B. Hobbs, J. E. Reynolds, P. C. C. Freire and F. Camilo, *Science* **314** (2006) 97.

164. M. Kramer and I. H. Stairs, *Ann. Rev. Astr. Ap.* **46** (2008) 541.
165. M. Kramer and N. Wex, *Classical and Quantum Gravity* **26** (2009) 073001.
166. T. Damour and N. Deruelle, *Ann. Inst. H. Poincaré (Physique Théorique)* **44** (1986) 263.
167. G. Boerner, J. Ehlers and E. Rudolph, *A&A* **44** (1975) 417.
168. T. Damour and R. Ruffini, *Academie des Sciences Paris Comptes Rendus Ser. Scie. Math.* **279** (1974) 971.
169. R. D. Ferdman, I. H. Stairs, M. Kramer, R. P. Breton, M. A. McLaughlin, P. C. C. Freire, A. Possenti, B. W. Stappers, V. M. Kaspi, R. N. Manchester and A. G. Lyne, *ApJ* **767** (2013) 85.
170. M. Burgay, A. Possenti, R. N. Manchester, M. Kramer, M. A. McLaughlin, D. R. Lorimer, I. H. Stairs, B. C. Joshi, A. G. Lyne, F. Camilo, N. D'Amico, P. C. C. Freire, J. M. Sarkissian, A. W. Hotan and G. B. Hobbs, *ApJ* **624** (2005) L113.
171. B. B. P. Perera, M. A. McLaughlin, M. Kramer, I. H. Stairs, R. D. Ferdman, P. C. C. Freire, A. Possenti, R. P. Breton, R. N. Manchester, M. Burgay, A. G. Lyne and F. Camilo, *ApJ* **721** (2010) 1193.
172. T. Damour and G. Esposito-Farese, *Phys. Rev. D* **54** (1996) 1474.
173. B. Bertotti, L. Iess and P. Tortora, *Nature* **425** (2003) 374.
174. J. D. Bekenstein, *Phys. Rev. D* **70** (2004) 083509.
175. R. W. Hellings and G. S. Downs, *ApJ* **265** (1983) L39.
176. G. H. Janssen, G. Hobbs, M. McLaughlin, C. G. Bassa, A. T. Deller, M. Kramer, K. J. Lee, C. M. F. Mingarelli, P. A. Rosado, S. Sanidas, A. Sesana, L. Shao, I. H. Stairs, B. W. Stappers and J. P. W. Verbiest, *ArXiv/1501.00127* (2015).
177. R. N. Manchester and IPTA, *Class. Quant Grav.* **30** (2013) 224010.
178. R. M. Shannon, V. Ravi, W. A. Coles, G. Hobbs, M. J. Keith, R. N. Manchester, J. S. B. Wyithe, M. Bailes, N. D. R. Bhat, S. Burke-Spolaor, J. Khoo, Y. Levin, S. Oslowski, J. M. Sarkissian, W. van Straten, J. P. W. Verbiest and J.-B. Wang, *Science* **342** (2013) 334.
179. A. Sesana, A. Vecchio and C. N. Colacino, *MNRAS* **390** (2008) 192.
180. A. Sesana and A. Vecchio, *Class. Quant Grav.* **27** (2010) 084016.
181. A. Sesana, *MNRAS* **433** (2013) L1.
182. A. Sesana, *Astrophysics and Space Science Proceedings* **40** (2015) 147.
183. C. Roedig, M. Dotti, A. Sesana, J. Cuadra and M. Colpi, *MNRAS* **415** (2011) 3033.
184. C. Roedig and A. Sesana, *Journal of Physics Conference Series* **363** (2012) 012035.
185. L. Sampson, N. J. Cornish and S. T. McWilliams, *ArXiv/1503.02662* (2015).
186. K. J. Lee, N. Wex, M. Kramer, B. W. Stappers, C. G. Bassa, G. H. Janssen, R. Karuppusamy and R. Smits, *MNRAS* **414** (2011) 3251.
187. C. M. F. Mingarelli, T. Sidery, I. Mandel and A. Vecchio, *Phys. Rev. D* **88** (2013) 062005.
188. D. J. Champion, G. B. Hobbs, R. N. Manchester, R. T. Edwards, D. C. Backer, M. Bailes, N. D. R. Bhat, S. Burke-Spolaor, W. Coles, P. B. Demorest, R. D. Ferdman, W. M. Folkner, A. W. Hotan, M. Kramer, A. N. Lommen, D. J. Nice, M. B. Purver, J. M. Sarkissian, I. H. Stairs, W. van Straten, J. P. W. Verbiest and D. R. B. Yardley, *ApJ* **720** (2010) L201.
189. K. J. Lee, F. A. Jenet and R. H. Price, *ApJ* **685** (2008) 1304.
190. S. J. Chamberlin and X. Siemens, *Phys. Rev. D* **85** (2012) 082001.
191. C. M. F. Mingarelli, K. Grover, T. Sidery, R. J. E. Smith and A. Vecchio, *Physical Review Letters* **109** (2012) 081104.

192. K. Lee, F. A. Jenet, R. H. Price, N. Wex and M. Kramer, *ApJ* **722** (2010) 1589.
193. T. Damour and G. Esposito-Farèse, *Phys. Rev. D* **58** (1998) 1.
194. N. Wex and S. Kopeikin, *ApJ* **513** (1999) 388.
195. M. Kramer, D. C. Backer, J. M. Cordes, T. J. W. Lazio, B. W. Stappers and S. . Johnston, *New Astronomy Reviews* **48** (2004) 993.
196. K. Liu, R. P. Eatough, N. Wex and M. Kramer, *MNRAS* (2014).
197. R. Nan, D. Li, C. Jin, Q. Wang, L. Zhu, W. Zhu, H. Zhang, Y. Yue and L. Qian, *International Journal of Modern Physics D* **20** (2011) 989.
198. E. Pfahl and A. Loeb, *ApJ* **615** (2004) 253.
199. G. C. Bower, A. Deller, P. Demorest, A. Brunthaler, R. Eatough, H. Falcke, M. Kramer, K. J. Lee and L. Spitler, *ApJ* **780** (2014) L2.
200. L. G. Spitler, K. J. Lee, R. P. Eatough, M. Kramer, R. Karuppusamy, C. G. Bassa, I. Cognard, G. Desvignes, A. G. Lyne, B. W. Stappers, G. C. Bower, J. M. Cordes, D. J. Champion and H. Falcke, *ApJ* **780** (2014) L3.
201. R. S. Lynch, R. F. Archibald, V. M. Kaspi and P. Scholz, *ApJ* .
202. R. S. Wharton, S. Chatterjee, J. M. Cordes, J. S. Deneva and T. J. W. Lazio, *ApJ* **753** (2012) 108.
203. A. M. Ghez, S. Salim, N. N. Weinberg, J. R. Lu, T. Do, J. K. Dunn, K. Matthews, M. R. Morris, S. Yelda, E. E. Becklin, T. Kremenek, M. Milosavljevic and J. Naiman, *ApJ* **689** (December 2008) 1044.
204. S. Gillessen, F. Eisenhauer, T. K. Fritz, H. Bartko, K. Dodds-Eden, O. Pfuhl, T. Ott and R. Genzel, *ApJ* **707** (2009) L114.
205. H. Falcke and S. B. Markoff, *Class. Quant Grav.* **30** (2013) 244003.
206. D. Psaltis, F. Özel and D. Chakrabarty, *ApJ* **787** (2014) 136.
207. J. M. Lattimer, *General Relativity and Gravitation* **46** (2014) 1713.
208. P. Podsiadlowski, J. D. M. Dewi, P. Lesaffre, J. C. Miller, W. G. Newton and J. R. Stone, *MNRAS* **361** (2005) 1243.
209. T. M. Tauris, N. Langer, T. J. Moriya, P. Podsiadlowski, S.-C. Yoon and S. I. Blinnikov, *ApJ* **778** (2013) L23.
210. J. M. Lattimer and B. F. Schutz, *ApJ* **629** (2005) 979.
211. I. A. Morrison, T. W. Baumgarte, S. L. Shapiro and V. R. Pandharipande, *ApJ* **617** (2004) L135.
212. M. Bejger, T. Bulik and P. Haensel, *MNRAS* **364** (2005) 635.
213. A. Loeb, Y. Shvartzvald and D. Maoz, *MNRAS* **439** (2014) L46.
214. J. Luan and P. Goldreich, *ApJL* **785** (2014) L26.
215. B. Dennison, *MNRAS* **443** (2014) L11.
216. S. R. Kulkarni, E. O. Ofek, J. D. Neill, Z. Zheng and M. Juric, *ApJ* **797** (2014) 70.
217. J. M. Cordes and I. Wasserman, *ArXiv e-prints/1501.00753* (2015).
218. M. Mannarelli, G. Pagliaroli, A. Parisi and L. Pilo, *Phys. Rev. D* **89** (2014) 103014.
219. A. Barrau, C. Rovelli and F. Vidotto, *Phys. Rev. D* **90** (2014) 127503.
220. Y.-W. Yu, K.-S. Cheng, G. Shiu and H. Tye, *J. of Cosm. & Astrop. Phys.* **11** (2014) 40.
221. Y. Lyubarsky, *MNRAS* **442** (2014) L9.
222. J.-P. Macquart, E. Keane, K. Grainge, M. McQuinn, R. P. Fender, J. Hessels, A. Deller, R. Bhat, R. Breton, S. Chatterjee, C. Law, D. Lorimer, E. O. Ofek, M. Pietka, L. Spitler, B. Stappers and C. Trott, *ArXiv/1501.07535* (2015).

Discussion

E. van den Heuvel Thank you very much. I would now like to start the discussion. Now probably the question of fundamental tests of gravity theories mentioned by Michael is still fresh in your mind. Since this is an important question for cosmology, I would be very pleased if not only the neutron-star people jump into the discussion but maybe also people interested in gravity theories may address questions to Michael, and consider the fact that you can already use double neutron stars to put very strong constraints on gravity theories. Already some of the scalar type theories have been refuted, as Michael mentioned. Neutron stars provide a very important test for any gravity theory, also quantum gravity will introduce differences in the strong and weak equivalence principles. So maybe we could start with considering this fundamental physical question concerning gravity. Who would like to start?

U. Pen I thank you for the fascinating talks. I have a very general question. I think we have a dilemma both in cosmology and in all other tests of general relativity: Einstein's theory of 1915 seems to work so far. And in fact it may work all the way. I am not sure about any expectations that general relativity has to be wrong some day, except maybe, as you mentioned, for quantum gravity. So what should we do if nothing unexpected turns up in the next five, ten or even hundred years? I guess it has been almost 99 years since Einstein's theory so should we go on for one more hundred years to put five more digits of testing? What should we do?

E. van den Heuvel Who would like to comment on this?

S. Tremaine This is not a comment, but perhaps an expansion on the question. A lot of effort is going into direct detection of gravitational waves, which we will hear about later on. Given the success of the quadrupole theory of gravitational waves predicting the behaviour of the pulsars, is there any conceivable alternative to relativity that would produce that agreement and not produce standard gravitational waves?

T. Damour No, I do not think so. Furthermore, when analyzing the tests that Michael Kramer told about in the context of space theories you find that even long-term observations from gravitational waves will not do better than pulsars tests. So there is really an issue today that the simplest alternatives to Einstein's theory with scalar fields are not really refuted, but there is only a small corner of parameter space, which is still available. There is now on the market new attempts for alternative theories of gravity, I mean there is a motivation that the graviton has a mass. There are also brain models of gravity. But these theories are not on a good theoretical bases in the sense that it is not clear that they are consistent and it is not clear that they are not already ruled out. More theoretical efforts are needed. But there is no strong incentive to find classically different theories of gravity. This is true.

S. Phinney The one thing I would like to say is that the direct detection of gravitational waves would, in principle, allow you to see the merging of two black holes, the formation of a new horizon and strong field gravity and also the scalar fields allowed for standard black holes. There are other theories. For example, we heard a lot about firewalls for black hole do not exist at all. Actually directly seen black holes and testing whether relativity really describes those, I think, will be the most interesting part of the tests of the gravitational waves and scalar theories.

D. Spergel This is in someway a response to Ue-Li's question about how long do we keep looking. I thought about the history of general relativity a bit and the precession of the perihelion of Mercury. I could imagine being at a meeting in the 1890s asking: we know Newton gravitational laws work remarkably well, why should one do precision work measuring the orbit of Mercury, let alone the very complicated perturbative theory estimates to compute what the corrections are due to all the planets? I suspect that if someone would have put-in a grant proposal in the 1890s to fund that kind of work, it would have been extremely difficult to get funded. I think we do not know when the surprises will be... I would also argue that the cosmological constant, for instance, was certainly something that was forced upon us by nature. It was not a part of the theory we would have insisted on, if the data did not compel us to it.

E. van den Heuvel Thank you very much for this comment, which is very in place I would think. So anymore comments?

S. Kulkarni This is a response to Ue-Li and Spergel. As an experimentalist observer, it seems to me what you have to do is always undertake precise observations whether they are meaningful or not. In this case, it is meaningful because the theory is there and we do not know how well it goes. If you improve any of these parameters by another factor of ten or hundred, it is a good thing to do. The problem I see is that theory, at this level, should be driven by experiments. Whereas we have a huge problem I would say in astronomy and far more in physics, that there are lots of theories and theoretical developments and that sort of keep the theorists busy. I find most of that work, however, to be marginal and even useless... I wonder how it is to get up in the morning as a theorist working on something with no guidance other than some numbers... and spend my life like that. So I am just actually very curious what those people think.

E. van den Heuvel I am afraid Shri that those people are not among us here. I think it is a very relevant comment but not for this collection of people here.

E. Silverstein Since the question of quantum gravity and even firewalls already came up, I just wanted to make a very brief comment. This is a problem where there is no theory and lots of thought experimental constraints to the extent that there is actually no model that fits all of even thought experimental constraints. Because of that, there is room for some dramatic effect that might eventually have observational consequences. Even if the firewalls exist at all, nobody is saying we need to set in very different physics as yet(?) because the timescale is way beyond the timescale that you would see in real astrophysical black holes. We could hope for a shorter timescale, there are reasons to believe that this might happen. The sharp arguments involve a timescale that is half the Hawking evaporation timescale of the black hole, which is way beyond anything that you would practically see.

E. van den Heuvel Thank you very much for this comment.

R. Wijers Just because it is so much fun to disagree with Shri, I would like to comment on the theory. If you look around, even for pulsars and black holes, Nature at the moment is throwing new mysteries at us at a sufficiently fast rate that, in fact, I would argue we are overdoing the observing and underdoing the theory. You may ask whether adding more digits to the precision measurements of general relativity is the first thing to do... You might dare say why not go looking for troubles. If you look at fast radio bursts and problems like that, I think there is plenty of room for basic imaginative theory to get a first approximation of what physics is involved there, let alone a precise theory.

E. van den Heuvel And we know of course that many theories which apparently did not have any connection to observations later turned out to be very valuable. I would like to stop the discussion at this point, and move to another fundamental question. We heard about these very massive neutron stars that recently have turned up. I have asked Jim Lattimer to prepare a small comment on what these very massive neutron stars might mean in terms of the equation of state.

Prepared comment

J. M. Lattimer: What Massive Stars Imply for the Equation of State

Within the past four years two relatively precise pulsar timing measurements of neutron stars with masses near $2\ M_\odot$ have been made. This establishes a lower limit M_m to the maximum mass of neutron stars. The actual limit is likely greater, as there are several additional, but less precise, measurements of even larger masses. Assuming the correctness of general relativity and that causality is not violated, this minimum value for the maximum mass is a powerful factor in determining the equation of state of dense matter. This discussion is a brief summary of points made in a recent review.[a]

Because of the relatively large value of M_m, theoretical mass-radius $(M-R)$ curves predicted by realistic equations of state must have the property that

---

[a]J.M. Lattimer, ARNPS 62, 485 (2012).

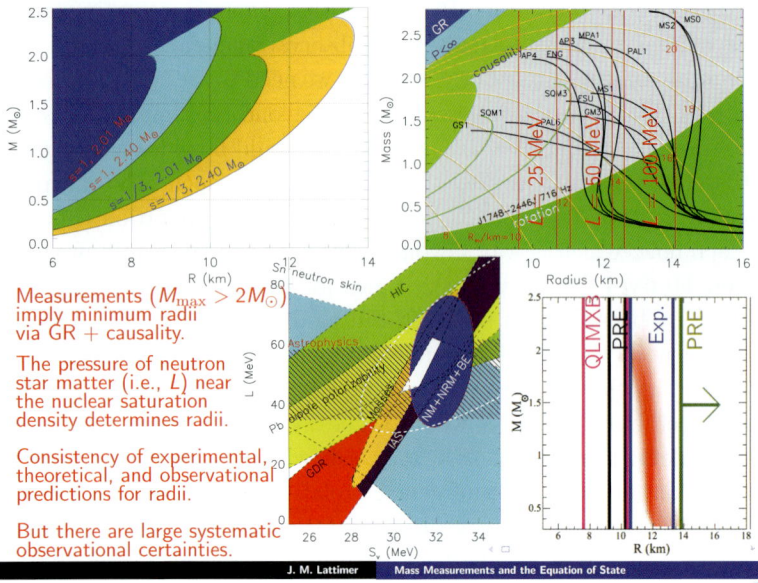

Fig. 1.   The slide presented at the meeting.

stars in the observed range of $1 - 2M_\odot$ have little radius variation. This is illustrated in the upper-right figure of my slide. Black curves are $M - R$ trajectories for several published equations of state. The colored regions in the upper left corner of the $M - R$ diagram are excluded by general relativity and causality; the green region in the lower right is excluded by observations of a 716Hz pulsar, which is spinning too rapidly for such configurations to be dynamically stable. Half of the equations of state illustrated can now be excluded because they cannot produce stars more massive than M_m. The viable $M - R$ curves have the constant radius property referred to above. Large values of M_m also disallow significant regions in $M - R$ space (upper-left figure). The contours labeled $s = 1$ assume causality and general relativity and for the indicated value of M_m forbid configurations to the left. With current precision measurements ($M_m > 2.01 M_\odot$), neutron stars with 1.4 $M_\odot$ cannot have radii smaller than about 8 km. If future measurements raise M_m to $2.4 M_\odot$, this radius limit increases to 11.1 km.

However, there are theoretical reasons for expecting that the causal limit is too extreme for realistic matter. In the high-density limit, it is believed that hadronic matter converts to asymptotically free quark matter in which the speed of sound is $c/\sqrt{3}$. If this lower value for the sound speed is utilized together with general relativity, larger radii limits are established for given values of M_m, as indicated in this figure with the $s = 1/3$ contours. Current measurements would then limit the minimum radii of neutron stars with

$1.4M_\odot$ to 11 km. Such a value conflicts with some recent small estimates of radii deduced from X-ray observations of both bursting neutron stars (PRE; black-bounded band in the lower-right figure) and cooling stars in quiescent low-mass X-ray binaries (QLMXB; magenta-bounded band).

While the maximum neutron star mass is an indicator of the pressure-density relation at high densities, 4 or more times the nuclear saturation density ($\rho_s \simeq 3 \times 10^{14}$ g cm^{-3}), the relatively constant neutron star radius for intermediate masses is an indicator of the equation of state at lower densities, $1 - 2\rho_s$. Specifically, it has been shown there is a high degree of correlation between the neutron star radius for $1.4M_\odot$ stars and the pressure of neutron star matter at ρ_s. If the symmetry energy E_{sym} of the nuclear force is defined as the difference between the energies of uniform cold neutron matter and uniform cold symmetric matter (i.e. having equal fractions of neutrons and protons), then a parameter L can be defined as $L = 3\rho_s(dE_{sym}/d\rho)_{\rho_s}$. In fact, since the proton fraction of neutron star matter near ρ_s is just a few percent, the pressure of cold neutron star matter at ρ_s is approximately $L\rho_s$. The resulting correlation between the radii of $1.4M_\odot$ neutron stars and the parameter L is shown by the vertical red bands in the upper-right figure. Therefore, there is a direct connection between the nuclear physics parameter L and neutron star radii.

There are several nuclear physics experiments which estimate L, including nuclear masses, giant dipole resonances, dipole polarizabilities, flows in heavy-ion collisions, energies of isobaric analog states, and the thicknesses of the neutron skin on neutron-rich nuclei. These experiments cannot individually establish L, but they do provide correlations between L and another symmetry parameter, $S_v = E_{sym}(\rho_s)$, as shown in the lower-left figure. The white trapezoidal region in this figure shows the consensus experimental values of S_v and L. The small range of permitted L values translates into a narrow region of permitted neutron star radii, indicated by the blue band in the lower-right figure.

Additionally, there have been recent advances in the theory of pure neutron matter from both quantum Monte Carlo and chiral Lagrangian approaches, constrained by scattering data and energies of light nuclei. The energy and pressure of pure neutron matter at ρ_s directly translate into ranges for S_v and L (the blue ellipsoid in the lower-left figure). The theoretical and experimental estimates agree well.

Some observational estimates of neutron star radii are relatively small (as mentioned above) and others are quite large (the olive-bounded region for PRE burst sources in the lower-right figure, for example), indicating there are still large systematic uncertainties in the interpretation of observational data for neutron stars. Nevertheless, the combined analysis of PRE and QLMXB sources, coupled with constraints provided by the general relativistic stellar structure equations, causality, the observed value of M_m, and the apparent existence of a well-understood neutron star crust, has led to intermediate radii estimates with an error of order ± 1 km (shown by the reddish band in the lower-right figure). Although this estimate did not rely on information from nuclear experiments or theoretical studies of neutron matter, it is rather consistent with experimental and theoretical predictions.

If the $s = 1/3$ sound speed constraint is valid for high-density matter, observations of neutron star radii and measurements of L from nuclear experiments and neutron matter theory are consistent with a neutron star maximum mass near $2M_\odot$. But as the upper-left figure indicates, should M_m increase to $2.4M_\odot$ or larger, tension between observations and experiment will exist.

Discussion

E. van den Heuvel Thank you very much. Lots of other types of observations could help us to constrain things further, spin rates perhaps.

V. Kaspi I just wanted to comment on some part of the plot. The maximum rate of rotation of neutron stars can also potentially be very constraining and if we can find neutron stars rotating much faster than a millisecond (currently the record holder is over 700 Hz) that could be quite constraining. I think it is important to recognize that in the past there has been huge selection biases against finding such objects. You might say that we haven't seen one yet so maybe it does not exist. But I think it is really only the modern present-day surveys that are sensitive to these objects. Perhaps, in the next few years we will find one and if not, it will be interesting to see what this will imply for the equation of state.

R. Blandford Another handle that you have involves the thermal properties and evolution of neutron stars, in particular the pion condensates that facilitate cooling. I would like to add that this sort of measurements can be competitive with the nuclear physics and mechanical deductions that have followed from pulsar timing.

J. Lattimer I think that the cooling measurements are giving more information about the internal composition of the star rather than the radii or global structure. So we are learning about superconductivity in the interior, the presence or absence of quark material and so on.

J. Ostriker I just have a question to Jim Lattimer. Are the observations of moments of inertia accurate enough to give any information?

J. Lattimer I refer this question to Michael Kramer. B. Schutz and I wrote a paper about this a few years ago. Measurement with ten percent accuracy would give a radius estimate on the order of five percent. That would be extremely important. We estimated that this could come perhaps in a decade but the paper was written almost a decade ago. Michael can perhaps give us an update about when this might be possible.

M. Kramer Yes, it is a challenge to extract the moment of inertia. We still work on it. The idea is that you measure the second post-Newtonian contribution to the periastron advance. If you isolate this contribution, it contains a term that has the moment of inertia in it. But in order to isolate it, we need to have two other post-Keplerian parameters added to the relativistic effects to the same accuracy. We have two, we need a third. It will take a few more years, and it will need bigger telescopes. But it is definitively doable. We published the paper in 2009 but we predicted when we reach there, I think it is about another five years or so depending on the accuracy of the observations that we have.

C. Aerts A suggestion from a stellar physicist. There is a real boost in this area by determining very precise masses and radii from seismology. It would seem to me that for the neutron stars, if you could just have seismic measurements, I know it is very hard and we need a space mission for that, we could deduce masses and radii either for single neutron stars or for binary neutron stars in the same way using the same methodology. Of course, we have to take out the variability due to the magnetism and the rotation. But that is similar to the things we have to do these days for exoplanet host stars to get their masses and radii up to a typical one-to-two percent precision. For anyone working on future space missions in X-rays, my suggestion would be that you try to give seismic detection capabilities because that will give you not only a handle on mass and radius but also on the properties of the equation of state. That is a suggestion I would like to give to the neutron star community.

E. van den Heuvel Thank you very much Conny. It is not so clear whether this is doable.

R. Romani I just want to return briefly to Jerry's question about the moment of inertia. There is a second path other than the railroad of pulsar timing to do this. And that is doing bolometric measurement of the total energy deposition. Of course from timing you get Ω and $\dot{\Omega}$. If you measure $\dot{E}$ with high precision, you get I. We showed there is a way of doing this using the pulsar $H\alpha$ bow shocks and we have made the first attempts. I have some observations coming out to improve such measurements. Yes there is some model dependence effects, but this is the kind of things you can do on a timescale of a year not ten years. Whether this will be five percent precision measurement or twenty percent remains to be seen. But one very attractive aspect of it, is that bow shocks are rare, but they occur both around binary neutron stars and around single millisecond pulsars. It would thus be possible, I believe, to make at least approximate estimates of moments of inertia for a wide variety of pulsars, even though the number will be small.

E. van den Heuvel Thank you Roger.

M. Rees The one topic where there has been little progress has been the understanding of why pulsars pulse. I would like to ask whether the experts have any comments on what is needed. Is it an issue of plasma physics? Or would it help to understand better the surface of the neutron star? the albedo? the work function? and topics like that?

V. Kaspi I could comment. It depends. You mean why they pulse, we think it is rotation. It is why they shine. I think it is what you mean. It depends on what waveband you are talking about. In the radio, it is quite complicated. Clearly there is plasma physics although there are different instabilities that have been proposed to explain the origin of the radio emission. But it is important to recognize that the radio emission represents only a tiny fraction of the spin-down luminosity. Although it is a very rich phenomenology, as an observer I would love to talk about what that pulsar does and what another one does. At the end, it is very little energy but the bulk of the energy is coming out in gamma rays and to a lesser degree in X-rays. For those emission mechanisms, I think there has been quite a bit of progress thanks to the Fermi telescope that has been detecting lots of millisecond pulsars. There has been certainly tremendous progress in modeling the geometry of the magnetosphere and probably Roger could say more words about progress.

E. van den Heuvel I think Roger is going to do that in his contribution.

T. Damour A comment. From gravitational waves emitted by binary neutron stars, we will be able to measure tidal polarizability of neutron stars and Love numbers, especially if their radii are larger than twelve kilometers for $1.4M_\odot$. Recent progress in the modeling of the waveform from binary neutron stars allows to see timing effects in the waveform up to the coalescence of the two neutron stars. And this is sensitive to the tidal polarizability and the fifth power of the radius of the neutron star.

J. Ostriker There is a tiny effect and I wonder whether this is included in the analysis, Vicky can say. If you look at the most distant pulsars in binary pulsars, the radiation coming from it will graze the surface near the neutron star and you will get the Shapiro delay. This will give you the size of the neutron star. Is that in the analysis?

M. Kramer It is not. It is just one post-Newtonian treatment of the Shapiro delay right now. It does not take this higher order effect into account yet. We do not have the precision yet that this requires. So in short, this is not taken into account, but we do not need to take it into account because we cannot measure it.

J. Ostriker In principle, this will give you the radius of the neutron star.

M. Kramer Sure, but it is not taken into account right now.

E. van den Heuvel Because of time and because there are lots of other subjects we would like to cover, I would like to close this discussion on the equation of state. We heard from Vicky that pulsars are born with large velocities, on average about 400 kilometres a second: they get a kick when they are born. But in recent years an important new development has been that some neutron stars are apparently born almost without a kick. That was found in a group of high mass X-ray binaries with almost circular orbits. So there seems to be another group of neutron stars, which hardly get any kick when they are born. Philipp Podsiadlowski has put forward some very interesting ideas about how this could relate to the different possible theoretical scenarios of neutron-star formation. I would like to ask Philipp to comment on this.

Prepared comment

Ph. Podsiadlowski: Neutron-Star Formation and Supernova Kicks

While there has been significant progress in recent years in modelling core-collapse supernovae and the formation of neutron stars, even the most sophisticated 3-dimensional hydrodynamical simulations are still unable to produce supernova explosions as are observed. Nevertheless, the most promising model is still that of *delayed neutrino-driven explosions*, where a fraction (typically a few per cent) of the neutrinos generated in the collapse are deposited just outside the accretion shock until enough energy has been accumulated there to reverse the infall and produce an outflow, i.e. a supernova explosion. In this paradigm, the supernova mechanism is a threshold process where the explosion energy is set by the binding energy of the outer regions of the core. This naturally explains why most neutron-star forming supernovae have a characteristic explosion energy of 10^{51} ergs.

There has also been dramatic progress in understanding the *supernova kicks* neutron stars receive when they are born in a supernova, as had long been established observationally from the high space motion of young radio pulsars. The most promising model involves a *standing accretion shock instability (SASI)*. As first shown in 2-d simulations, the surface of the accretion shock that surrounds the proto-neutron star in the infall phase is unstable and sloshes around, thereby imparting momentum to the proto-neutron star. This can give a net kick velocity of several 100 km/s to the neutron star, consistent with the observations of pulsar velocities, provided the instability has enough time to grow. As the timescale of the instability is set by the convective turn-over time behind the shock (tens of ms), it typically requires that this accretion phase lasts more than ~ 500 ms, i.e. hundreds of dynamical times, before the explosion is initiated.

One important implication of this paradigm is that it predicts that the kicks should be different depending on whether a neutron star forms from the collapse of an iron core in a massive star (i.e. in an *iron-core collapse supernova*) or is triggered by the capture of electrons onto Ne and Mg in an ONeMg core (i.e. in a so-called *electron-capture supernova*). The latter is believed to occur in stars with an initial mass around $10\,M_\odot$. As the binding energy of the envelope in these systems is much lower than in an iron core, these are expected to explode much more easily and with much lower energies than is the case for an iron core collapse. As these also produce neutron stars with a very characteristic mass of $1.25\,M_\odot$, we theoretically expect two classes of neutron stars: those formed by electron capture with a mass of $1.25\,M_\odot$ and low kick velocities and those formed by iron core collapse with a broader mass distribution $(1.34 \pm 0.03\,M_\odot)$ and large kick velocities. Present observations, including the famous double pulsar PSR J0737-3039, appear to confirm these expectations.

However, as a caveat one needs to note that, in the most recent 3-dimensional simulations, the nature of the SASI instability and the geometry of the explosion appear to be quite different from the 2-dimensional case. Finally, in the case of the formation of a black hole, a similar kick mechanism may be responsible for the observed kicks in some black-hole binaries, provided the black hole forms in a two-step process, passing through an intermediate neutron-star phase with a weak neutrino-driven explosion.

Discussion

E. van den Heuvel Thank you very much Philipp. Who would like to address a question to Philipp?

T. Piran I want to mention that the binary pulsars in which two pulsars are seen have a very peculiar orbit, almost circular and it is in the Galactic plane. If you combine just the orbital parameters of this system, including the direction of rotation (which has been measured from the pulsar observations) and the fact that the system is almost standing still (peculiar velocities are only ten kilometres per second), one can get extremely precise determination of the initial condition of the system before the last collapse. And you find that the progenitor of pulsar B was actually a star which had to be about $1.5 \pm 0.05 M_\odot$. This star collapsed almost without mass ejection, between $0.02 M_\odot$ and $0.1 M_\odot$ only. This is a unique progenitor, which, I think, does not fit any of the current formation mechanisms. One has to keep in mind that such system does exist and wonder what was really happening in the formation process.

P. Podsiadlowski I am slightly surprised by that statement, because I would actually use that system as a particular example for what I was talking about. Because the pulsar B with its mass $1.25 M_\odot$, I think, was from electron-capture and that is very much consistent with the progenitor. If you look at binary evolution models, a progenitor of $1.5 M_\odot$ will eject $0.1 M_\odot$ or $0.2 M_\odot$. We have models like that. So I would actually use that in strong support of this picture.

E. van den Heuvel Maybe one could add here that neutron stars forming with small kicks produce a large selection effect in the production of binary pulsars because those are the ones that survive in binaries. So you would expect many of the ones in binaries to have these small kicks actually. And the same hold for globular clusters. So maybe someone could also comment on that.

P. Podsiadlowski I would like to comment on that because this is a very important selection factor. It is also particularly important for the direct detection of gravitational waves from neutron star-neutron star and black hole mergers because these systems all have to survive two supernovae. If you only have high kicks, 99 percent of the systems get unbound after the first supernova. This is why those rates are so uncertain.

S. Kulkarni Back to Philipp's suggestions of electron capture, there is some development which may have some bearing on this particular model. There is a class of transients, which we see now in reasonable abundance. They shine primarily in the near infrared. These luminous red novae, which occur in spiral arms, may in fact be the supernovae associated with stars anywhere in the range $8 - 10 M_\odot$, where as you know, there is a longstanding uncertainty on what exactly happens to them. However there is less dispute on what happens in their AGB evolution. The general agreement is that they go though an extreme AGB phase, which will then provide a big cocoon that then explains why these supernovae, unlike ordinary supernovae, shine primarily in the mid-infrared. There are many selection effects in finding these things. But the fact that we have found a few of them accidentally suggests to me that the rates are in accord with this idea of electron capture. There is a very nice paper, which combines all this by Kochanek and it is worth reading for those interested in this topic.

E. van den Heuvel That is a great comment. Thank you very much for this.

C. Aerts I have a question for Philipp. As far as I am aware from the literature, the simulations from the supernova explosions do not yet take into account the latest news on the internal rotation and mixing of the progenitor. My question is: how much dependent are you on what is for us the outcome of stellar evolution while this is for you the initial condition?

P. Podsiadlowski In principle, that could completely change the picture of course because if you start with really rapidly rotating neutron stars you can extract a significant amount of the rotational energy. Now we do not believe that most stars are rapidly rotating at the end of their evolution because they spin down rapidly mainly by magnetohydrodynamical processes and lose that angular momentum in a wind. In fact, the observations you are referring to, suggests that we underestimate the amount of angular momentum loss. These cores, which have been measured by Kepler, are rotating more slowly than expected. So at least for the normal neutron stars, we do not expect this effect to be important. That could be different for a subset, most or almost all massive stars are in binary systems and if you have a late interaction, you could easily end up with a rapidly rotating core. Then the picture may be quite different.

C. Aerts So the Kepler data only tell us something on what you are referring to on red giants. These are much lower masses. I am talking about OB stars. So I guess we could talk off-line on a coffee break about that.

M. Kamionkowski I have a question about the SASI. Few years ago, there was some analytical work suggesting that there might be physical ingredients not included in the simulations and that could quench the SASI. Could you remind me what the status is?

P. Podsiadlowski I think the status is still quite confusing. I presented it in a very simplified way to make it more understandable. But the actual fact is that in three dimensions things look quite different. Now the simulators tell you that you can still get high kicks of a few hundred kilometres per second. But the description is quite different. I have not really answered your question actually, but I think the details still need to be sorted out. Let me recall that these simulations do not produce explosions that we actually see.

E. van den Heuvel Now I would like to continue with other aspects of the formation of neutron stars and their evolution. We heard about the millisecond pulsars, that they are recycled; other people call them "reborn" because they were pulsars first and then later in the binary system they were spun up for a long time and after the companion switches off accretion they became radio pulsars again. Now I would like to call two prepared presentations, one by Roger Romani and one by Sterl Phinney, which both have to do with the latest millisecond pulsar discoveries. The Fermi satellite came up with very surprising discoveries, namely that the millisecond pulsars are very strong high-energy gamma-ray sources. Actually it was predicted by Srinivasan in the 1980's that they should be, but I think that prediction was completely forgotten. So when Fermi discovered that, this was a great surprise. Many new millisecond pulsars turned up just by looking at the gamma rays. I would like to ask Roger to tell us a little bit about these discoveries.

Prepared comment

R.W. Romani: Fermi Pulsars – Lessons Learned

In the study of neutron stars astrophysics osculates fundamental physics. We heard nice reviews of the zoo of astrophysical neutron star manifestations and of the precision tests of gravity that are accessible through pulsar timing. Historically, neutron star discovery and characterization is centered in the radio and X-ray bands. However, the *Fermi* γ-ray space telescope has opened up a new high energy window on the neutron star population, revealing a variety of pulsars that offer some novel physics opportunities. I would like to summarize this progress.

Before *Fermi*, we had clear γ-ray pulsations from a half-dozen radio pulsars and from one pulsar not seen in the radio, 'Geminga'. Only hints of a few additional sources had been collected. Five years into the *Fermi* mission the (ever increasing) GeV-band detections stand at:

- 49 young radio pulsars
- 41 young γ-selected pulsars, not detected in the radio
- 68 Millisecond pulsars (MSP), *all* radio- *and* γ-detected, including
  - 10 'Redbacks' or pulsar/LMXB transition objects with $\sim 0.1 - 0.3 M_\odot$ binary companions experiencing strong mass loss
  - 18 'Black Widows' with $\sim 0.01 - 0.03 M_\odot$ companions

Although the basic properties of the energetic GeV pulsar population were, I would argue, discernible even in the pre-*Fermi* era, this harvest has caused a fairly robust convergence on a few basic characteristics. The principal conclusions from the overall sample are:

1) **Spin-powered neutron star magnetospheres are copious producers of multi-TeV** e^+/e^- **pairs.** In transiting the large scale dipole field, these produce γ-ray pulsations, likely principally curvature radiation.[Thereafter the pairs power 'Pulsar Wind Nebula' beam dumps, bright in the X-ray (synchrotron) and TeV (Compton). When, as for the Crab, the synchrotron component is *Fermi*-detected, this indicates PeV particles that challenge acceleration schemes, as presently understood] Remarkably, the pulsed GeV νF_ν peak represents the dominant photon product of the pulsar machine, taping up to half the spindown power. The clear lesson is that, at least for typical pulsar magnetic moments, the dominant radiative output of a spinning magnet in vacuum is few-GeV photons. Details remain to be deciphered, but since $(\text{GeV·keV})^{1/2} \approx m_e c^2$, we suspect that pair production on thermal or synchrotron photons may be relevant. In any event the GeV band is the place to study the pulsar machine. Accreting sources and magnetars do not appear to set up a similar particle accelerator.

2) **The GeV emission arises at high altitude**, approaching, or even slightly exceeding, the light cylinder radius $r_{LC} = cP/2\pi$. This means that the γ-ray beams cover much of the sky, in contrast to the more tightly beamed radio pulse, which arises below $10 - 100 r_{NS}$. Of course, for MSP with small P, r_{LC} lies *within* $10 - 100 r_{NS}$, so radio and gamma-rays should be co-extensive, and both visible.

So, γ-ray beams are wide and penetrating. Accordingly *Fermi* can discover objects with important physics implications. For example:

A) For young pulsars, *Fermi* finds Geminga-type objects beamed away from Earth in the radio. Many are quite nearby and their e^+/e^- winds may well be a significant contributor to the local cosmic ray positron excess. Had we missed these, we might be more likely to attribute this signal to exotica such as annihilating dark matter particles.

B) For old pulsars (MSP), the GeV γ-rays can penetrate dense companion winds that scatter or absorb the corresponding radio beams. Thus *Fermi* has increased the number of known Galactic black widows and redbacks ten-fold. In compact orbits with few-hour periods, these systems provide important insights into X-ray binary evolution and the origin of the strongly recycled single and binary MSP so useful for gravitational wave searches. Intriguingly, the neutron stars in these systems appear to have experienced extensive accretion, raising the prospect that careful follow-on mass measurements can constrain QCD via the dense matter equation of state.

We conclude that the *Fermi* γ-ray sky provides a unique treasure map, pointing to powerful neutron star accelerators. Multi-wavelength study of these gems promises new probes of extreme physics.

S. Phinney: Black Widows

Discussion

R. Sunyaev I just want to speak about different little things. This is great to hear how diverse are neutron stars in our Galaxy. But there is also a very important class of objects: low mass X-ray binaries. Three hundred objects have been found in our Galaxy and maybe thousands or more may have been seen by Chandra in elliptical galaxies. These objects, we believe, are responsible for the spin-up of neutron stars. This is the connection to previous discussions. It was very interesting to see today during the talk of Vicky Kaspi that maybe half of millisecond pulsars have magnetic fields less than 10^8 G, some of them may even have magnetic field of order 10^7 G or less. In the case of so small magnetic fields and very high rate of accretion, the accretion disk comes to the surface of neutron stars and we have invaluably important boundary layers. This is hydrodynamics, which I think we will never be able to create in our labs because accretion disk plasma is moving around neutron stars with a velocity of the order of half of the velocity of light. This is a real hydrodynamical flow and we are observing now a lot of instabilities in these flows. These flows are radiation dominated. The pressure of radiation exceeds many tens or even sometimes hundred of times the pressure of matter, which we cannot create in the lab either, except for laser plasma. We made with Igumensahchev, ten years ago the theory of this radiation dominated boundary layer. There are two bright belts around the neutron star, one above and one below the equator, where the radiation flux is very close to the local Eddington flux and everything is in balance and levitating. It was possible to compute how these belts will be radiating: their radiation is much harder than the radiation from the accretion disk. Also it was possible to show that all instabilities produced there exhibit a much more rapid variability than the instabilities in the extended accretion disk. What is interesting is that all these key predictions are now observed. Really there is a very rapidly variable component, which has a temperature of the order of 7- 8 keV, and another component which is much softer coming from the accretion disk whose characteristic variability timescale is much longer. For me it is a surprise that nobody can simulate this boundary layer, which is enormously interesting because the velocities that we are able to reach in the labs are four orders of magnitude smaller than those in these accretion disks. It is beautiful physics and with a lot of consequences. It gives the possibility by just observing the variability (for instance the ratio of the luminosities of the accretion disk and of the boundary layer) to obtain a lot of information about the moment of inertia, general relativity, and many properties of the accretion disk. This is a new window and I think it is good to ask numerical people to continue to think in this direction.

E. van den Heuvel I would like to shift to another subject now, and discuss about the relation between neutron stars and gamma-ray bursts. In the past, e.g. T. Piran in the late eighties, it was suggested that the coalescence of double neutron stars could account for gamma-ray bursts and we will come to that a little later. Recently, it has been also suggested that the long gamma-ray bursts are related to neutron stars. That would mean, given that some of the long gamma-ray bursts have very high redshifts (up to 8 or even 9) that they may come from the epoch of re-ionisation. This means that there could even be a connection between neutron stars and the epoch of re-ionisation. I would like to ask now Ralph Wijers to give a short presentation on the connection between long gamma-ray burst, neutron stars and black holes.

Prepared comment

R. Wijers: Causes and Uses of (Stellar) Explosions

We have known since ancient times that occasionally the sky presents us with new objects, whose arrival generally comes unpredicted. Only in the last decades, however, have we learnt that in some wavelengths (in particular X and gamma rays), sources appearing at random, or varying greatly, are the rule rather than the exception. And now that in the present decade new instruments and computer techniques have made sensitive, all-sky searches for such transients possible, we find they exist at all wavelengths, and new types of transient are found every year. These discoveries have opened up new frontiers, and we have every indication that the time-domain era has only just started.

A good example of the new lessons of the time domain era comes from supernovae and gamma-ray bursts, and their comparison. Classical studies of supernovae suggested that these are relatively standard events, with a single mechanism controlled by only a few key parameters (notably, the explosion energy and ejected Ni mass). For a subset, the Type Ia supernovae, this seems to work: they can be so well standardized that we can test and even discover cosmological models with them (even though we do not yet agree on their cause!). All the others, thought to arise from various types of core collapse of massive stars, have however been displaying an ever greater variety, with peak luminosities now ranging from 1/100,000th to 100 times the former semi-standard value. At the same time, a somewhat less old mystery, the long gamma-ray bursts, has also been traced to the deaths of very massive stars. The main difference is not in explosion energy, which seems to cover the upper range of supernovae or perhaps a bit more, but the fact that ultrarelativistic ejecta are produced by a highly anisotropic inner engine. While the leading model for this is some type of magnetized outflow from a newly born, rapidly spinning black hole, very few models have really been excluded. For example, rapidly spinning magnetars might also work, albeit that we understand even less how to make those. And who is to say that we should not consider a mechanism similar to GRBs for regular supernovae, with the difference that the jet engine ceases before jets emerge from the stellar envelope? Some supernovae, after all, are known from spectropolarimetry to be quite asymmetric. And meanwhile, the range of explosion energies of gamma-ray bursts has also expanded greatly, with so-called 'low-luminosity GRBs' being up to 100,000 times fainter and potentially much more common. I think all this should encourage us to take a fresh look at the problem of stellar explosion mechanisms, and perhaps leave some oldish dogma's about them behind us. Although I would still bet for black-hole births as the most likely cause for bright, long GRBs, I also think that Nature will slowly fill in the parameter space of observed phenomena, showing us that 'the' mechanism for them does not exist, but a few different physical mechanisms are at play to various degrees in stellar death, producing a wide and continuous range of phenomena.

Since long GRBs can be observed to very high redshift (the current record is 9.4), the question has been posed whether, like type Ia supernovae, they could be cosmological probes, or even be used for precision cosmology. Despite many attempts at the latter, I personally believe using them as precison cosmology tools is very premature. Claimed intrinsic relationships between GRB parameters are wrought with strong and yet poorly understood selection effects, making conclusions about the geometry of our Universe from using such relationships quite unreliable. However, this does not mean that they cannot be used as tools to understand stellar evolution and the star formation history of our universe better. Each line of sight to a

GRB is a unique probe of cosmic history, revealing details such as metallicity, star formation rate, and other parameters of a sample of galaxies that is selected very differently from normal galaxy and star formation rate surveys. Some of these indicate that the star formation rate measured by GRBs differs from that measured by other types of indicator. That could teach us important clues on what type of star or stellar evolution path more likely leads to a GRB, and also possibly clues to how star formation in the early universe might have differed from that at the present time. Here too, however, one should be aware that all these star formation rate measures have biases and assumptions built in, which we do not yet understand well enough to draw definitive conclusions. This makes it all the more beneficial that we have several very differently biased probes and samples of the star formation rate. What is certain, however, is that the unique view of the early universe provided by these bright transients has been quite valuable. Lastly, it is interesting to note that in the past few years, more types of new transient have been discovered that could be even rarer or stranger than GRBs. An example are the so-called Fast Radio Bursts: millisecond pulses of radio radiation that show a very strong dependence of pulse arrival time with frequency, characteristic of propagation through a very large column density of free electrons. So large, in fact, that these are likely extragalactic and of very speculative origin (the merger of two neutron stars has been suggested, in which case coincidence with gravity-wave bursts is predicted). If true, these objects might be the best probes ever of the extragalactic matter density outside large cosmological mass concentrations, besides being another very interesting riddle in high-energy astrophysics.

In short, I predict that the next Solvay meeting with an astrophysical bent will have time domain astronomy as one of its key topics, and that in it weirder monsters will be shown than the ones I just spoke of.

Discussion

E. van den Heuvel Thank you very much. Still I would like to discuss the connection to neutron stars. I remember that in the very distant past, Jerry Ostriker and Gunn put the question that maybe very fast spinning neutron stars could make supernovae and this idea was revived with millisecond magnetars which could be driving long gamma-ray burst. Lars Bildsten and collaborators made some models for that, but unfortunately he is not here so we cannot ask him.

R. Wijers I think it is not excluded as the difference in energy between a gamma-ray burst and a supernova is only a factor of a few. I am not sure whether from a neutron star you could make things as strongly asymmetric as gamma-ray bursts need to be and that is why I have some slight preference from making them via black holes, where we know you could make very sharply collimated jets. But, you know, to some extent this remains to be proven.

E. van den Heuvel Who would like to comment? If not, I would like to ask Tsvi Piran to present the relation between merging double neutron stars and effects in the gamma ray range and other types of emission, which they could produce (including gravitational waves). This will be immediately followed by the contribution from Chryssa Kouveliotou, who will tell us about the relation between double neutron star mergers, short gamma-ray bursts and gravitational wave signals. I would like to recall that Chryssa was the discoverer of short gamma-ray bursts in 1993, as a separate category.

Prepared comment

T. Piran: Short Gamma-Ray Bursts, Macronovae, Gold and Radio Flares

The discovery of a faint IR signal following the short GRB (sGRB) 130603B[a,b] has opened the possibility to resolve several long standing puzzles in Astronomy. The tentative identification of this signal as a Li-Paczynski[c] Macronova (also referred to as kilonova) indicated that a few hundredths of a solar mass of neutron star matter were ejected and have radioactively decayed into heavy r-process elements. If correct, this confirms the long standing prediction[d] that on the one hand, sGRBs are produced in compact binary mergers and on the other hand that these events are significant and possibly the dominant sources of heavy r-process nuclei.

r-Process Nucleosynthesis: The decompression of cold nuclear matter ejected during a compact binary merger would lead to the conditions suitable for r-process nucleosynthesis.[e,4] Numerical simulations[f] revealed that $\sim 10^{-2} M_\odot$ ejected a merger event. Nuclear network calculations based on these hydrodynamic simulations[g] showed that the resulting abundance pattern agrees well with the observed solar system abundances for $A > 130$. These results were later refined and confirmed by numerous authors.

Gamma-Ray Bursts: Eichler et al.[4] placed compact binary mergers in a broader context. They suggested that in addition to bursts of gravitational waves and neutrinos these mergers are the engines of GRBs. Later on it turned out that they are the likely origin of the subclass of short GRBs.

---

[a] N. R. Tanvir, et al., Nature 500, 547, 2013.

[b] E. Berger, W. Fong, and R. Chornock, ApJL. 774, L23, 2013.

[c] L. Li and B. Paczyński, ApJL. 507, L59, 1998.

[d] D. Eichler, M. Livio, T. Piran, and D. N. Schramm, Nature 340, 126, 1989.

[e] J. M. Lattimer and D. N. Schramm, ApJL, 192, L145, 1974.

[f] S. Rosswog, M. Liebendörfer, F.-K. Thielemann, M. Davies, W. Benz, and T. Piran, A&A 341, 499, 1999.

[g] C. Freiburghaus, S. Rosswog, and F.-K. Thielemann, ApJL, 525, L121, 1999.

Estimating the rate of mergers they also suggested that these events could be a major source of r-process material.

Macronova: Li and Paczynski[3] suggested the third link in this chain. They pointed out that the radioactive decay of the neutron-rich nuclei in this ejecta would produce a Macronova: a short lived optical - IR weak supernova-like signal. A critical point in current estimates of Macronova emission is the realization[h] that that the opacity of the ejecta is very large as it is dominated by Lanthanides. This has led to a qualitative shift in the expected light curve. Earlier estimates, based on iron group opacities, predicted a UV-optical signal peaking at around half a day after the burst. The huge opacities of the Lanthanides imply a weaker IR signal peaking at a week.[i,j,k]

GRB 130603B: Remarkably these revisions in the predicted light curve were done in the spring of 2013 just a few weeks before the light of GRB 130603B reached Earth. Hubble observations of the late afterglow of this burst have shown a nIR point source with an apparent magnitude of $H_{160,AB} = 25.73 \pm 0.2$ ($M_{J,AB} \approx -15.35$), corresponding to an intrinsic luminosity of $\approx 10^{41}$erg/sec. The upper limit on the R band emission at the same time, $R_{606,AB} > 28.5$, suggests that the regular afterglow has already decayed. This nIR source is consistent with the emission of $0.02 - 0.04 M_\odot$ Macronovae. Unfortunately this was observed only once ($\sim$9 days after the burst) before it decayed as well. The interpretation hinges on just this single data point.

GRB 060614: This discovery motivated a detailed study of other afterglows. A recent re-analysis of the afterglow data of GRB 060614 provided a tentative identification of a Macronova signal in that event[l] as well. Unfortunately also in that case there is only a single data point showing an IR excess.

Cosmochemical evolution: Assuming that this interpretation is correct and using the current estimates of the rate of sGRBs[m] and a beaming factor of 50, mergers associated with sGRBs can produce all the observed heavy r-process material in the Universe. However, r-process elements have been observed in halo and disk stars covering a metallicity range [Fe/H] $\approx$ -3.1-0.5.[n] This requires a significant fraction of r-process nucleosynthesis to take place within a very short time after the onset of star formation. One may wonder whether mergers, that depend on orbital decay due to gravitational

[h]D. Kasen, N. R. Badnell, and J. Barnes , ApJ, 774, 25, 2013.

[i]J. Barnes and D. Kasen, ApJ., 775, 18, 2013.

[j]M. Tanaka and K. Hotokezaka K., 2013, ApJ, 775, 113, 2013.

[k]D. Grossman, O. Korobkin, S. Rosswog and T. Piran, MNRAS, 439, 757, 2014.

[l]B. Yang, et al., Nature communication, in press, 2015, arXiv:1503.07761.

[m]D. Wermanand and T. Piran, MNRAS, 448, 3026, 2015.

[n]e.g. V. Hill, et al., A. & A. 387, 560, 2002.

radiation emission, can take place so early. Current data involving either sGRBs or binary neutron stars in the Galaxy is insufficient to determine whether this is indeed the case[12]. Interestingly, the abundance of the short lived ^{244}Pu in the early solar system compared with its abundance at the current ISM[o] suggests that a merger has enriched the pre-solar system material with a heavy r-process elements just a few hundred million years before its formation.

Radio Flares: Can we find further support for the Macronova interpretation? Luckily the answer is yes. Nakar and Piran[p] pointed out that the interaction of the merger's ejecta with the surrounding matter will produce a Radio Flare lasting a few years. The identification of such a flare is still possible in both GRB 130603B and GRB 060614 and searches for such signals are in place now. If detected it would clearly confirm this interpretation.

Implications: If the interpretation is correct this discovery has several far reaching implications:

- It provides the first *direct*[q] evidence that sGRBs arise from compact binary mergers.
- The late time IR emission arises only if the opacity is dominated by Lanthanides. Thus, it indicates that a substantial amount of heavy r-process nuclei was produced in this event.
- Macronovae and Radio Flares are the electromagnetic counterparts of the mergers' gravitational waves signals that the advanced detectors, like LIGO Virgo and KAGRA, aim to detect. They could be used to localize and identify these sources and enhance the sensitivity of these detectors.[r]

C. Kouveliotou: Short Gamma-Ray Bursts, Mergers, Gravitational Waves, and *ISS-Lobster*

In this era of extreme specialization and scientific stove-piping, it is extremely gratifying to identify areas that bring together a broad range of disciplines. Gamma-Ray Bursts (GRBs) are a prime example of such an area. Since their discovery in 1967 as explosive transients in γ-rays, the field has expanded to multiple *wavelengths* (X-rays, UV, optical, IR, mm- and radio waves), *messengers* (Ultra High Energy Cosmic Rays, neutrinos, gravitational waves (GWs)), and *fields* (galaxy formation and evolution,

---

[o]A. Wallner, et al., Nature Communication, 6, 5956, 2015.
[p]E. Nakar and T. Piran, Nature 478, 82, 2011.
[q]So far there is only indirect evidence that short GRBs arise from mergers: nature of hosts, position within the hosts and overall rate.
[r]C.S. Kochanek, and T. Piran, ApJL, 417, L17, 1993.

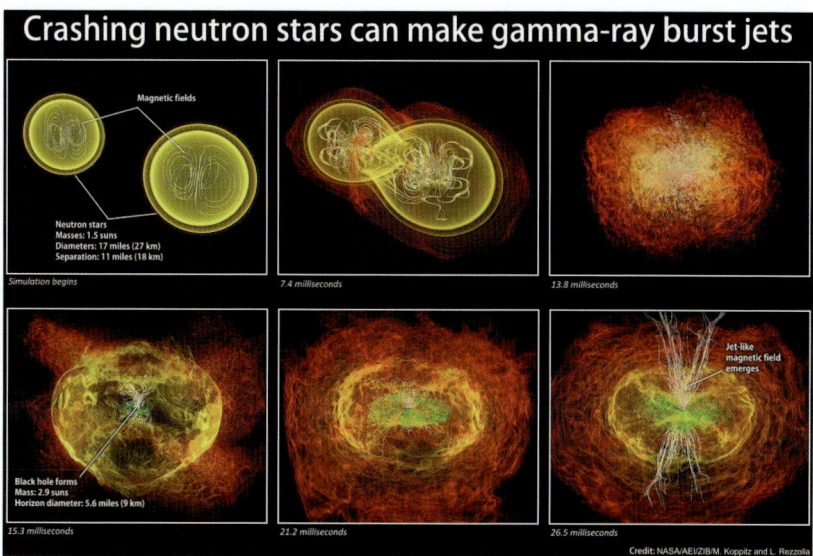

Fig. 1. Snapshots[x] at 6 representative times of the evolution of a binary (NS-NS) merger and the formation of a large-scale ordered magnetic field.

star formation and metallicities, dynamic cosmic chemical evolution, warm hot intergalactic medium, population III stars and the re-ionization era, dust properties, jet dynamics and energetics, stellar collapses/mergers and their physics, extragalactic background light, quantum gravity).[s] Here I was asked to comment on the connection among short GRBs, compact object mergers, and GWs.

GRBs are the most luminous electromagnetic transients in our Universe. Their durations span ms to thousands of seconds with very diverse light curves, their sky distribution is isotropic, and their currently known cosmological distances occupy the redshift range $0.0085 < z < 9.6$. The only classification that has withstood time so far is their apparent duration and hardness bimodality,[t] separating events at roughly 2 s into short–soft and long–hard GRBs. Although this is a phenomenological classification, it has provided the basis for dividing the GRB progenitors in two prevalent models: *collapsars* (mostly long GRBs resulting from the catastrophic collapse of rapidly rotating Wolf-Rayet stars[u]) and *mergers* (short GRBs resulting from the merging of two neutron stars or a neutron star and a black hole[v]). Short GRBs constitute $< 25\%$ of the total population and they release on

[s]GRBs, Kouveliotou, C., Wijers R., Woosley, S. 2012, Cambridge University Press.
[t]Kouveliotou, C. et al. 1993, APJL 413, 101.
[u]Woosley, S. 1993, ApJ 405, 273.
[v]Eichler, D. et al. 1989, Nature 340, 126.

average 10 − 1000 times less energy (isotropic equivalent) than long ones. Their afterglow counterparts are also fainter and shorter, thus presenting an observational challenge until the launch of NASA's *Swift* satellite in 2004 which utilized its rapid repointing capability to enable the first accurate locations of short GRBs and their multi-wavelength follow up observations. A solid confirmation of their progenitors, however, is still lacking, despite strong circumstantial evidence linking them to coalescing compact objects. Among these, is the unambiguous identification of an elliptical galaxy with an old (> 1 Gyr) stellar population as the host of the short GRB 050724,[w] pointing to mergers as the progenitors of short GRBs.

From the theoretical point of view, state of the art simulations[x] have shown using a fully general relativistic and ideal-magnetohydrodynamic (MHD) framework that the merger of two modestly magnetized NSs (10^{12} G) can produce an ultra-strong poloidal field (10^{15} G) along the merger spin axis that can help launch a relativistic jet (Figure 1) with properties broadly consistent with γ- and X-ray observations. This and follow up work[y] exemplify that mergers may indeed be short GRB progenitors. Furthermore, the merging process produces a distinct GW signal until the actual merger and the formation of a black hole (Figure 2, left panel). The beauty of short GRBs is that they produce transient electromagnetic (EM) and gravitational signals whose simultaneous detections would be a groundbreaking discovery.

Enter *ISS-Lobster*. This proposed Mission of Opportunity experiment operates on the simple concept of locating the EM counterpart of the GW signal of a merger. It can rapidly repoint its 900 sq. deg. field of view and identify the source of aLIGO transients locating them to $< 1'$ accuracy for ground-based follow up observations, which will in turn accurately determine their source distances. Thus the powerful combination of GW and multi-wavelength EM observations will revolutionize multi-messenger astrophysics, determine in unprecedented detail the physics of mergers and GRBs, test theoretical models and state-of-the-art simulations,[z] and, most importantly, provide astrophysical context to the GW sources, associating their now known distances with their physical properties.

Discussion

E. van den Heuvel Thank you very much. We have still some 5 minutes left for discussions about the last three presentations.

[w]Berger, E. et al. 2005, Nature 438, 988.
[x]Rezzolla, L. et al. 2011, APJL, 732, 6.
[y]e.g., Giacomazzo, B. et al. 2011, PhRvD, 83, 44014.
[z]Noble, S. et al. 2012 ApJ, 755, 51.

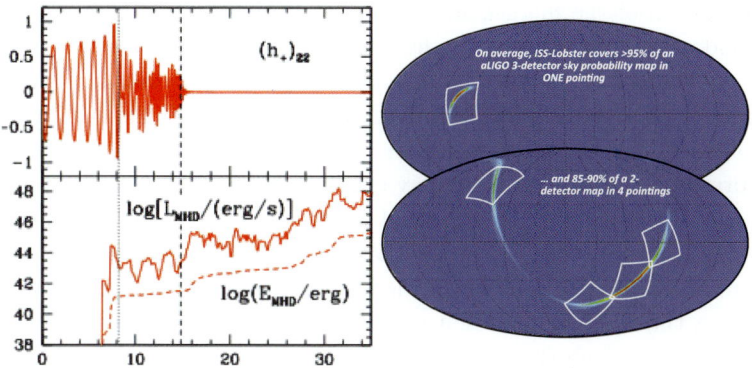

Fig. 2. Left: Evolution of the GW signal during a merger (top); and of the MHD luminosity (bottom) (from Figure 2 in Rezzolla et al. (2010)). Right: Locating the electromagnetic counterparts of GW signals: the white box represents the *ISS-Lobster* field of view; the colored contours are the aLIGO 3-detector (top) and 2-detector sky probability maps (bottom).

W. Hoffman Just a point of clarification: can you understand or intuitively describe what leads to the magnetic field build-up on that timescale and what determines the opening angle of the jet?

C. Kouveliotou The assumption is that the torus is very conductive and it rotates differentially hence it produces a shearing mode and therefore it creates the alignment into a conical form, this vessel type of alignment. The simulations show that both the toroidal and the poloidal fields can grow up to 10^{15} G. So it is the combination of the high conductivity and the differential rotation of the torus which is responsible for the creation of this.

M. Begelman There is a real controversy now about the nature of the poloidal field that is necessary to produce these jets. The general picture where you have a rotating body with a magnetic field producing a jet usually assumes that the power of the jet is proportional to the square of the magnetic flux. The magnetic flux is conserved, and is limited to value originally contained in the neutron star. Without violating total flux conservation, a dynamo might create large fluctuations in the flux threading the jet. But the question is whether enough flux can thread the crank at a given time to create the required power or whether the magnetic field reconnects and dissipates before that happens.

C. Kouveliotou Yes, the original magnetic field, to my recollection, is about 10^{12} G in each star and the abrupt rise of the magnetic field power leads to a magneto-rotational instability in the simulation. I am pretty sure that this has been taken into account.

T. Piran I just want to remark that one has to be careful about the comparison of what comes out of the central engine and the observations. Right now, there is the realisation that the matter which is ejected in the process of the formation of the merger and also matter which has been ejected from the wind of the precursor of the hypermassive neutron star before it collapsed to a black hole, is surrounding the system. Whatever is formed by the central engine, has to pass through this material, so the eventual opening angle of the jet, will depend on the interaction of this outflow with the surrounding material. This may change significantly the observed opening angle of the jet compared to what is actually being produced in the central core. In some sense, it means that even the short gamma-ray bursts are similar to collapsars in which the jets have to cross the whole star. Now, it is not the whole star that has to be crossed, but still, the previous ejecta are influential on the observed features.

E. van den Heuvel Thank you for this last remark. I would like to thank not only our two rapporteurs but also those who presented prepared comments and all who have contributed to the discussions.

Session 2

Black Holes

Chair: *Scott Tremaine*, IAS, Princeton, USA
Rapporteurs: *Reinhard Genzel*, MPE, Garching, Germany and *Mitchell Begelman*, University of Colorado, USA
Scientific secretaries: *Geoffrey Compère* and *Stephane Detournay*, Université Libre de Bruxelles, Belgium

Rapporteur Talk by R. Genzel: Massive Black Holes: Evidence, Demographics and Cosmic Evolution

Abstract

The article summarizes the observational evidence for the existence of massive black holes, as well as the current knowledge about their abundance, their mass and spin distributions, and their cosmic evolution within and together with their galactic hosts. We finish with a discussion of how massive black holes may in the future serve as laboratories for testing the theory of gravitation in the extreme curvature regimes near the event horizon.

1. Introduction

In 1784 Rev. John Michell was the first to note that a sufficiently compact star may have a surface escape velocity exceeding the speed of light. He argued that an object of the mass of the Sun (or larger) but with a radius of 3 km (instead of the Sun's radius of 700,000 km) would thus be invisible. A proper mathematical treatment of this problem then had to await Albert Einstein's General Relativity ("GR", 1916). Karl Schwarzschild's (1916) solution of the vacuum field equations in spherical symmetry demonstrated the existence of a characteristic event horizon, the Schwarzschild radius $R_s = 2GM/c^2$, within which no communication is possible with external observers. Roy Kerr (1963) generalized this solution to spinning black holes. The mathematical concept of a black hole was established (although the term itself was coined only later by John Wheeler in 1968). In GR, all matter within the event horizon is predicted to be inexorably pulled toward the center where all gravitational energy density (matter) is located in a density singularity. From considerations of the information content of black holes, there is significant tension between the predictions of GR and Quantum theory (e.g. Susskind 1995, Maldacena 1998, Bousso 2002). It is generally thought that a proper quantum theory of gravity will modify the concepts of GR on scales comparable to or smaller than the Planck length, $\ell_{Pl} \propto 1.6 \times 10^{-35}$ m, remove the concept of a central singularity, and potentially challenge the interpretation of the GR event horizon (Almheiri et al. 2013).

But are these objects of GR realized in Nature?

2. First Evidence

Astronomical evidence for the existence of black holes started to emerge in the 1960s with the discovery of distant luminous 'quasi-stellar-radio-sources/objects' (QSOs, Schmidt 1963) and variable X-ray emitting binaries in the Milky Way (Giacconi et al. 1962). It became clear from simple energetic arguments that the enormous luminosities and energy densities of QSOs (up to several 10^{14} times the luminosity of the Sun, and several 10^4 times the entire energy output of the Milky Way Galaxy), as well as their strong UV-, X-ray and radio emission can most plausibly be explained by accretion of matter onto massive black holes (e.g. Lynden-Bell 1969, Shakura & Sunyaev 1973, Rees 1984, Blandford 1999). Simple theoretical considerations show that between 7% (for a non-rotating Schwarzschild hole) and 40% (for a maximally rotating Kerr hole) of the rest energy of an infalling particle can in principle be converted to radiation outside the event horizon, a factor 10 to 100 more than in stellar fusion from hydrogen to helium. To explain powerful quasars by this mechanism, black hole masses of 10^8 to 10^9 solar masses and accretion flows between 0.1 to 10 solar masses per year are required. QSOs are located (without exception) at the nuclei of large, massive galaxies (e.g. Osmer 2004). QSOs just represent the most extreme and spectacular among the general nuclear activity of most galaxies. This includes variable X- and γ-ray emission and highly collimated, relativistic radio jets, all of which cannot be accounted for by stellar activity.

The 1960s and 1970s brought also the discovery of X-ray stellar binary systems (see Giacconi 2003 for an historic account). For about 20 of these compact and highly variable X-ray sources dynamical mass determinations from Doppler spectroscopy of the visible primary star established that the mass of the X-ray emitting secondary is significantly larger than the maximum stable neutron star mass, $\sim$3 solar masses (McClintock & Remillard 2004, Remillard & McClintock 2006, Özel et al. 2010).

The binary X-ray sources thus are excellent candidates for stellar black holes (SBH). They are probably formed when a massive star explodes as a supernova at the end of its fusion lifetime and the compact remnant collapses to a stellar hole.

An unambiguous proof of the existence of a stellar or massive black hole, as defined by GR, requires the determination of the gravitational potential to the scale of the event horizon. This proof can in principle be obtained from spatially resolved measurements of the motions of test particles (interstellar gas or stars) in close orbit around the black hole. In practice it is not possible (yet) to probe the scale of an event horizon of any black hole candidate (SBH as well as MBH) with spatially resolved dynamical measurements. A more modest goal then is to show that the gravitational potential of a galaxy nucleus is dominated by a compact non-stellar mass and that this central mass concentration cannot be anything but a black hole because all other conceivable configurations are more extended, are not stable, or produce more light (e.g. Maoz 1995, 1998). Even this test cannot be conducted yet in distant QSOs from dynamical measurements. It has become feasible over the last decades in nearby galaxy nuclei, however, including the Center of our Milky Way.

3. NGC 4258

Solid evidence for central 'dark' (i.e. non-stellar) mass concentrations in about 80 nearby galaxies has emerged over the past two decades (e.g. Magorrian 1998, Kormendy 2004, Gültekin et al. 2009, Kormendy & Ho 2013, McConnell & Ma 2013) from optical/infrared imaging and spectroscopy on the Hubble Space Telescope (HST) and large ground-based telescopes, as well as from Very Long Baseline radio Interferometry (VLBI).

The first truly compelling case that such a dark mass concentration cannot just be a dense nuclear cluster of white dwarfs, neutron stars and perhaps stellar black holes emerged in the mid-1990s from spectacular VLBI observations of the nucleus of NGC 4258, a mildly active galaxy at a distance of 7 Mpc (Miyoshi et al. 1995, Moran 2008, Figure 1). The VLBI observations show that the galaxy nucleus contains a thin, slightly warped disk of H2O masers (viewed almost edge on) in Keplerian rotation around an unresolved mass of 40 million solar masses (Figure 1). The inferred density of this mass exceeds a few 109 solar masses pc-3 and thus cannot be a long-lived cluster of 'dark' astrophysical objects of the type mentioned above (Maoz 1995). As we will discuss below, a still more compelling case can be made in the case of the Galactic Center.

4. The Galactic Center Black Hole

The central light years of our Galaxy contain a dense and luminous star cluster, as well as several components of neutral, ionized and extremely hot gas (Genzel, Hollenbach & Townes 1994, Genzel, Eisenhauer & Gillessen 2010). The central dark mass concentration discussed above is associated with the compact radio source SgrA*, which has a size of about 10 light minutes and is located at the center of

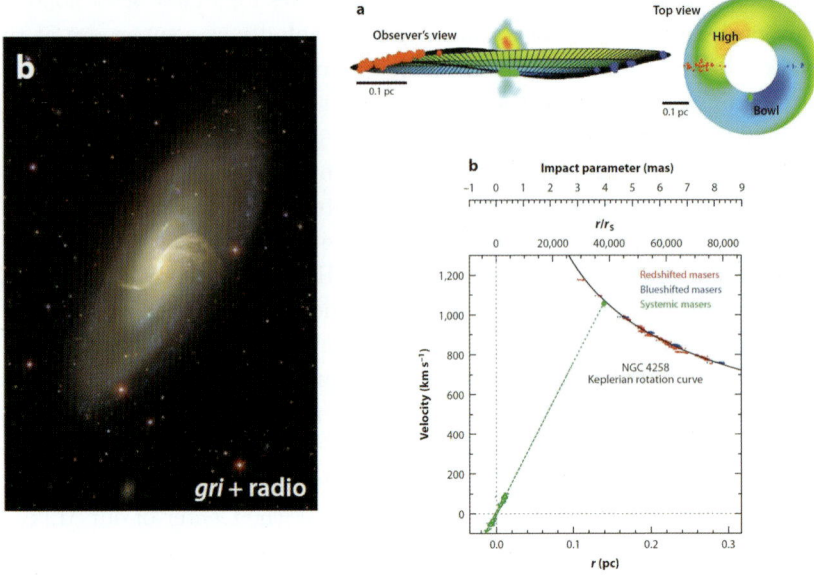

Fig. 1.   Left: Optical and radio image of the active galaxy NGC4258. This disk galaxy exhibits a spectacular curved twin radio and X-ray jet, visible in orange in this picture. Right: (top) Schematic edge-on (left) and face-on (right) views of the almost-edge-on, warped maser disk of NGC 4258 (from Moran 2008) with warp parameters from Herrnstein et al. (2005) and including the inner contours of the radio jet. The relative positions of the receding, near-systemic, and approaching H_2O masers are indicated by red, green, and blue spots, respectively. Differences in line-of-sight projection corrections to the slightly tilted maser velocities account for the departures in the high-velocity masers from exact Keplerian rotation. The near-systemic masers are seen tangent to the bottom of the maser disk bowl along the line of sight. They drift from right to left in ~ 12 years across the green areas where amplification of the background radio continuum is sufficient for detection. (b) NGC 4258 rotation velocity versus radius in units of parsec (bottom axis), Schwarzschild radii (top axis), and milliarcsec (extra axis). The black curve is a Keplerian fit to 4255 velocities of red- and blue-shifted masers (red and blue dots). The small green points and line show 10036 velocities of near-systemic masers and a linear fit to them. The green filled circle is the corresponding mean velocity point. The maser data are taken from Argon et al. (2007) (adapted from Kormendy & Ho 2013).

the nuclear star cluster. SgrA* thus may be a MBH analogous to QSOs, albeit with orders of magnitude lower mass and luminosity. Because of its proximity – the distance to the Galactic Center is about 8.3 kilo-parsecs (kpc), about 10^5 time closer than the nearest QSOs – high resolution observations of the Milky Way nucleus offer the unique opportunity of carrying out a stringent test of the MBH-paradigm and of studying stars and gas in the immediate vicinity of a MBH, at a level of detail that will not be accessible in any other galactic nucleus for the foreseeable future. Since the Center of the Milky Way is highly obscured by interstellar dust particles in the plane of the Galactic disk, observations in the visible part of the electromagnetic spectrum are not possible. The veil of dust, however, becomes transparent at longer wavelengths (the infrared, microwave and radio bands), as well as at shorter wavelengths (hard X-ray and γ-ray bands), where observations of the Galactic Center thus become feasible.

The key obviously lies in very high angular resolution observations. The Schwarzschild radius of a 4 million solar mass black hole at the Galactic Center subtends a mere 10^{-5} arc-seconds.[a] For high resolution imaging from the ground an important technical hurdle is the correction of the distortions of an incoming electromagnetic wave by the refractive Earth atmosphere. For some time radio astronomers have been able to achieve sub-milli-arcsecond resolution VLBI at millimeter wavelengths, with the help of phase-referencing to nearby compact radio sources. In the optical/near-infrared waveband the atmosphere distorts the incoming electromagnetic waves on time scales of milliseconds and smears out long-exposure images to a diameter of more than an order of magnitude greater than the diffraction limited resolution of large ground-based telescopes (Figure 2). From the early 1990s onward initially 'speckle imaging' (recording short exposure images, which are subsequently processed and co-added to retrieve the diffraction limited resolution) and then later 'adaptive optics' (AO: correcting the wave distortions on-line) became available, which have since allowed increasingly precise high resolution near-infrared observations with the currently largest (10 m diameter) ground-based telescopes of the Galactic Center (and nearby galaxy nuclei).

Early evidence for the presence of a non-stellar mass concentration of 2-4 million times the mass of the Sun ($M_\odot$) came from mid-infrared imaging spectroscopy of the 12.8μm [NeII] line, which traces emission from ionized gas clouds in the central parsec region (Wollman et al. 1977, Lacy et al. 1980, Serabyn & Lacy 1985). However, many considered this dynamical evidence not compelling because of the possibility of the ionized gas being affected by non-gravitational forces (shocks, winds, magnetic fields). A far better probe of the gravitational field are stellar motions, which started to become available from Doppler spectroscopy in the late 1980s. They confirmed the gas motions (Rieke & Rieke 1988, McGinn et al. 1989, Sellgren et al. 1990, Krabbe et al. 1995, Haller et al. 1996, Genzel et al. 1996). The ultimate breakthrough came from the combination of AO techniques with advanced imaging and spectroscopic instruments (e.g. 'integral field' imaging spectroscopy, Eisenhauer et al. 2005) that allowed diffraction limited near-infrared spectroscopy and imaging astrometry with a precision initially at the few milli-arcsecond scale, and improving to a few hundred micro-arcseconds in the next decade (c.f. Ghez et al. 2008, Gillessen et al. 2009). With diffraction limited imagery starting in 1992 on the 3.5m New Technology Telescope (NTT) of the European Southern Observatory (ESO) in La Silla/Chile, and continuing since 2002 on ESO's Very Large Telescope (VLT) on Paranal, a group at MPE was able to determine proper motions of stars as close as ~0.1" from SgrA* (Eckart & Genzel 1996, 1997). In 1995 a group at the University of California, Los Angeles started a similar program with the 10m

---

[a]10 μarc-seconds correspond to about 2cm at the distance of the Moon.

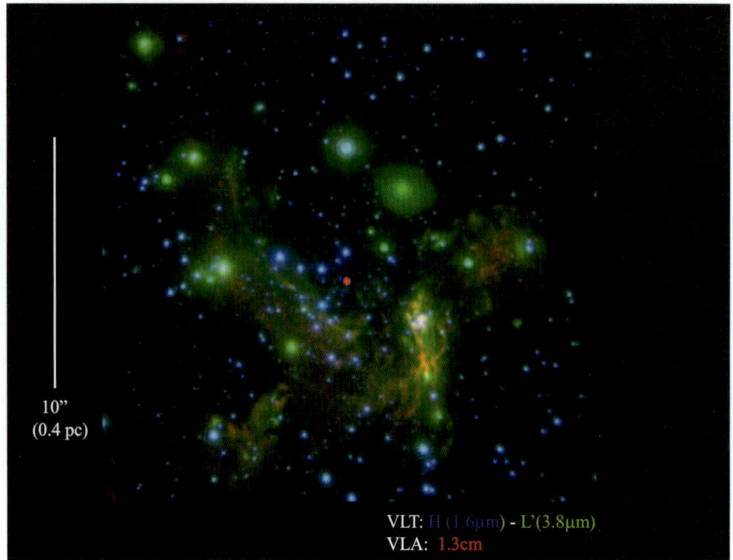

10"
(0.4 pc)

VLT: H (1.6μm) - L'(3.8μm)
VLA: 1.3cm

Fig. 2.   Near-infrared/radio, color-composite image of the central light years of Galactic Center. The blue and green colors represent the 1.6 and 3.8μm broad band near-infrared emission, at the diffraction limit ($\sim$ 0.05") of the 8m Very Large Telescope (VLT) of the European Southern Observatory (ESO), and taken with the 'NACO' AO-camera and an infrared wavefront sensor (adapted from Genzel et al. 2003). Similar work has been carried out at the 10 m Keck telescope (Ghez et al. 2003, 2005). The red color image is the 1.3cm radio continuum emission taken with the Very Large Array (VLA) of the US National Radio Astronomy Observatory (NRAO). The compact red dot in the center of the image is the compact, non-thermal radio source SgrA*. Many of the bright blue stars are young, massive O/B- and Wolf-Rayet stars that have formed recently. Other bright stars are old, giants and asymptotic giant branch stars in the old nuclear star cluster. The extended streamers/wisps of 3.8μm emission and radio emission are dusty filaments of ionized gas orbiting in the central light years (adapted from Genzel, Eisenhauer & Gillessen 2010).

diameter Keck telescope in Hawaii (Ghez et al. 1998). Both groups independently found that the stellar velocities follow a 'Kepler' law ($v \propto 1/\sqrt{R}$) as a function of distance from SgrA* and reach $\geq 10^3$ km/s within the central light month.

Only a few years later both groups achieved the next and crucial steps. Ghez et al. (2000) detected accelerations for three of the 'S'-stars, Schödel et al. (2002) and Ghez et al. (2003) showed that the star S2/S02 is in a highly elliptical orbit around the position of the radio source SgrA*, and Schödel et al. (2003) and Ghez et al. (2005) determined the orbits of 6 additional stars. In addition to the proper motion/astrometric studies, they obtained diffraction limited Doppler spectroscopy of the same stars (Ghez et al.2003, Eisenhauer et al. 2003, 2005), allowing precision measurement of the three-dimensional structure of the orbits, as well as the distance to the Galactic Center. Figure 3 shows the data and best fitting Kepler orbit for S2/S02, the most spectacular of these stars with a 16 year orbital period (Ghez et al. 2008, Gillessen et al. 2009, 2009a). At the time of writing, the two groups have determined individual orbits for more than 40 stars in the central light month. These orbits show that the gravitational potential indeed is that of a point mass centered on SgrA*. These stars orbit the position of the radio source SgrA*like planets around the Sun. The point mass must be concentrated well within the peri-approaches of the innermost stars, ~10-17 light hours, or 70 times the Earth orbit radius and about 1000 times the event horizon of a 4 million solar mass black hole. There is presently no indication for an extended mass greater than about 2 % of the point mass.

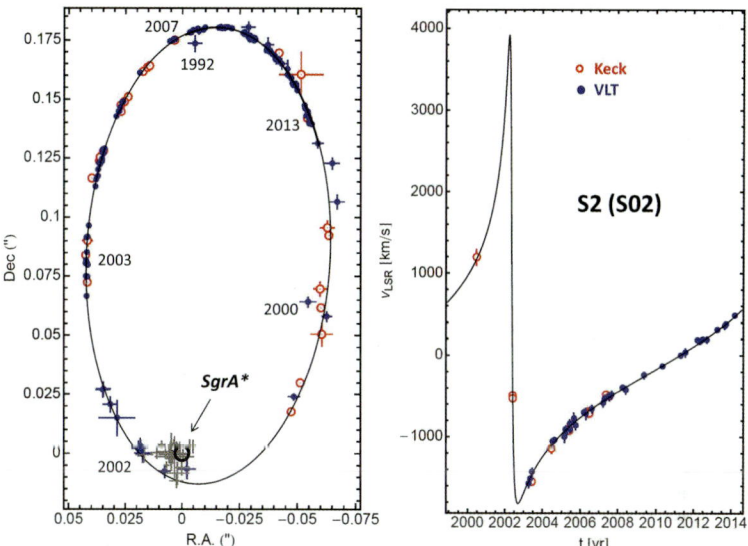

Fig. 3. Position on the sky as a function of time (left) and Doppler velocity (relative to the Local Standard of Rest) as a function of time (right) of the star 'S2 (or S02)' orbiting the compact radio source SgrA*. Blue filled circles denote data taken with the ESO NTT and VLT as part of the MPE Galactic Center monitoring project (Schödel et al. 2002, 2005, Gillessen et al. 2009), and red open circles denote data taken with the Keck telescope as part of the UCLA monitoring project (Ghez et al. 2003, 2008, see Gillessen et al. 2009a for the calibration to a common reference frame). Superposed is the best fitting elliptical orbit (continuous curve: central mass $4.26(\pm0.14)_{statistical}(\pm0.2)_{systematics}$ million solar masses, distance $8.36(\pm0.1)_{stat}(\pm0.15)_{syst}$ kpc) with its focus at (0,0) in the left inset (including the independent distance constraints of Reid et al. 2014, Chatzopoulos et al. 2014). The astrometric position of SgrA* is denoted by a circle, grey crosses mark the locations of infrared flares (of typical duration 1-3 hours) that are believed to originate from within the immediate vicinity of the event horizon. The radio source is coincident within the 2 milli-arcsecond errors with the gravitational centroid of the stellar orbit. Since the beginning of the MPE monitoring project (1991/1992), the star has completed its first full orbit in 2007, and it passed its peri-center position 17 light hours from SgrA* in spring 2002 (and again in spring 2018).

VLBI observations have set an upper limit of about 20 km/s and 2 km/s to the motion of SgrA* itself, along and perpendicular to the plane of the Milky

Way, respectively (Reid & Brunthaler 2004). When compared to the two orders of magnitude greater velocities of the stars in the immediate vicinity of SgrA*, this demonstrates that the radio source must indeed be massive, with simulations giving a lower limit to the mass of SgrA* of $\sim 10^5$ solar masses (Chatterjee, Hernquist & Loeb 2002). The intrinsic size of the radio source at about 1mm is only about 4 times the event horizon diameter of a 4 million solar mass black hole (Bower et al. 2004, Shen et al. 2005, Doeleman et al. 2008). Combining radio size and proper motion limit of SgrA* with the dynamical measurements of the nearby orbiting stars leads to the conclusion that SgrA* can only be a massive black hole, beyond any reasonable doubt (Genzel et al. 2010).

The current Galactic Center evidence eliminates all plausible astrophysical plausible alternatives to a massive black hole. These include astrophysical clusters of neutron stars, stellar black holes, brown dwarfs and stellar remnants (e.g., Maoz 1995, 1998; Genzel et al. 1997, 2000; Ghez et al. 1998, 2005), and even fermion balls (Viollier, Trautmann & Tupper 1993, Munyaneza, Tsiklauri & Viollier 1998, Ghez et al. 2005; Genzel, Eisenhauer & Gillessen 2010). Clusters of a very large number of mini-black holes and boson balls (Torres, Capozziello & Lambiase 2000; Schunck & Mielke 2003; Liebling & Palenzuela 2012) are harder to exclude. The former have a large relaxation and collapse time, the latter have no hard surfaces that could exclude them from luminosity arguments (Broderick, Loeb & Narayan 2009), and they are consistent with the dynamical mass and size constraints. However, such a boson 'star' would be unstable to collapse to a MBH when continuously accreting baryons (as in the Galactic Center), and it is very unclear how it could have formed. Under the assumption of the validity of General Relativity the Galactic Center is now the best quantitative evidence that MBH do indeed exist.

5. Massive Black Holes in the Local Universe

Beyond the "gold standards" in the Galactic Center and NGC 4258, evidence for the presence of central mass concentrations (which we will henceforth assume to be MBH even though this conclusion can be challenged in most of the individual cases), and a census of their abundance and mass spectrum comes from a number of independent methods:

- robust evidence for MBH in about 10 galaxies comes from VLBI studies of H_2O maser spots in circum-nuclear Keplerian disks of megamaser galaxies akin to NGC4258 (the NRAO "megamaser cosmology project", https://safe.nrao.edu/wiki/bin/view/Main/MegamaserCosmologyProject, Braatz et al. 2010, Kuo et al. 2011, Reid et al. 2013);
- robust evidence for MBH for about 80 galaxies comes from modeling of the spatially resolved, line-of-sight integrated stellar Doppler-velocity distributions with the Hubble Space Telescope (HST) and large ground based telescopes with AO (see the recent reviews of Kormendy & Ho 2013, McConnell & Ma 2013 and references therein). Among the latter, a particu-

larly impressive case is the nucleus of M31, the Andromeda galaxy, where a $10^8 \, M_\odot$ central mass is identified from the rapid (~ 900 km/s) rotation of a compact circum-nuclear stellar disk (Bender et al. 2005);

– for a number of galaxies, observations of the spatially resolved motions of ionized gas also provide valuable evidence for central mass concentrations, which, however, can be challenged, as mentioned for the Galactic Center, by the possibility of non-gravitational motions (Macchetto et al. 1997, van der Marel & van den Bosch 1998, Barth et al. 2001, Marconi et al. 2003, 2006, Neumayer et al. 2006);

– most recently, high resolution interferometric observations of CO emission have become available as a promising new tool for determining robust central masses (Davis et al. 2013);

– qualitative evidence for the presence of accreting black holes naturally comes for all bona-fide AGN from their IR-optical-UV- and X-ray spectral signatures. In the case of type 1 AGN with broad permitted lines coming from the central light days to light years around the black hole (c.f. Netzer 2013 and references therein), it is possible to derive the size of the broad line region (BLR) from correlating the time variability of the (extended) BLR line emission with that of the (compact) ionizing UV continuum. This reverberation technique (Blandford & McKee 1982) has been successfully applied to derive the BLR sizes (Peterson 1993, 2003, Netzer & Peterson 1997, Kaspi et al. 2000) in several dozen AGN, and has yielded spatially resolved imaging of the BLR in a few (e.g. Bentz et al. 2011, Kaspi et al. 2000). These observations show that the size of the BLR is correlated with the AGN optical luminosity, $R_{BLR} \sim [(\nu L_\nu)_{5100\mathring{A}}]^{0.7}$. After empirical calibration of the zero points of the correlation measurements of the line width of the BLR and the rest frame optical luminosity of the AGN are sufficient to make an estimate of the MBH mass. As this requires only spectro-photometric data, the technique can be applied even for distant (high redshift) type 1 AGNs (Vestergaard 2004, Netzer et al. 2006, Traktenbrot & Netzer 2012), as well as for low-luminosity AGN in late type and dwarf galaxies (Filippenko & Sargent 1989, Ho, Filippenko & Sargent 1997, Greene & Ho 2004, 2007, Ho 2008, Reines et al. 2011, Greene 2012, Reines, Greene & Geha 2013).

6. Demographics and MBH-galaxy "Co-evolution"

These data give a fairly detailed census of the incidence and of the mass spectrum of the local (and less so, also of the distant) MBH population. MBH masses span a range at least five orders of magnitudes from $10^5 \, M_\odot$ in dwarf galaxies to $10^{10} \, M_\odot$ in the most massive central cluster galaxies. Most massive spheroidal/bulged galaxies appear to have a central MBH. The occupation fraction drops in bulgeless systems with decreasing galaxy mass (Greene 2012). It is not clear yet whether the lack of observational evidence below 10^5 $M_\odot$ is real, or driven by observational detectability. The inferred black hole mass and the mass of the galaxy's spheroidal component (but not its disk, or dark matter halo) are strongly correlated (Magorrian et al. 1998, Häring & Rix 2004). The most recent analyses of Kormendy & Ho (2013) and McConnell & Ma (2013) find that between 0.3 and 0.5% of the bulge/spheroid mass is in the central MBH. The scatter of this relation is between ±0.3 and ±0.5 dex, depending on sample and analysis method (McConnell & Ma 2013). A correlation of comparable scatter exists between the black hole mass and the bulge/spheroid velocity dispersion σ (MBH $\sim \sigma^\beta$, with $\beta \sim 4.2 - 5.5$, Ferrarese & Merritt 2000, Gebhardt et al. 2000, Tremaine et al. 2002, Kormendy & Ho 2013, McConnell & Ma 2013, Figure 4).

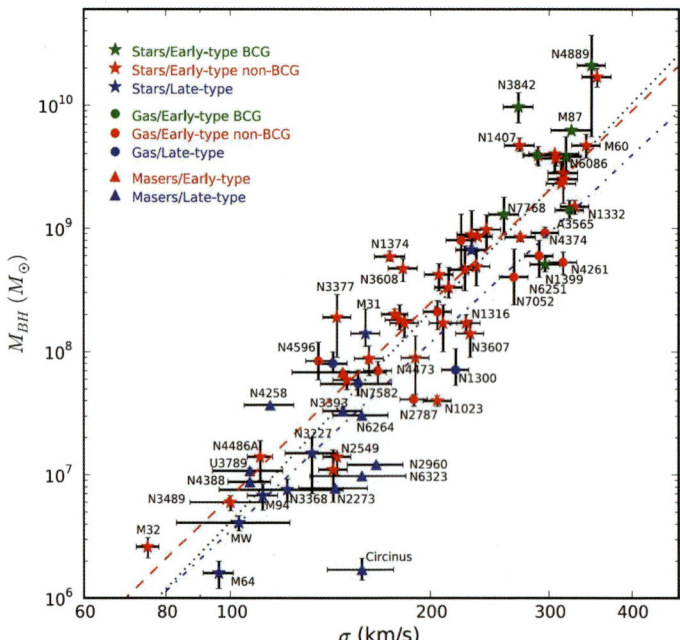

Fig. 4.   Black hole mass MBH (vertical axis) as a function of galaxy velocity dispersion σ (horizontal axis), for all 72 galaxies in the compendium of McConnell & Ma (2013). Asterisks, filled circles and filled triangles denote the technique that was used to determine the MBH mass (stellar kinematics, gas kinematics, or masers), and red, green and blue colors denote the type of host galaxy (spheroidal galaxy, very massive spheroidal galaxy at the center of a galaxy cluster (BCG), and late type (disk/ irregular) galaxy). The black dotted line shows the best-fitting power law for the entire sample: log10 $(M_{BH}/M_\odot) = 8.32 + 5.64 \log(\sigma/200\text{km/s})$. When early-type and late-type galaxies are fitted separately, the resulting power laws are $\log(M_{BH}/M_\odot) = 8.39 + 5.20 \log(\sigma/200\text{km/s})$ for the early-type (red dashed line), and $\log (M_{BH}/M_\odot) = 8.07 + 5.06 \log (\sigma/200\text{km/s})$ for the late-type galaxies (blue dot-dashed line). The plotted values of σ are derived using kinematic data within the effective radius of the spheroidal galaxy component (adapted from McConnell & Ma 2013).

Ever since this correlation between central black hole mass and galaxy host spheroidal mass (or velocity dispersion) component has been established, the interpretation has been that there must be an underlying connection between the formation paths of the galaxies' stellar components and their embedded central MBHs. This underlying connection points back to the peak formation epoch of massive galaxies about 6-10 Gyrs ago (e.g. Madau et al. 1996, Haehnelt 2004). The fact that the correlation is between the black hole mass and the bulge/spheroidal component, and not the total galaxy or dark matter mass, has been taken as evidence that most of the MBH's growth, following an early evolution from a lower mass seed, is triggered by a violent dissipative process at this early epoch. The most obvious candidate are major mergers between early gas rich galaxies, which are widely thought to form bulges in the process (Barnes & Hernquist 1996, Kauffmann & Haehnelt 2000, Haiman & Quataert 2004, Hopkins et al. 2006, Heckman et al. 2004). Compelling support for the AGN - merger model comes from the empirical evidence that dusty ultra-luminous infrared galaxies (ULIRGs, $LIR_{\textrm{¿}}10^{12}L_{\odot}$) in the local Universe are invariably major mergers of gas-rich disk galaxies (Sanders et al. 1988); the majority of the most luminous late stage ULIRGs are powered by obscured AGN (Veilleux et al. 1999, 2009).

This 'strong' co-evolution model is further supported by the fact that the peak of cosmic star formation 10 Gyrs ago is approximately coeval with the peak of cosmic QSO activity (Boyle et al. 2000), and that the amount of radiation produced during this QSO era is consistent with the mass present in MBHs locally for a 10-20% radiation efficiency during MBH mass growth (Soltan 1982, Yu et al. 2002, Marconi et al. 2004, Shankar et al. 2009). There is an intense ongoing discussion whether or not MBHs and their hosts galaxies formed coevally and grew on average in lock-step (Figure 5, Marconi et al. 2004, Shankar et al. 2009, Alexander & Hickox 2012, Mullaney et al. 2012, del Vecchio et al. 2014), or whether MBHs started slightly earlier or grew more efficiently (Jahnke et al. 2009, Merloni et al. 2010, Bennett et al. 2011). The fact that the correlation appears to be quite tight suggests that feedback between the accreting and rapidly growing black holes during that era and the host galaxy may have been an important contributor to the universal shutdown of star formation and mass growth in galaxies above the Schechter mass, $M_S \geq 10^{10.9} \, M_{\odot}$ (Baldry et al. 2008, Conroy & Wechsler 2009, Peng et al. 2010, Moster et al. 2013, Behroozi et al. 2013).

7. AGN-MBH Feedback

Throughout the last 10 billion years galaxies have been fairly inefficient in incorporating the cosmic baryons available to them into their stellar components. At a dark matter halo mass near $10^{12} \, M_{\odot}$ this baryon fraction is only about 20% (of the cosmic baryon abundance), and the efficiency drops to even lower values on either side of this mass (e.g. Madau et al. 1996; Baldry et al. 2008, Conroy & Wechsler 2009, Guo et al. 2010, Moster et al. 2013, Behroozi et al. 2013). Galactic winds driven by

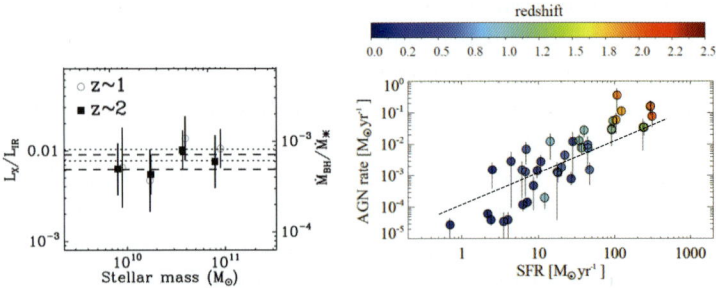

Fig. 5. Evidence for average MBH-galaxy growth co-evolution by stacking deep X-ray data (as quantitative indicators of AGN growth) on multi-wavelength images of star forming galaxies (including mid- and far-IR emission as extinction- and (nearly) AGN-independent tracers of star formation rates) in GOODS-S (left, Mullaney et al. 2012) and GOODS-S and COSMOS (right, delVecchio et al. 2014). The left plot shows that the ratio of inferred black hole growth to star formation rate is the same in several mass bins and at z~1 and z~2. The right plot shows the integrated MBH growth rate as a function of star formation and redshift (colors). The dotted line has slope unity and is not a fit to the data. However, once the dependence on mass and redshift are de-coupled, the best fitting correlation does have unity slope, suggesting average co-evolution.

supernovae and massive stars have long been proposed to explain the low baryon content of halos much below $\log(M_h/M_\odot)\sim12$ (e.g. Dekel & Silk 1986, Efstathiou 2000). The decreasing efficiency of galaxy formation above $\log(M_h/M_\odot) \sim 12$ may be caused by less efficient cooling and accretion of baryons in massive halos (Rees & Ostriker 1977, Dekel & Birnboim 2006). Alternatively or additionally efficient outflows driven by accreting MBH may quench star formation at the high mass tail, at and above the Schechter stellar mass, $M_S \sim 10^{10.9}\,M_\odot$ (di Matteo, Springel & Hernquist 2005, Croton et al. 2006, Bower et al. 2006, Hopkins et al. 2006, Cattaneo et al. 2007, Somerville et al. 2008, Fabian 2012).

In the local Universe, such 'AGN or MBH feedback' has been observed in the so called 'radio mode' in very massive, central cluster galaxies driving jets into the intra-cluster medium. In these cases the central MBHs typically is in a fairly low or quiescent radiative state. Considerations of energetics suggest that radio mode feedback plausibly prevents cooling cluster gas to fall onto these massive galaxies that would otherwise lead to substantial further star formation and mass growth (McNamara & Nulsen 2007, Fabian 2012, Heckman & Best 2014). A second MBH feedback mode (termed 'QSO mode'), in which the MBH is active (i.e. the AGN is luminous) is detected as ionized winds from AGN (e.g. Cecil, Bland, & Tully 1990, Veilleux, Cecil & Bland-Hawthorn 2005, Westmoquette et al. 2012, Rupke & Veilleux 2013, Harrison et al. 2014) and from obscured QSOs (Zakamska & Greene 2014). The QSO mode feedback in form of powerful neutral and ionized gas outflows has also been found in late stage, gas rich mergers (Fischer et al. 2010, Feruglio et al. 2010, Sturm et al. 2011, Rupke & Veilleux 2013, Veilleux et al. 2013), which however are rare in the local Universe.

At high-z AGN QSO mode feedback has been seen in broad absorption line quasars (Arav et al. 2001, 2008, 2013, Korista et al. 2008), in type 2 AGN (Alexander et al. 2010, Nesvadba et al. 2011, Cano Díiaz et al. 2012, Harrison et al. 2012), and in radio galaxies (Nesvadba et al.2008). However, luminous AGNs near the Eddington limit are again rare, constituting less than 1% of the star forming population in the same mass range (e.g. Boyle et al. 2000). QSOs have short lifetimes relative to the Hubble time ($t_{QSO} \sim 10^7 - 10^8$ yr $\ll t_H$, Martini 2004) and thus have low duty cycles compared to galactic star formation processes ($t_{SF} \sim 10^9$ yr, Hickox et al. 2014). It is thus not clear whether the radiatively efficient QSO mode can have much effect in regulating galaxy growth and star formation shutdown, as postulated in the theoretical work cited above (Heckman 2010, Fabian 2012).

From deep adaptive optics assisted integral field spectroscopy at the ESO VLT, Förster Schreiber et al. (2014) and Genzel et al. (2014) have recently reported the discovery of broad ($\sim 10^3$ km/s), spatially resolved (a few kpc) ionized gas emission associated with the nuclear regions of very massive ($\log(M_*/M_\odot) > 10.9$) z$\sim$1-2 star forming galaxies (SFGs). While active AGN do exhibit similar outflows, as stated above, the key breakthrough of this study is that it provides compelling evidence for wide-spread and powerful nuclear outflows in most ($\sim$70%) normal massive star forming galaxies at the peak of galaxy formation activity. The fraction of active, luminous AGN among this sample is 10-30%, suggesting that the nuclear outflow phenomenon has a significantly higher duty cycle than the AGN activity. If so, MBHs may indeed be capable to contribute to the quenching of star formation near the Schechter mass, as proposed by the theoretical work mentioned above.

8. Non-Merger Evolution Paths of MBHs

The most recent data on MBH demographics (Figure 4) suggest that the simple scenario of early MBH-galaxy formation through mergers and strong "co-evolution" might be too simplistic. Kormendy & Ho (2013, see also Kormendy, Bender & Cornell 2011) as well as McConnell & Ma (2013) find that MBHs in late type galaxies tend to fall below the best correlation of the pure spheroidal systems. The "pseudo"-bulges in these disk galaxies (including the Milky Way itself) typically rotate rapidly and may have partially been formed by radial transport of disk stars to the nucleus mediated through slow, secular angular momentum transport, rather than by rapid merger events. In these systems the efficiency and growth processes of MBHs appears to be lower than in the very massive spheroids that formed a long time ago. In the local Universe, the Sloan Digital Sky Survey has shown that most AGN are not involved in active mergers or galaxy interactions (Li et al. 2008). Most lower luminosity AGN are in massive early type hosts that are not actively fed. Most of the lower-mass MBH growth at low redshift happens in lower mass galaxies (Kauffmann et al. 2003, Heckman et al. 2004).

Lower mass ($< 10^{5.3..7} M_\odot$) MBHs have been found in bulge-less disks and even dwarf galaxies (Filippenko & Ho 2003, Barth et al. 2001, Barth, Greene & Ho

2005, Greene & Ho 2004, 2007, Reines et al. 2011, 2013), in which there appears to be no or little correlation between the properties of the galaxy and its central MBH, in contrast to the bulged/spheroid systems (Greene 2012). These MBHs must have formed more through an entirely different path. MBH growth in these cases is more likely to be controlled by local processes, such gas infall from local molecular clouds (Sanders 1998, Genzel et al. 2010 and references therein) and stellar mass loss following a nuclear 'starburst' (Scoville & Norman 1988, Heckman et al.2004, Davies et al. 2007, Wild et al. 2010).

At the peak of the galaxy formation epoch (redshifts z∼1-2) imaging studies show little evidence for the average AGNs to be in ongoing mergers (Cisternas et al. 2011, Schawinski et al. 2011, Kocevski et al. 2012). Instead most AGNs at this epoch are active star forming galaxies, including large disks, near the 'main-sequence' of star formation (Shao et al. 2010, Rosario et al. 2012, 2013). For active MBH, AGN luminosity and star formation rates are not or poorly correlated, excepting at the most extreme AGN luminosities (Netzer 2009, Rosario et al. 2012), yet the average MBH and galaxy growth rates are (Mullaney et al. 2012, del Vecchio et al. 2014). The empirical evidence for the AGN-merger model based on luminosity functions and spatial correlations (Hopkins et al. 2006) has been shown to not be a unique interpretation (Conroy & White 2013).

All these findings suggest that the concept of co-evolution between MBH growth and galaxy growth may most of the time be applicable only on average, or merely as a non-causal, statistical 'central limit' (Jahnke & Maccio 2011). One might call this 'weak' co-evolution. The instantaneous MBH growth rate at any given time exhibits large amplitude fluctuations (Hopkins et al.2005, Novak et al. 2011, Rosario et al. 2012, Hickox et al. 2014). Relatively rare gas rich mergers may be able to stimulate phases of strong co-evolution at all redshifts. At other times, radial transport of gas (and stars) in galaxy disks may be an alternative channel of MBH growth, at least at the peak of galaxy-MBH formation, since galaxies 10 billion years were gas rich (Tacconi et al. 2013), resulting in efficient radial transport from the outer disk to the nucleus (a few hundred million years, Bournaud et al. 2011, Alexander & Hickox 2012). These inferences from the empirical data are in good agreement with the most recent hydrodynamical simulations (Sijacki et al. 2014).

9. MBH Spin

X-ray spectroscopy of the 6.4-6.7 keV Fe K-complex finds relativistic Doppler motions in several tens of AGNs, following the initial discovery in the iconic Seyfert galaxy MCG-6-30-15 (Tanaka et al. 1995, Figure 5). The Fe-K profiles can be modelled as a rotating disk on a scale of few to 20 RS that reflects a power law, hard X-ray continuum emission component likely located above the disk (Tanaka et al. 1995, Nandra et al. 1997, 2007, Fabian et al. 2000, 2002, Fabian & Ross 2010, Reynolds 2013). While the X-ray spectroscopy by itself does not yield black hole masses, it provides strong support for the black hole interpretation. In addition, reverberation techniques of the time variable spectral properties are beginning to deliver interesting constraints on the spatial structure of the continuum and line components (Fabian et al. 2009, Uttley et al. 2014). From the modeling of the spectral profiles it is possible to derive unique constraints on the MBH spin, assuming that the basic modeling assumptions are applicable. The inferred spin for MCG-6-30-15 is near maximal (Figure 6). In a sample of 20 MBHs investigated in this way at least half have a spin parameter $a > 0.8$, providing tantalizing, exciting evidence for a frequent occurrence of high-spin MBH (Figure 6, Reynolds 2013 and references therein). These measurements promise to yield important information on the growth processes of MBH.

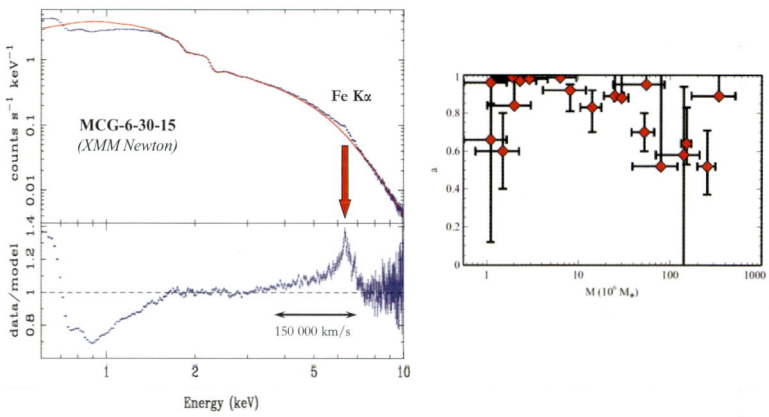

Fig. 6. Left: Fe-K profile of the iconic Seyfert galaxy MCG-6-30-15 (Tanaka et al. 1995) obtained with XMM-Newton (Fabian et al. 2002, Fabian & Vaughan 2003). The blue extension of the relativistic emission extends as low as ~ 3 keV. In the framework of a rotating accretion disk reflecting a hard X-ray power law component this means that there is emission at $\sim 2 \times R_{grav}$. If the MBH has a low angular momentum parameter, this would require significant reflection occurring within the last stable orbit. Alternatively, the MBH in MCG-6-3-15 is a near maximum Kerr hole with $a > 0.97$ (Brenneman & Reynolds 2006). Right: Inferred spin parameter a as a function of black hole mass, for 20 MBHs (Reynolds 2013)

10. Early Growth

The formation and evolution of MBHs faces two basic problems. One is angular momentum. To make it into the MBH event horizon from the outer disk of a galaxy, a particle has to lose all but 10^{-9} of its original angular momentum, a truly daunting task (c.f. Alexander & Hickox 2012). For this reason, major mergers have been considered natural candidate for being the sites of rapid MBH growth (Hopkins et al. 2006), since the mutual gravitational torques in a galaxy-galaxy interaction can reduce more than 90% of the angular momentum of a significant fraction of the total interstellar gas (e.g. Barnes & Hernquist 1996). However, this is by far not sufficient. It is likely that several stages of additional angular momentum loss much closer to the nucleus are involved in growing MBH, plausibly including star formation events at different 'way-points' including the nucleus itself (Scoville & Norman 1988, Davies et al. 2007, Hopkins & Quataert 2010, Meyer et al. 2010, Wild, Heckman & Charlot 2010, Alexander & Hickox 2012 and references therein). For this reason, black hole growth, accretion and radiation are probably highly time variable and strongly influenced by the properties of the gas and stellar environment in the sphere of influence around the black hole (Genzel et al. 2010, Hickox et al. 2014).

The second major obstacle is the time needed to grow to a final mass M from an initial seed of much lower mass M_0 (Volonteri 2010, 2012). This time is given by,

$$\frac{t_m}{t_{Salpeter}} = \frac{\eta}{1-\eta} \times \frac{1}{L/L_{Edd}} \ln\left(\frac{M}{M_0}\right),$$

where $t_{Salpeter} = 4 \times 10^8$ yr, η is the radiative efficiency, L_{Edd} is the Eddington luminosity $(3.4 \times 10^4 M(M_\odot))$ where the accreting MBH's radiation pressure equals its gravity. To grow to $M = 10^9 M_\odot$ at z~6 (~1 billion yr after the Big Bang) with $\eta \sim 0.1$ at the Eddington rate requires $4 \times 10^7 \ln(M/M_0)$ yr. If the initial seed formed in the re-ionization epoch at z~10, the seed mass has to exceed $\sim 10^4 M_\odot$. While $10^9 M_\odot$ MBHs as early as z~6 are rare (10-10 Mpc-3, Fan et al. 2006), and most very massive MBHs could have reached their final masses later, this example does show that standard Eddington accretion from a relative low mass seed, such as a super massive star $(M_0 \sim 10^2 M_\odot)$, cannot account for the oldest MBHs. Possibilities include fairly massive seeds $(\geq 10^4 M_\odot)$ formed from direct collapse of a dense gas cloud (Silk & Rees 1998), perhaps including a phase of super-Eddington accretion (see the more detailed account in Mitch Begelman's contribution in this volume, as well as the discussion in Begelman, Volonteri & Rees 2006, Volonteri 2010, 2012).

11. Zooming in on the Event Horizon

Looking forward to the next decade, there are several avenues to get still firmer constraints on the black hole paradigm, and determine the gravitational field still closer to the event horizon, in particular in the Galactic Center. Infrared spectroscopy of S2 during the next peri-approach in 2018 will have a good chance of detecting post-Newtonian parameters (Roemer effect, gravitational redshift, longitudinal Doppler effect, e.g. Zucker et al. 2006). Within the next decade it should be possible with current astrometric capabilities to detect S2's Schwarzschild precession angle,

$$\Delta\Phi_S = \frac{3\pi}{1-e^2}\left(\frac{R_s}{a}\right) \sim 12'. \tag{1}$$

The Schwarzschild precession and perhaps even the Lense-Thirring precession (due to the spin and quadrupole moment of the MBH) are obviously more easily detectable for stars with smaller semi-major axes and shorter orbital periods than S2.

One such star, S102/S55 has been reported by Meyer et al. (2012), but the current confusion limited imagery on 10m-class telescopes prevents further progress.

This barrier will be broken in the next few years by the GRAVITY near-IR interferometric experiment (Eisenhauer et al. 2005a, 2011, Gillessen et al. 2006), and with the next generation 30m-class telescopes a decade later (Weinberg et al. 2005). GRAVITY will combine the four VLT telescopes interferometrically, with the goal of 10μarcsec precision, near-infrared imaging interferometry (angular resolution a few milli-arcseconds). GRAVITY will also be able to search for dynamical signatures of the variable infrared emission from SgrA* itself (Genzel et al. 2003, Eckart et al. 2006, Do et al. 2009, Dodds-Eden et al. 2009, c.f. Baganoff et al. 2001). These 'flares' originate from within a few milli-arcseconds of the radio position of SgrA* and probably occur when relativistic electrons in the innermost accretion zone of the black hole are substantially accelerated so that they are able to produce infrared synchrotron emission and X-ray synchrotron or inverse Compton radiation (Markoff et al. 2001). As such the infrared variable emission as well as the millimeter and submillimeter emission from SgrA* probe the inner accretion zone between a few to 100 R_S. If orbital motion (of hot spots) could be detected the space time very close to the event horizon could potentially be probed (Paumard et al. 2005, Broderick & Loeb 2006, 2009, Hamaus et al. 2009).

VLBI at short millimeter or submillimeter wavelengths may be able to map out the strong light bending ('shadow') region inside the photon orbit of the MBH (Bardeen 1973, Falcke et al. 2000). This 'Event Horizon Telescope' Project (http://www.eventhorizontelescope.org/) will soon benefit the observations of the Galactic Center (and the relatively nearby $\sim 6 \times 10^9 M_\odot$ MBH in M87) from the much enhanced sensitivity and additional u-v-coverage with the ALMA interferometer in Chile (Lu et al. 2014). It is hoped that the shadow signature can be extracted fairly easily even from data with a sparse coverage of the UV-plane (e.g. Doeleman 2010). As in the case of the GRAVITY observations of the infrared flares, it is not clear, however, how the potentially complex emission structure from the inner accretion zone, including a possible radio jet, may compromise the interpretation of EHT maps in terms of GR effects (Dexter & Fragile 2013, Moscibrodzka et al. 2014).

Given the current presence of ~ 200 OB stars in the central parsec (Genzel et al. 2010) and extrapolating to earlier star formation episodes, there should be 100-1000 neutron stars, and thus potentially many pulsars within the parsec-scale sphere of influence of the Galactic Center MBH (Pfahl & Loeb 2004, Wharton et al. 2012). Until recently none have been found, despite many radio searches. The blame was placed on the large dispersion of the radio pulses by large columns of electrons in front of the Galactic Center sight line. In 2013 SWIFT and NuSTAR discovered a magnetar, 1745-2900, within 3" of SgrA*, with a pulse period of 3.7 s, whose radio pulse characteristics have since been studied in detail (Kennea et al. 2013, Mori et al. 2013, Eatough et al. 2013, Spitler et al. 2014, Bower et al. 2014). While the magnetar itself cannot be used for timing studies, its detection renews the hope that radio pulsars can be detected with sufficient sensitivity but also increases the suspicion that there are physical germane to the Galactic Center region suppressing the formation of normal radio pulsars. If radio pulsars can be detected, however, precision timing, especially with the future capabilities of the Square Kilometer Array (SKA), has the potential to detect not only post-Newtonian parameters (including the Shapiro delay and the Schwarzschild precession term), but also the Lense-Thirring and quadrupole terms (Liu et al. 2012).

If one or several of these efforts is successful, it may be ultimately possible to test GR in the strong curvature limit, and test the no-hair theorem (Will 2008, Merritt et al. 2010, Psaltis & Johanssen 2011).

Acknowledgments

I would like to thank Stefan Gillessen, David Rosario, Amiel Sternberg, Scott Tremaine and Stijn Wuyts for helpful comments on this manuscript.

References

1. Alexander, D. M., Swinbank, A. M., Smail, I., McDermid, R., & Nesvadba, N. P. H. 2010, MNRAS, 402, 2211.
2. Alexander, D.M. & Hickox, R.C., 2012, New.Astr.Rev., 56, 92.
3. Almheiri, A. Marolf, D., Polchinski, J. & Sully, J. 2013. JHEP 02. 62.
4. Arav, N., de Kool, M., Korista, K. T., et al. 2001, ApJ, 561, 118.
5. Arav, N., Moe, M., Costantini, E., Korista, K. T., Benn, C. & Ellison, S. 2008, ApJ, 681, 954.
6. Arav, N., Borguet, B., Chamberlain, C., Edmonds, D. & Danforth, C. 2013, MNRAS, 436, 3286.
7. Argon, A.L., Greenhill, L.J., Reid, M.J., Moran, J.M. & Humphreys, E.M.L. 2007, ApJ, 659, 1040.
8. Baganoff, F. et al. 2001, Nature, 413, 45.
9. Baldry, I. K., Glazebrook, K., & Driver, S. P. 2008, MNRAS 388, 945.
10. Bardeen J. M., 1973, in Black holes (Les astres occlus), DeWitt B. S., DeWitt C., eds., New York: Gordon and Breach, p. 215.
11. Barnes, J.E. & Hernquist, L. 1996, ApJ, 471, 115.
12. Barth, A.J. Sarzi, M., Rix, H.-W., Ho, L. C., Filippenko, A.V. & Sargent, W. L. W. 2001, ApJ, 555, 685.
13. Barth, A.J., Greene, J.E. & Ho, L.C. 2005, ApJ, 619, L151.
14. Begelman, M.C., Volonteri, M. & Rees, M.J. 2006, MNRAS, 370, 289.
15. Behroozi, P.S., Wechsler, R.H. & Conroy, C. 2013, ApJ 770, 57.
16. Bender, R. et al. 2005, ApJ 631, 280.
17. Bennert, V. N., Auger, M.W., Treu, T., Woo, J.-H. & Malkan, M.A. 2011, ApJ, 742, 107.
18. Bentz, M.C. et al. 2010, ApJ, 720, L46.
19. Blandford, R.D. & McKee, C.F. 1982, ApJ, 255, 419.
20. Blandford, R.D. 1999, ASPC, 160, 265.
21. Bournaud, F., Dekel, A., Teyssier, R., Cacciato, M., Daddi, E., Juneau, S. & Shankar, F. 2011, ApJ, 741, L33.
22. Bousso, R. 2002, Rev.Mod.Phys. 74, 825.
23. Bower, G.C. et al. 2004, Science 304, 704.

24. Bower, G.C. et al. 2014, ApJ, 780, L2.
25. Bower, R.G, Benson, A. J., Malbon, R., Helly, J. C., Frenk, C. S., Baugh, C. M., Cole, S. & Lacey, C. G. 2006, MNRAS, 370, 645.
26. Boyle, B. J., Shanks, T., Croom, S. M., Smith, R. J., Miller, L., Loaring, N. & Heymans, C. 2000, MNRAS, 317, 1014.
27. Braatz, J.A., Reid, M.J., Humphreys, E.M.L., Henkel, C., Condon, J.J. & Lo, K.Y. 2010, ApJ, 718, 657.
28. Brenneman, L.W. & Reynolds, C.S. 2006, ApJ, 652, 1028.
29. Broderick, A. E. & Loeb, A. 2006, MNRAS, 367, 905.
30. Broderick, A., Loeb, A. & Narayan, R. 2009, ApJ, 701, 1357.
31. Broderick, A. E., Fish, V. L., Doeleman, S. S. & Loeb, A. 2009, ApJ, 697, 45.
32. Cano-Diaz, M., Maiolino, R., Marconi, A., et al. 2012, A&A, 537, L8.
33. Cattaneo, A., Dekel, A., Devriendt, J., Guiderdoni, B. & Blaizot, J. 2006, MNRAS, 370, 1651.
34. Cecil, G., Bland, J. & Tully, B. R. 1990, ApJ, 355, 70.
35. Chatterjee, P., Hernquist, L. & Loeb, A. 2002, ApJ, 572, 371.
36. Chatzopoulos, S., Fritz, T., Gerhard, O., Gillessen, S., Wegg, C., Genzel, R. & Pfuhl, O. 2014, arXiv1403.5266.
37. Cisternas, M. et al. 2011, ApJ, 726, 57.
38. Conroy, C. & Wechsler, R.H. 2009, ApJ, 696, 620.
39. Conroy, C. & White, M. 2013, ApJ, 762, 70.
40. Croton, D.J. et al. 2006, Mon.Not.Roy.Astr.Soc. 365, 11.
41. Davies, R.I., Müller Sánchez, F., Genzel, R., Tacconi, L. J., Hicks, E. K. S., Friedrich, S. & Sternberg, A. 2007, ApJ, 671, 1388.
42. Davis, T. A., Bureau, M., Cappellari, M., Sarzi, M. & Blitz, L. 2013, Nature, 494, 328.
43. Dekel, A. & Silk, J. 1986, ApJ 303, 39.
44. Dekel, A. & Birnboim, Y. 2006, MNRAS 368, 2.
45. DelVecchio, I. Lutz, D., Berta, S. et al. 2014, in prep.
46. Dexter, J. & Fragile, P.C. 2013, MNRAS, 432, 2252.
47. Di Matteo, T., Springel, V. & Hernquist, L. 2005, Nature, 433, 604.
48. Do, T., Ghez, A. M., Morris, M. R., Yelda, S., Meyer, L., Lu, J. R., Hornstein, S. D. & Matthews, K. 2009, ApJ, 691, 1021.
49. Dodds-Eden, K. et al. 2009, ApJ, 698, 676.
50. Doeleman, S.S. et al. 2008, Nature, 455, 78.
51. Doeleman, S.S. 2010, in Proceedings of the 10th European VLBI Network Symposium and EVN Users Meeting: VLBI and the new generation of radio arrays. September 20-24, 2010. Manchester, UK. Published online at http://pos.sissa.it/cgi-bin/reader/conf.cgi?confid=125, id.53.
52. Eatough, R.P. et al. 2013, Nature, 501, 391.
53. Eckart, A. & Genzel, R. 1996, Nature 383, 415.
54. Eckart, A. & Genzel, R. 1997, MNRAS, 284, 576.
55. Eckart, A. et al., 2006, A&A 450, 535.
56. Efstathiou, G. 2000, MNRAS, 317, 697.
57. Einstein, A. 1916, Ann.Phys. 49, 50.
58. Eisenhauer, F. et al. 2003, ApJ 597, L121.
59. Eisenhauer, F. et al. 2005, ApJ 628, 246.
60. Eisenhauer, F., Perrin, G., Rabien, S., Eckart, A., Lena, P., Genzel, R., Abuter, R., & Paumard, T. 2005a, Astr.Nachrichten 326, 561.
61. Eisenhauer, F. et al. 2011, ESO Msngr. 143, 16.

62. Fabian, A.C., Iwasawa, K., Reynolds, C.S. & Young, A.I., 2000, PASP, 112, 1145.
63. Fabian, A. C., Vaughan, S., Nandra, K., Iwasawa, K., Ballantyne, D. R., Lee, J. C., De Rosa, A., Turner, A. & Young, A. J. 2002, MNRAS, 335, L1.
64. Fabian, A.C. & Vaughan, S. 2003, MNRAS 340, L28.
65. Fabian, A.C. et al. 2009, Nature, 459, 540.
66. Fabian, A.C. & Ross, R.R., 2010, Sp. Sc. Rev. 157, 167.
67. Fabian, A. C. 2012, ARA&A, 50, 455.
68. Falcke, H., Melia, F. & Algol, E. 2000, ApJ 528, L13.
69. Fan, X. et al. 2006, AJ, 132, 117.
70. Feruglio, C., Maiolino, R., Piconcelli, E., et al. 2010, A&A, 518, L155.
71. Ferrarese, L. & Merritt, D. 2000, ApJ 539, L9.
72. Filippenko, A.V. & Sargent, W.L.W. 1989, ApJ, 342, L11.
73. Filippenko, A.V. & Ho, L.C. 2003, ApJ, 588, L13.
74. Fischer, J., Sturm, E., González-Alfonso, E., et al. 2010, A&A 518, L41.
75. FÖrster Schreiber, N. M., Genzel, R., Newman, S. F., et al. 2014, ApJ, 787, 38.
76. Gebhardt, K. et al. 2000, ApJ, 539, L13.
77. Genzel, R., Hollenbach, D., & Townes, C. H., 1994, Rep. Prog. Phys., 57, 417.
78. Genzel, R., Thatte, N., Krabbe, A., Kroker, H. & Tacconi-Garman, L.E. 1996, ApJ, 472, 153.
79. Genzel, R., Eckart, A., Ott, T. & Eisenhauer, F. 1997, MNRAS, 291, 219.
80. Genzel, R., Pichon, C., Eckart, A., Gerhard, O.E. & Ott, T. 2000, MNRAS, 317, 348.
81. Genzel, R. et al. 2003, Nature, 425, 934.
82. Genzel, R., Eisenhauer, F. & Gillessen, S. 2010, Rev.Mod.Phys. 82, 3121.
83. Genzel, R. et al. 2014, ApJ in press (arXiv1406.0183).
84. Ghez, A.M., Klein, B.L., Morris, M. & Becklin, E.E. 1998, ApJ 509, 678.
85. Ghez, A.M. et al. 2000, Nature, 407, 349.
86. Ghez, A.M. et al. 2003, ApJ 586, L127.
87. Ghez, A. et al. 2005, ApJ, 620, 744.
88. Ghez, A. et al. 2008, ApJ, 689, 1044.
89. Giacconi, R., Gursky, H., Paolini, F. & Rossi, B.B. 1962, Phys.Rev.Lett. 9, 439.
90. Giacconi, R. 2003, Rev.Mod.Phys. 75, 995.
91. Gillessen, S. 2006, SPIE, 6268, E11.
92. Gillessen, S. et al. 2009, ApJ, 692, 1075.
93. Gillessen, S. et al. 2009a, ApJ 707, L114.
94. Greene, J.E. & Ho, L.C. 2004, ApJ, 610, 722.
95. Greene, J.E. & Ho, L.C. 2007, ApJ 667, 131.
96. Greene, J.E., 2012, Nature Comm., 10.1038, 2314.
97. Gültekin, K. et al. 2009, ApJ 698, 198.
98. Guo, Q., White, S., Li, C., & Boylan-Kolchin, M. 2010, MNRAS 404, 1111.
99. Haehnelt, M. 2004 in 'Coevolution of Black Holes and Galaxies', Carnegie Observatories Centennial Symposia. Cambridge University Press, Ed. L.C. Ho, p. 405.
100. Häring, N. & Rix, H.-W. 2004, ApJ, 604, L89.
101. Haiman, Z. & Quataert, E. 2004, in Supermassive Black Holes in the Distant Universe. Edited by Amy J. Barger, Astrophysics and Space Science Library Volume 308. ISBN 1-4020-2470-3 (HB), ISBN 1-4020-2471-1 (e-book). Published by Kluwer Academic Publishers, Dordrecht, The Netherlands, p. 147.
102. Haller, J.W., Rieke, M.J., Rieke, G.H., Tamblyn, P., Close, L. & Melia, F. 1996, ApJ, 456, 194.
103. Hamaus, N., Paumard, T., Müller, T., Gillessen, S., Eisenhauer, F., Trippe, S. & Genzel, R. 2009, ApJ, 692, 902.

104. Harrison, C. M., Alexander, D. M., Swinbank, A. M., et al. 2012, MNRAS, 426, 1073.
105. Harrison, C. M., Alexander, D. M., Mullaney, J. R., & Swinbank, A. M. 2014, MN-RAS, 441, 3306.
106. Heckman, T. M., Kauffmann, G., Brinchmann, J., Charlot, S., Tremonti, C. & White, S. D. M. 2004, ApJ, 613, 109.
107. Heckman, T.M. 2010, in Co-Evolution of Central Black Holes and Galaxies, Proceedings of the International Astronomical Union, IAU Symposium, Volume 267, p. 3-14.
108. Heckman, T.M. & Best, P.N. 2014, ARAA in press (astro-ph 1403.4620).
109. Herrnstein, J.T., Moran, J.M., Greenhill, L.J. & Trotter, A.S. 2005, ApJ, 629, 719.
110. Hickox, R. C., Mullaney, J. R., Alexander, D. M., Chen, C.-T. J., Civano, F.M., Goulding, A. D., & Hainline, K. N. 2014, ApJ, 782, 9.
111. Ho, L.C., Filippenko, A.V. & Sargent, W.L.W., 1997, ApJS, 112, 315.
112. Ho, L.C. 2008, ARAA, 46, 475.
113. Hopkins, P.F. et al. 2005, ApJ, 630, 716.
114. Hopkins, P.F. 2006, ApJS, 163, 1.
115. Hopkins, P.F. & Quataert, E. 2010, MNRAS, 407, 1529.
116. Jahnke, K, et al. 2009, ApJ, 706, L215.
117. Jahnke, K. & Maccio, A.V., 2011, ApJ, 743, 92.
118. Kaspi, S., Smith, P. S., Netzer, H., Maoz, D., Jannuzi, B. T. & Giveon, U. 2000, ApJ, 533, 631.
119. Kauffmann, G. & Haehnelt, M. 2000, MNRAS, 311, 576.
120. Kauffmann, G. et al. 2003, MNRAS, 346, 1055.
121. Kennea, J.A. et al.2013, ApJ, 770, L24.
122. Kerr, R. 1963, Ph.Rev.Lett., 11, 237.
123. Kocevski, D.D. et al., 2012, ApJ, 744, 148.
124. Korista, K. T., Bautista, M. A., Arav, N., et al. 2008, ApJ, 688, 108
125. Kormendy, J. 2004, in 'Coevolution of Black Holes and Galaxies', Carnegie Observatories Centennial Symposia. Cambridge University Press, Ed. L.C. Ho, p. 1.
126. Kormendy, J., Bender, R. & Cornell, M.E. 2011 Nature 469, 374.
127. Kormendy, J. & Ho, L. 2013, ARAA 51, 511.
128. Krabbe, A. et al. 1995, ApJ, 447, L95.
129. Kuo, C.Y., Braatz, J.A., Condon, J.J. et al. 2011, ApJ, 727, 20.
130. Lacy, J.H., Townes, C.H., Geballe, T.R. & Hollenbach, D.J. 1980, ApJ 241, 132.
131. Lawson, P.R., Unwin, S.C. & Beichman, C.A. 2004, JPL publication 04-014.
132. Li, C., Kauffmann, G., Heckman, T. M., White, S. D. M. & Jing, Y. P. 2008, MNRAS, 385, 1915.
133. Liebling, S.L. & Palenzuela, C. 2012, LRR, 15, 6.
134. Liu, K., Wex, N., Kramer, M., Cordes, J. M. & Lazio, T. J. W. 2012, ApJ, 747, 1.
135. Lynden-Bell, D. 1969, Nature 223, 690.
136. Lu, R.-S. et al. 2014, ApJ 788, L120.
137. Macchetto, F., Marconi, A., Axon, D.J., Capetti, A., Sparks, W. & Crane, P. 1997, ApJ, 489, 579.
138. Madau, P. et al. 1996, MNRAS, 283, 1388.
139. Magorrian, J. et al. 1998, AJ, 115, 2285.
140. Maldacena, J. 1998, Ad. Th.Math.Phys. 2, 231.
141. Maoz, E. 1995, ApJ, 447, L91.
142. Maoz, E., 1998, ApJ 494, L181.
143. Marconi, A. et al. 2003, ApJ, 586, 868.

144. Marconi, A., Risaliti, G., Gilli, R., Hunt, L. K., Maiolino, R. & Salvati, M. 2004, MNRAS, 351, 169.

145. Marconi, A., Pastorini, G., Pacini, F., Axon, D. J., Capetti, A., Macchetto, D., Koekemoer, A. M. & Schreier, E. J. 2006, A&A, 448, 921.

146. Markoff, S., Falcke, H., Yuan, F. & Biermann, P.L. 2001, Astr.&Ap. 379, L13.

147. Martini, P., 2004, in Coevolution of Black Holes and Galaxies, from the Carnegie Observatories Centennial Symposia, Cambridge University Press, as part of the Carnegie Observatories Astrophysics Series, edited by L. C. Ho, p. 169.

148. Mayer, L., Kazantzidis, S., Escala, A. & Callegari, S. 2010, Nature, 466, 1082.

149. McClintock, J. & R. Remillard 2004, in Compact Stellar X-ray sources , eds. W.Lewin and M.van der Klis, Cambirdge Univ. Press (astro-ph 0306123).

150. McConnell, N. & Ma, C.-P. 2013, ApJ 764, 184.

151. McGinn, M.T., Sellgren, K., Becklin, E.E. & Hall, D.N.B. 1989, ApJ, 338, 824.

152. McNamara, B.R. & Nulsen, P.E.J. 2007, ARAA, 45, 117.

153. Merloni, A. et al. 2010, ApJ, 708, 137.

154. Merritt, D., Alexander, T., Mikkola, S. & Will, C.M. 2010, PhRev D, 81, 2002.

155. Meyer, L., Ghez, A. M., SchÖdel, R., Yelda, S., Boehle, A., Lu, J. R., Do, T., Morris, M. R., Becklin, E. E. & Matthews, K. 2012, Sci, 338, 84.

156. Mori, K. et al. 2013, ApJ, 770, L23.

157. Moscibrodzka, M., Falcke, H., Shiokawa, H. & Gammie, C. F. 2014, arXiv:1408.4743.

158. Moster, B. P., Naab, T., & White, S. D. M. 2013, MNRAS 428, 3121.

159. Miyoshi, M. et al. 1995, Nature 373, 127.

160. Moran, J.M. 2008, ASPC, 395, 87.

161. Mullaney, J.R , Daddi, E., Béthermin, M., Elbaz, D., Juneau, S., Pannella, M., Sargent, M. T., Alexander, D. M., Hickox, R. C. 2012, ApJ, 753, L30.

162. Munyaneza, F, Tsiklauri, D. & Viollier, R.D. 1998, ApJ, 509, L105.

163. Nandra, K., George, I. M., Mushotzky, R. F., Turner, T. J. & Yaqoob, T. 1997, ApJ, 477, 602.

164. Nandra, K., O'Neill, P. M., George, I. M. & Reeves, J. N. 2007, MNRAS, 382, 194.

165. Nesvadba, N. P. H., Polletta, M., Lehnert, M. D., et al. 2011, MNRAS, 415, 2359.

166. Nesvadba, N. P. H., Lehnert, M. D., De Breuck, C., Gilbert, A. M. & van Breugel, W. 2008, A&A, 491, 407.

167. Netzer, H. & Peterson, B.M. 1997, ASSL, 218, 85.

168. Netzer, H., Mainieri, V., Rosati, P. & Trakhtenbrot, B. 2006, A&A, 453, 525.

169. Netzer, H. 2009, MNRAS, 399, 1907.

170. Netzer, H. 2013, The Physics and Evolution of Active Galactic Nuclei,Cambridge University Press.

171. Neumayer, N., Cappellari, M., Rix, H.-W., Hartung, M., Prieto, M. A., Meisenheimer, K. & Lenzen, R. 2006, ApJ, 643, 226.

172. Novak,G.S., Ostriker, J.P. & Ciotti, L. 2011, ApJ, 737, 26.

173. Osmer. P.S. 2004, in Coevolution of Black Holes and Galaxies, from the Carnegie Observatories Centennial Symposia. Published by Cambridge University Press, as part of the Carnegie Observatories Astrophysics Series. Edited by L. C. Ho, p. 324.

174. Özel, F., Psaltis, D., Narayan, R. & McClintock, J. E. 2010, ApJ 725, 1918.

175. Paumard, T., Perrin, G., Eckart, A., Genzel, R., Lena, P., Schoedel, R., Eisenhauer, F., Mueller, T. & Gillessen, S. 2005, Astr.Nachrichten 326, 568.

176. Peng, Y. et al. 2010, ApJ, 721, 193.

177. Peterson, B.M. 1993, PASP, 105, 247.

178. Peterson, B.M. 2003, ASPC, 290, 43.

179. Pfahl, E. & Loeb, A. 2004, ApJ, 615, 253.

180. Psaltis,D. & Johanssen, T. 2011, JPhCS, 283, 2030.
181. Rees, M.J. & Ostriker, J.P. 1977 MNRAS 179, 541.
182. Rees, M. 1984, Ann.Rev Astr.Ap. 22, 471.
183. Reid, M.J. & Brunthaler, A. 2004, ApJ 616, 872.
184. Reid, M.J. Braatz, J. A., Condon, J. J., Lo, K. Y., Kuo, C. Y., Impellizzeri, C. M. V. & Henkel, C. 2013, ApJ, 767, 154.
185. Reid, M.J. et al. 2014, ApJ, 783, 130.
186. Reines, A. E., Sivakoff, G. R., Johnson, K. E. & Brogan, C. L.2011, Nature, 470, 66.
187. Reines, A.E., Greene, J.E. & Geha, M. 2013, ApJ, 775, 116.
188. Reynolds, C.S. 2013, CQGra., 30, 4004.
189. Remillard, R.A. & McClintock, J.E. 2006, ARAA 44, 49.
190. Rieke, G.H. & Rieke, M.J. 1988, ApJ, 330, L33.
191. Rosario, D. et al. 2012, A&A, 545, 45.
192. Rosario, D. et al. 2013, ApJ, 771, 63.
193. Rupke, D. S. N. & Veilleux, S. 2013, ApJ, 768, 75.
194. Sanders, D. B. et al. 1988, ApJ, 325, 74.
195. Sanders, R.H. 1998, MNRAS, 294, 35.
196. SchÖdel, R. et al. 2002, Nature 419, 694.
197. SchÖdel, R. et al. 2003, ApJ 596, 1015.
198. Schmidt, M. 1963, Nature 197, 1040.
199. Schunck, F.E. & Mielke, E.W. 2003, CQGra, 20, R301.
200. Schawinski, K.,Treister, E., Urry, C. M., Cardamone, C. N., Simmons, B., Yi, S. K. 2011, ApJ, 727, L31.
201. Schwarzschild, K., 1916, Sitzungsber. Preuss. Akad.Wiss., 424.
202. Scoville, N.Z. & Norman, C.A. 1988, ApJ, 332, 163.
203. Sellgren, K., McGinn, M.T., Becklin, E.E. & Hall, D.N. 1990, ApJ 359, 112.
204. Serabyn, E. & Lacy, J.H. 1985, ApJ 293, 445.
205. Shakura, N.I. & Sunyaev, R.A. 1973, A&A 24, 337.
206. Shankar, F., Weinberg, D.H. & Miralda-Escude, J. 2009, ApJ, 690, 20.
207. Shao, L. et al. 2010, A&A, 518, L26.
208. Shen, Z.Q., Lo, K.Y., Liang, M.C., Ho, P.T.P. & Zhao, J.H. 2005, Nature 438, 62
209. Sijacki, D. Vogelsberger, M., Genel, S., Springel, V., Torrey, P., Snyder, G., Nelson, D. & Hernquist, L. 2014, arXiv1408.6842.
210. Soltan, A. 1982, MNRAS, 200, 115
211. Somerville, R., Hopkins, P. F., Cox, T. J., Robertson, B. E. & Hernquist, L. 2008, MNRAS, 391, 481.
212. Spitler, L.G. et al. 2014, ApJ, 780, L3.
213. Sturm, E., González-Alfonso, E., Veilleux, S., et al. 2011, ApJ, 733, L16.
214. Susskind, L. 1995, JMP 36, 6377.
215. Tacconi, L.J., Neri, R., Genzel, R., et al. 2013, ApJ, 768. 74.
216. Tanaka, Y. et al. 1995, Nature, 375, 659.
217. Torres,D.F., Capoziello, S. & Lambiase, G. 2000, PhRv D, 62, 4012.
218. Townes, C. H., Lacy, J. H., Geballe, T. R. & Hollenbach, D. J. 1982, Nature 301, 661.
219. Trakhtenbrot, B. & Netzer, H. 2012, MNRAS, 427, 3081.
220. Tremaine, S. et al. 2002, ApJ, 574, 740.
221. Uttley, P. et al. 2014, A&Arev, 22, 72.
222. van der Marel, R.P. & vn den Bosch, F.C. 1998, AJ, 116, 2220.
223. Vestergaard, M. 2004, ApJ 601, 676.
224. Veilleux,S., Kim, D.-C. & Sanders, D.B. 1999, ApJ, 522, 113.

225. Veilleux, S., Cecil, G. & Bland-Hawthorn, J. 2005, ARAA, 43, 769.
226. Veilleux, S. et al. 2009, ApJS, 182, 628.
227. Veilleux, S., Meléndez, M., Sturm, E., et al. 2013, ApJ, 776, 27.
228. Viollier, R.D, Trautmann, D. & Tupper 1993, PhLB, 306, 79.
229. Volonteri, M. 2010, A&AR, 18, 279.
230. Volonteri, M. 2012, Sci, 337, 544.
231. Weinberg, N. N., Milosavljevic, M. & Ghez, A. M. 2005, ApJ 622, 878.
232. Westmoquette, M. S., Clements, D. L., Bendo, G. J., & Khan, S. A. 2012, MNRAS, 424, 416.
233. Wharton, R. S., Chatterjee, S., Cordes, J. M., Deneva, J. S. & Lazio, T. J. W. 2012, ApJ, 753, 108.
234. Wheeler, J.A. 1968, Amer.Scient. 56, 1.
235. Wild, V., Heckman, T. & Charlot, S. 2010, MNRAS, 405, 933.
236. Will, C.M. 2008, ApJ, 674, L25.
237. Wollman, E. R., Geballe, T. R., Lacy, J. H., Townes, C. H., Rank, D. M. 1977, ApJ 218, L103.
238. Yu, Q & Tremaine, S. 2002, MNRAS, 335, 965.
239. Zakamska, N.L. & Greene, J.E. 2013, MNRAS, 442, 784.
240. Zucker, S., Alexander, T., Gillessen, S., Eisenhauer, F. & Genzel, R. 2006, ApJ 639, L21.

Discussion

S. Tremaine We have time for a couple of questions or comments while Mitch sets up. Several of the themes that Reinhard mentioned, in particular black hole spin and feedback will have contributions later, so please try to focus on other issues.

S. White You focus mostly on large black holes; but stellar mass black holes are also interesting. Although the angular sizes are much smaller, the time scales are much shorter so you can use variability phenomena to investigate the properties of black holes close in, so would you like to say a couple of things about this.

R. Genzel I do not think so, I'm looking at Andy maybe, Andy would you...

A. Fabian The instructions were for Reinhard to focus on dynamical measurements in inactive black holes and for Mitch to focus on active ones, so you will hear something from Mitch on this.

M. Rees Just a comment about Scott Tremaine's suggestion that the Galactic center could be filled with billions of Jupiters rather than a black hole. I suspect you rule that out because gas would settle in that potential well and would form a pretty massive star pretty quickly, so you probably end up with a black hole there inside your cluster of Jupiters anyway.

M. Kamionkowski So just a question: if there was anybody who still believes in this Jupiter explanation for supermassive black holes, would observations of tidal disruption events, or the evidence that we have for tidal disruption events rule out that scenario?

S. Tremaine Depending on what Reinhard chooses to say, one could argue that we have not seen a tidal disruption event in the Galactic center. If you want to assume that what's in the middle of our Galaxy is the same as what's in the middle of other galaxies, then perhaps.

V. Kaspi Just a very quick comment about just the magnetar in the galactic center, just to point out that the one thing it did teach us, I totally agree, that dynamically it's so noisy that as a timer it's not so useful, but the radio emission from it is far less scattered than models had predicted, which actually tells you that there is a good hope of one day finding a pulsar in the Galactic center.

M. Kramer Actually it is changing the scattering, it is now scattering much more than it was a year ago. So it is very interesting.

S. Tremaine Let's move on to Mitch Begelman from University of Colorado who's speaking on accreting black holes.

Rapporteur Talk by M. Begelman: Accreting Massive Black Holes

Abstract

I outline the theory of accretion onto black holes, and its application to observed phenomena such as X-ray binaries, active galactic nuclei, tidal disruption events, and gamma-ray bursts. The dynamics as well as radiative signatures of black hole accretion depend on interactions between the relatively simple black-hole spacetime and complex radiation, plasma and magnetohydrodynamical processes in the surrounding gas. I will show how transient accretion processes could provide clues to these interactions. Larger global magnetohydrodynamic simulations as well as simulations incorporating plasma microphysics and full radiation hydrodynamics will be needed to unravel some of the current mysteries of black hole accretion.

1. The Spacetime Context

Accretion of gas onto a black hole provides the most efficient means known for liberating energy in the nearby universe. This is good for observational astronomers, since it allows regions close to the event horizon to be studied directly through emitted radiation. But it is also crucial for understanding the influence of black holes on galaxy formation and evolution. The huge amounts of energy released close to accreting black holes, particularly in the form of winds and jets, but also in the form of energetic radiation, can affect the thermal and dynamical states of matter out to large distances, making black holes important agents of change on cosmological scales.

Astrophysical black holes are described by two parameters that effectively set the inner boundary conditions for accretion. The mass, M, basically determines the characteristic length and time scales close to the horizon, whereas the Kerr spin parameter a/M (where $a = J/M$ is the specific angular momentum of the hole, in geometric units $G = c = 1$), with $0 \leq a/M \leq 1$, determines the efficiency of energy release and, coupled to the magnetic and radiative properties of the infalling gas, the forms in which energy is liberated.

Although the horizon (at R_H) marks the point of invisibility and no return for matter being accreted by a black hole, the energy efficiency of accretion is determined somewhat farther out, near the *innermost stable circular orbit* (ISCO). A test particle orbiting inside this radius will be swallowed by the black hole without giving up any additional energy or angular momentum. Because R_{ISCO} decreases from $6M$ ($= 3R_H$) for a Schwarzschild black hole to M ($= R_H$) for a corotating orbit around an extreme ($a/M = 1$) Kerr hole, accretion is more efficient for a rotating hole than for a stationary one. (Note, however, that R_{ISCO} increases with a/M for counter-rotating orbits and approaches $9M$ for extreme Kerr, affording much lower efficiency.) The specific angular momentum corresponding to the ISCO decreases from $2\sqrt{3}M$ to $2M/\sqrt{3}$ as a/M increases from 0 to 1, and the efficiency of energy release increases from about 6% of Mc^2 to about 42%. Hartle's undergraduate textbook on relativity[1] provides a very readable discussion of these key features.

2. The Gaseous Environment

It is important to keep in mind that elements of gas in an accretion flow do not behave exactly as test particles close to the ISCO, and in some cases their dynamics can be influenced strongly by pressure and magnetic forces. Gas (or radiation) pressure forces, directed inward, can allow gas to remain in orbit slightly closer to the black hole than the ISCO, and with somewhat higher angular momentum. Gas plunging into the black hole from such an orbit would have a lower binding energy and therefore a lower accretion efficiency. In the limiting case where gas orbits a Schwarzschild black hole down to $4M$ (the *marginally bound orbit*), the binding energy of the accreted gas approaches zero and so does the accretion efficiency.[2]

Likewise, net magnetic flux, accumulating in the innermost regions of an accretion flow, could hold back the gas, creating a *magnetically arrested disk*.[3] The angular momentum close to the black hole might then be lower than that of any stable test particle orbit, but infall could be regulated by interactions between the gas and the magnetic field, such as interchange instabilities and reconnection.

In light of these considerations, we can identify at least four factors that must play an important role in governing black hole accretion flows. The first three of these may be regarded as outer boundary conditions for the problem.

– *Angular momentum.* The specific angular momentum at the marginally bound orbit, somewhat larger than ℓ_{ISCO} but still of order a few GM/c, represents the largest angular momentum per unit mass that can be accreted by a black hole. Given that the radius of the gas reservoir supplying black hole accretion is usually at least a few hundred times M, and often much more, accretion is seldom possible without the loss of some angular momentum. Angular momentum is thought to be transferred outward through the flow via the magnetorotational instability (MRI),[4] which works in the limit of sufficiently *weak* magnetic field. Angular momentum can also be lost through winds or electromagnetic torques; the latter process depends on net magnetic flux, the third factor below.
– *Radiative efficiency.* Energy liberated during the accretion process is thought to be transferred outward by the same processes that wick away the excess angular momentum. Some of this energy can go into driving coherent motions such as circulations or outflows, or can be removed by electromagnetic torques.[5] But much (or most) of it is likely to be dissipated as heat or radiation. If much of this energy is retained by the flow, the associated pressure can partially support the flow against gravity, reducing the relative importance of rotation. When the local rotation rate is still a large fraction of the Keplerian value, the flow resembles a disk but the added pressure force can drive circulation or even mass loss.[6,7] When

pressure support dominates rotational support, or even becomes comparable to it, the flow can inflate into a nearly spherical configuration and interesting stability questions arise.

– *Magnetic flux.* The MRI basically uses distortions of a poloidal magnetic field (i.e. parallel to the rotation axis) to extract free energy from differential rotation.[4] Therefore, it is not surprising that the vigor of angular momentum transport driven by MRI is sensitive to the presence of poloidal magnetic flux.[8] Recent numerical experiments[9] suggest that a poloidal magnetic pressure as small as 0.1% of the gas pressure, coherent over scales comparable to the disk thickness, is enough to enhance angular momentum transport by a substantial factor. Small patches of magnetic flux could arise through statistical fluctuations as a result of local MRI[10] but larger coherent fields probably need to be accumulated by advection of flux from larger distances. Whether efficient flux advection is possible is highly uncertain,[11] and probably depends sensitively on details of the distant outer boundary conditions[12] and vertical disk structure.[13] Sufficiently coherent poloidal magnetic flux, threading the innermost regions of the rotating accretion flow, is an essential element for extracting rotational energy from the black hole, and presumably also for producing a strong disk wind.

– *Black hole spin.* Part of the gravitating mass M of a spinning black hole, as perceived by a distant observer, is contributed by the spin energy. In principle, all of this energy — up to $0.29M$ for an extreme Kerr hole — can be extracted. In practice this can be done using coherent poloidal magnetic fields, through the Blandford–Znajek (BZ) process.[14] Since the power extracted is proportional to the square of the net magnetic flux threading the hole as well as the square of the spin parameter a, the efficiency of the BZ effect is sensitive to the amount of flux that can be accumulated and held in place against the black hole. Analytic calculations and relativistic magnetohydrodynamic simulations show that the efficiency can be appreciable, possibly even exceeding the efficiency of the accretion process.[15,16]

3. Modes of Accretion

The dynamical properties of black-hole accretion flows are affected most strongly by their angular momentum distributions and secondarily by the efficiency of radiative losses. In almost all astrophysically interesting circumstances the flow is expected to be *centrifugally choked*, meaning that the specific angular momentum of the gas supply exceeds a small multiple of GM/c. Various possible modes of accretion, discussed in more detail below, are summarized in Figure 1.

Hypothetical flows which are not centrifugally choked should resemble spherically symmetric Bondi accretion,[17] which may or may not have a high radiative efficiency, depending on the presence or absence of a dynamically significant magnetic field.

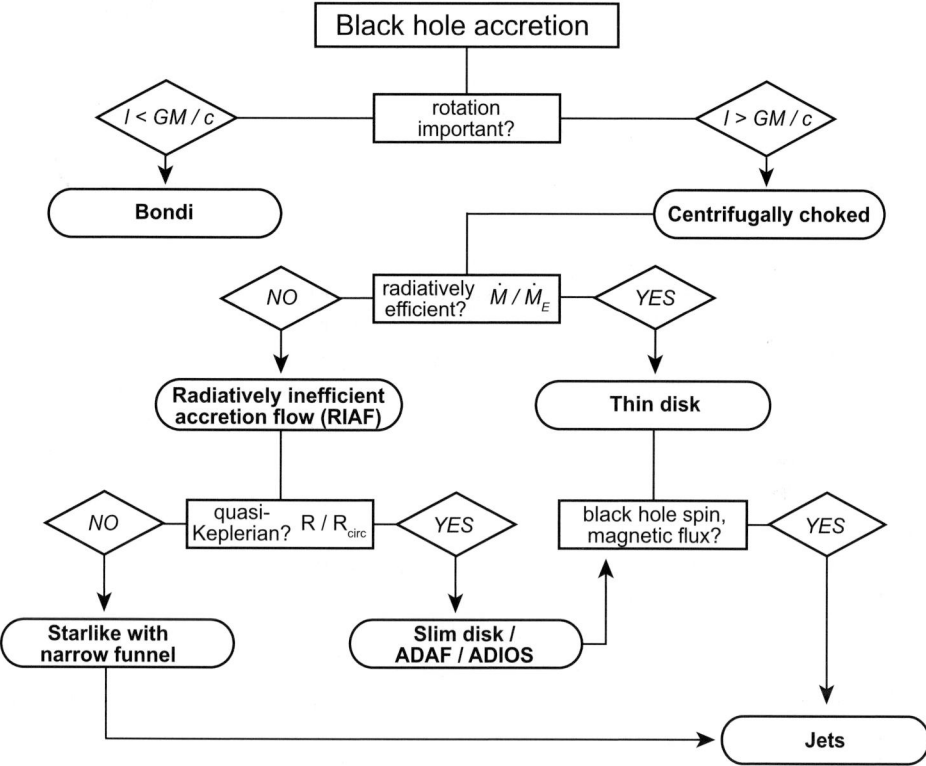

Fig. 1. Flow chart illustrating the main factors determining the dynamical state of black hole accretion flows. The specific angular momentum relative to GM/c determines whether the flow is centrifugally choked or not, and the same quantity, relative to the Keplerian angular momentum, determines whether the flow resembles a disk or a quasi-spherical ("starlike") envelope. Radiative inefficiency is a necessary but not sufficient condition for strongly sub-Keplerian rotation. Additional factors, such as net magnetic flux and the spin of the central black hole, mainly affect the rate of angular momentum transport and the production of jets and winds.

3.1. *Disklike Accretion*

Without significant pressure support, gas will follow approximately Keplerian orbits. Since energy is generally thought to be dissipated more quickly than angular momentum is transported away, a radiatively efficient flow will circularize to form a thin disk.[18] Gas then slowly spirals towards the black hole as it gives up angular momentum, until it reaches the ISCO and plunges into the hole with relatively little additional dissipation. The energy dissipated at every radius is radiated away locally, and is a well-defined function of mass accretion rate $\dot{M}$, radius and black hole spin that can be converted into a run of effective temperatures and (assuming a large disk optical depth, which usually applies) a continuum spectrum. A key strength of accretion disk theory is that these results do not depend on the angular momentum transport mechanism. Fitting the continuum spectrum close to the inner edge of the disk is the basis of the continuum fitting method for determining black hole spins.[19] However, portions of the spectrum originating farther out often seem hotter than predicted, probably due to irradiation of the disk, poorly understood optical depth effects, or nonlocal dissipation.

At lower radiative efficiency, the disk will thicken as more internal energy is retained, leading to a *slim disk*,[20] in which radial pressure forces depress the angular momentum at every radius below the Keplerian value and radial advection of energy becomes appreciable.

It is important to note that the energy dissipated locally in an accretion disk is *not* the same as the gravitational binding energy liberated locally. Far from the black hole where the angular momentum is close to the Keplerian value, the dissipation rate is three times the local rate of energy liberation because two-thirds of the dissipated energy is transported from closer in by the same torques that transport angular momentum outward. Overall energy conservation is maintained because the dissipation rate close to the ISCO is lower than the local rate at which energy is liberated. This means that the outer parts of an accretion disk will accumulate internal energy and become unbound unless at least two-thirds of the dissipated energy is radiated away.

This problem was noted by Narayan & Yi[21,22] in their models for *advection-dominated accretion flows* (ADAFs), which assume steady accretion with low radiative efficiency. Self-similar ADAF models exhibit positive Bernoulli function B (where B is the sum of gravitational potential energy, kinetic energy and specific gas enthalpy), implying that elements of gas in the flow are able to unbind neighboring elements by doing work on them. Eventually the entire flow should disperse. Possible resolutions that preserve disklike structure include *adiabatic inflow-outflow solutions* (ADIOS),[6] *convection-dominated accretion flows* (CDAFs),[23,24] and models with large-scale matter circulation.[7] These models operate by suppressing the accretion rate relative to the mass supply and/or providing an escape route for energy and angular momentum that avoids the accreting gas. All of them require some unspecified mechanism for separating accreting matter from outflowing mass, energy and angular momentum. Despite these uncertainties, various numerical simulations[25-27] suggest that such a separation can take place naturally.

3.2. *Starlike Accretion*

The dynamical significance of the Bernoulli parameter can be demonstrated by considering a sequence of two-dimensional, axisymmetric, self-similar models of quasi-Keplerian rotating flows with pressure. Quasi-Keplerian here means that the specific angular momentum scales with radius as $R^{1/2}$. If B is either constant or has a radial scaling $\propto 1/R$, then as B approaches zero from below the surface of the disklike flow closes up to the rotation axis, and the flow becomes *starlike*.[21,28] If B becomes positive, the flow becomes unbounded. This suggests that disklike ADAF models without some kind of escape valve (such as outflow or circulation) are untenable.

Could such flows alternatively resolve their energy crisis by expanding and becoming nearly spherical?

It turns out that the ratio of the specific angular momentum to the Keplerian angular momentum is the parameter that decides whether an accretion flow is disk-like or starlike. Any of the disklike, radiatively inefficient flows have relatively "flat" density distributions as a function of radius, $\rho \propto R^{-n}$ with $1/2 < n < 3/2$. But in order for such flows to remain in dynamical equilibrium with $B < 0$, the specific angular momentum has to maintain a fairly large fraction of the Keplerian value. For the interesting case of a radiatively inefficient accretion flow (RIAF) dominated by radiation pressure, meaning that the adiabatic index γ is $4/3$, the minimum specific angular momentum compatible with disklike flow ranges between 74% and 88% of Keplerian, as n ranges from $3/2$ to $1/2$. Any lower value of angular momentum leads to starlike flow. (This result is sensitive to γ, with the constraint relaxed for larger values of this parameter. For the limiting case of a gas pressure-dominated RIAF with $\gamma = 5/3$ and $n = 3/2$, disklike flow is formally possible for any specific angular momentum, but even in this limit an angular momentum 30% of Keplerian forces the disk surface to close up to within $5°$ of the rotation axis.)

The outer boundary conditions for the mass supply probably play the dominant role in determining whether the angular momentum is large enough to keep the flow disklike. Gas that is supplied from the outside, in the form of a thin disk, probably holds on to enough angular momentum to remain disklike all the way to the black hole. Accretion flows in X-ray binaries fall into this category. On the other hand, flows that start out with a finite supply of angular momentum, such as debris from tidal disruption events or the interiors of stellar envelopes, might become starved of angular momentum (particularly if the envelope expands as it absorbs energy released by accretion) and forced into a starlike state.

The presumption for disklike flows is that the accretion rate adjusts itself so that the circulation, outflow, convection — or whatever it is that relieves the energy crisis — is able to carry away any accretion energy that isn't lost to radiation. This is possible because the radial density and pressure gradients are rather flat, giving the outer flow a high "carrying capacity" for excess energy, compared to the inner flow where most of this energy is liberated. But this presumption fails in the case of starlike accretion flows, where the density and pressure profiles are forced to steepen in order to keep the gas bound while satisfying dynamical constraints. This means that matter is more centrally concentrated around the black hole in a starlike flow, and if this matter is swallowed with an energy efficiency typical for the ISCO it will liberate much more energy than can be carried away by the outer parts of the flow.

One can envisage several outcomes of this situation: (1) there is no equilibrium configuration, and the flow either blows itself up or becomes violently unsteady, with successive episodes of energy accumulation and release; (2) the gas close to the black hole is pushed inward by the pressure until it approaches the marginally bound orbit, in which case the large accretion rate releases very little energy; and (3) the inner flow finds a way to release most of the accretion energy locally, without forcing it to propagate through the outer flow. This third possibility, for example, could involve the production of powerful jets that escape through the rotational funnel. It is hard to decide by pure thought which, if any, of these possibilities is most likely, and simulations have not yet addressed this problem. Nevertheless, there are observational indications that some tidal disruption events, at least, choose the third option (Sec. 4.3).

3.3. *Causes of Radiative Inefficiency*

The dynamical properties of black hole accretion flows depend on whether the gas radiates efficiently or not, but do not depend on the mechanisms that determine radiative efficiency. From an observational point of view, however, these details are crucial because radiatively inefficient flows can be either very faint or, paradoxically, very luminous. The fiducial mass flux that governs radiative efficiency is related to the Eddington limit, which is the luminosity (assumed to be isotropic) at which radiation pressure force balances gravity for a gas with opacity κ: $L_E = 4\pi GMc/\kappa$. The accretion rate capable of producing such a luminosity is $\dot{M}_E = L_E/\varepsilon c^2$, where ε is the radiative efficiency of accretion. Radiation escaping from such a flow will exert outward pressure forces competitive with centrifugal force, thus creating the dynamical conditions prevalent in radiatively inefficient flow. In fact, flows with $\dot{M} > \dot{M}_E$ are literally radiatively inefficient because radiation is trapped and advected inward by the large optical depth at radii $R < (\dot{M}/\dot{M}_E)GM/c^2$ (to within a factor $\sim \varepsilon$).[29]

Thus, accretion flows with high accretion rates are radiatively inefficient because they liberate more energy than they can radiate, but they are also very luminous because they radiate at close to the Eddington limit. Such flows are strongly dominated by radiation pressure and should be modeled using an adiabatic index of 4/3.

Accretion flows with low accretion rates can also be radiatively inefficient, because their densities are so low that radiative processes (which typically scale as ρ^2) cannot keep up with dissipation (which scales as ρ). Because electrons cool much more rapidly than ions, such flows are expected to develop a "two-temperature" thermal structure, with $T_i \gg T_e$.[30] They would be optically thin, dominated by thermal gas pressure, and characterized by an adiabatic index 5/3. If thermal coupling between ions and electrons is provided by Coulomb interactions, such flows can exist for accretion rates $\dot{M} < \alpha^2 \dot{M}_E$ (very roughly), where $\alpha \ll 1$ is a widely-used parameter that describes the rate of angular momentum transport.[18]

At high ($> \dot{M}_E$) and intermediate ($\alpha^2 \dot{M}_E < \dot{M} < \dot{M}_E$) accretion rates, the thermal state of accretion is uniquely determined by $\dot{M}$. But flows with low accretion rates may exist in either a radiatively efficient (thin disk) or inefficient (two-temperature) state. It is not understand what would trigger a thin disk in this regime to transition to the radiatively inefficient state, or vice-versa, but there is evidence that such transitions do occur in X-ray binaries.

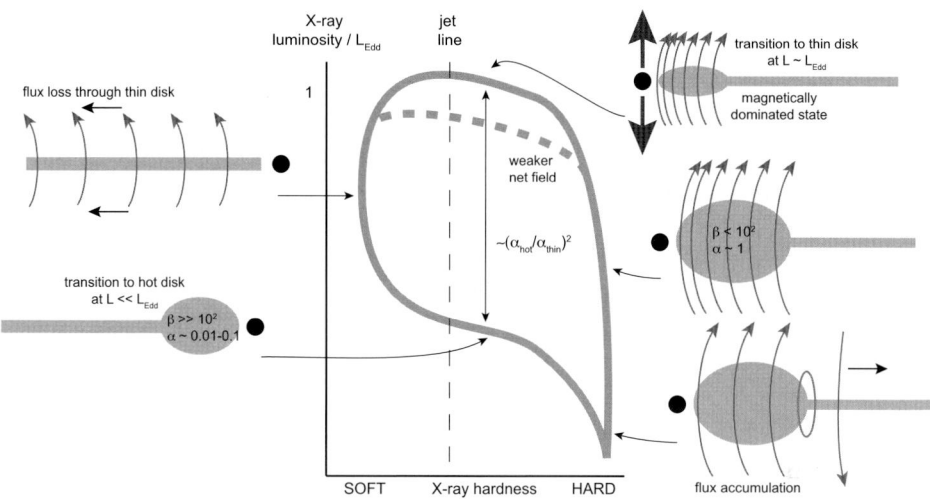

Fig. 2. X-ray binaries during outburst exhibit a temporal sequence of spectral states in the plane of X-ray hardness vs. luminosity. A similar sequence is seen in the plane of rms variability vs. luminosity, with variability and hardness strongly correlated. The sense of the cycle is counterclockwise, implying that a hard spectrum with increasing luminosity precedes a rapid transition to a soft spectrum at roughly constant luminosity, etc. The origin of the hysteresis exhibited by the cycle, in which the hard to soft transition occurs at a higher luminosity than soft to hard, is an unsolved problem. Cartoons illustrating different stages of the cycle show the geometric/thermal characteristics of each state (thin radiatively efficient disk vs. radiatively inefficient hot flow), which are widely agreed on, along with a more speculative proposal for the correlated behavior of the magnetic flux.[12] Figure is reproduced from Ref. 12.

4. Phenomenology of Black Hole Accretion

4.1. *X-Ray Binaries*

X-ray binaries (XRBs) are the observational manifestations of stellar-mass black holes undergoing disklike accretion of matter captured from a companion star. The mean rate of mass supply is typically far below $\dot{M}_E$, but the outer disks undergo occasional instabilities[31] that dump matter into the central regions at a rate that can approach or even exceed the Eddington limit. These outbursts can last months, and usually follow a specific sequence of thermal and spectral "states" that has never been adequately explained. A schematic representation of the sequence of states in the luminosity–spectral "hardness" plane is shown in Figure 2,[12,32] which is annotated to illustrate a possible model for the origin of the observed hysteresis.

Although the cyclic evolution of XRB outbursts is not well understood, there is reasonable confidence about the physical nature of each state.[33,34] As the luminosity increases at the start of the outburst, the observed X-rays have a very "hard" (i.e. biased towards high energies), "nonthermal" spectrum that looks nothing like that expected from a thin (radiatively efficient), optically thick accretion disk. This is usually interpreted as evidence for a radiatively inefficient, two-temperature flow occupying the region around the black hole. As the accretion rate increases, this hot region shrinks and collapses to form a radiatively efficient disk. Both the pre- and post-collapse states are highly variable, and show evidence for a relativistic jet that is thought to indicate strong magnetic activity. Whether this activity is tied to the accumulation of magnetic flux is a subject of current speculation.[12,35] A hard spectral "tail," combined with evidence of thermal disk emission, suggests that a highly magnetized corona surmounts the disk. Only at the highest luminosities, exceeding a few percent of the Eddington limit, does the magnetic activity calm down (often following a dramatic, final outburst) and the flow resembles a standard thin disk with a thermal spectrum peaking at moderate X-ray energies.

As the accretion rate declines, the flow does not retrace its previous evolution. Rather than developing signs of strong magnetic activity and evaporating to form a hot, two-temperature state, the disk remains thin and rather quiescent, decreasing in luminosity while remaining radiatively efficient. The transition back to a hot accretion state with strong magnetic activity does not occur until the luminosity has dropped to a fraction of the luminosity at which the forward transition took place. This hysteresis, illustrated in Fig. 2, also occurs in XRBs containing a neutron star rather than a black hole,[36] so it is apparently intrinsic to the accretion flow and not the properties of the central object.

4.2. *Active Galactic Nuclei*

An active galactic nucleus (AGN) is produced when a central supermassive ($M > 10^6 M_\odot$) black hole accretes matter from its host galaxy. The outer boundary conditions for AGN accretion flows are much less understood (and probably much more diverse) than those for XRBs. Any standard thin disk model set up to supply the requisite mass for a luminous AGN is predicted to become self-gravitating beyond a fraction of a parsec, suggesting that star formation would interrupt or at least modify the flow.[37,38] Attempts to resolve this problem — such as thermal regulation of the disk through stellar processes, magnetically supported disks,[39,40] and stochastic injection of matter in random orbits[41] — show varying degrees of promise, but none has been strongly supported by observations.

The diversity of boundary conditions, as well as the cooler temperatures that typically prevail in AGN accretion flows, presumably lead to the highly diverse observational manifestations of these objects.[42] Many AGN are apparently surrounded by a geometrically thick torus of opaque, dusty gas that obscures a direct view of the inner accretion flow and reprocesses its emission to longer wavelengths. Unobscured AGN often exhibit an apparently thermal emission component peaking in the ultraviolet or soft X-rays, as predicted for thin disks with luminosities of roughly a few percent Eddington around black holes in this mass range. Such AGN also exhibit Doppler broadened emission lines that result from reprocessing of the thermal UV and X-rays, as well as corresponding absorption lines whose large blueshifts indicate that they are produced in a cool, fast wind.

Other AGN appear to be rather radiatively inefficient, with little evidence for thermal disk emission and luminosities well below the Eddington limit. Such AGN are primarily detected through broadband emission from powerful, relativistic jets.[43] These are especially well-studied in the radio (mainly because radio telescopes are especially sensitive compared to telescopes operating in other bands), but also emit strongly in all other bands ranging up to X-rays and gamma-rays. The relativistic nature of these jets is established through clear signatures of Doppler beaming (e.g. one-sided structure on the sky and rapid variability) and apparent superluminal motion (an illusion due to light travel-time effects) in jets pointing close to the line of sight.

AGN exhibit variability but, unlike XRBs, seldom change their accretion state dramatically. It could be that the manner in which gas is supplied determines the long-term mode of accretion, e.g. disklike and radiatively efficient when interstellar clouds accumulate and settle into a disk, hot (two-temperature) and perhaps starlike when matter is supplied from a hot halo of gas surrounding the nucleus of the galaxy or from stellar winds, as is believed to be the case in the Galactic Center.[44] The fact that radiatively inefficient flows with powerful jets occur much more frequently in elliptical galaxies (which have predominantly hot interstellar matter) than in spirals (which have an abundant cool gas) seems to support this environmental interpretation. These differences in accretion mode could also affect the accumulation of magnetic flux, thus governing jet production indirectly.[45] But one cannot rule out the possibility that AGN undergo state transitions analogous to XRBs, as their vastly larger spatial scales suggest that such transitions would take place too slowly to detect.

4.3. *Tidal Disruption Events*

Tidal disruption events (TDEs) are transient episodes of accretion triggered when a star is partially or completely disrupted by the tidal gravitational field of a supermassive black hole.[46,47] In a typical event about half the mass escapes, but the other half falls back gradually over time, with a rate scaling roughly as $t^{-5/3}$ at late times.[48]

Only black holes less massive than about $10^8 M_\odot$ are capable of disrupting main-sequence stars; more massive holes would swallow such stars whole. A solar-type star would have to venture to within about 30 Schwarzschild radii of the Galactic Center's black hole (which has a mass of several million $M_\odot$) in order to be disrupted, implying that the mean specific angular momentum of the debris would only be a few times GM/c. This means that fallback cannot lead initially to a disklike accretion flow that extends to large radii, because there is not enough angular momentum compared to the Keplerian value. At later times, however, as matter is swallowed leaving behind most of its initial angular momentum, the mean angular momentum per unit mass of the remaining material increases and the flow can evolve toward a more disklike state.

Despite the relatively small cross-section for tidal disruption, TDEs are expected to be fairly common in nearby galaxies, occurring once every 10^4 years or so in a galaxy like the Milky Way. About two dozen candidate TDEs have been identified through the soft X-ray thermal spectra predicted to characterize their accretion disks and their characteristic light curves, which peak days to months after the disruption (depending mainly on the mass of the black hole) and then decline roughly according to $t^{-5/3}$.[49,50]

Disruptions of solar-type stars by black holes in the mass range $10^5 - 10^6 M_\odot$ are predicted to lead to fallback rates large enough to produce super-Eddington luminosities for $\sim 1 - 3$ yr, if the debris were accreted in real time.[51,52] Two TDE candidates have been discovered which appear to exceed the Eddington limit by about two orders of magnitude (for the estimated SMBH mass of $\sim 10^6 M_\odot$), even after correcting for beaming.[53–56] Their observed decay rates suggest that the luminosity tracks the fallback rate, and the presence of a radio afterglow in both cases suggests the production of a relativistic jet.

If these events represented disklike accretion, one would expect self-regulation of the mass flux reaching the black hole to a value that did not exceed $\dot{M}_E$ by a large factor.[57,58] But super-Eddington TDE accretion flows are probably starlike, given the low specific angular momentum of the accreting matter and the fact that it is probably pushed out to rather large distances by the pressure of trapped radiation. In Sec. 3.2 we argued that radiatively inefficient, starlike flows are unable to regulate the rate at which matter reaches the black hole, but were unable to decide the outcome of the energy crisis that likely ensues. Observations of super-Eddington X-ray luminosities and jets from TDEs suggest that, at least in these systems, the excess energy finds a relatively stable escape route through the rotational axis.[59]

4.4. *Gamma-Ray Bursts*

The close coincidence between long-duration gamma-ray bursts (GRBs) and core collapse supernovae supports the collapsar model, in which the burst results from the formation and rapid growth of a black hole or neutron star at the center of a massive stellar envelope.[60] The long duration of such bursts (minutes or more) implies sustained accretion at an extremely high rate,[61] while the large total energies involved favor a black hole engine, at least for the most luminous bursts. The inferred accretion rates initially can be as large as a tenth of a solar mass *per second*.[62] While enormously super-Eddington (by up to 14 orders of magnitude), the initial episode of accretion is not necessarily radiatively inefficient, because neutrinos can carry away most of the liberated energy; however, as the accretion rate declines and neutrino losses become insignificant, the flow must revert to an extremely radiatively inefficient state.

Our ignorance about the angular momentum distribution inside the stellar progenitor make it difficult to determine whether the late-time accretion flow during a long-duration burst is disklike or starlike. The weight and optical depth of the overlying envelope certainly prevent the Eddington limit from being a factor in regulating the rate at which mass reaches the black hole — radiation is too thoroughly trapped. However, we would expect a disklike flow to adjust so that some outward advection or circulation of energy suppresses the accretion by about a factor $(c_s/c)^2$ below the Bondi value, where the latter is calculated using the self-consistent value of the density and sound speed c_s at the "Bondi radius" GM/c_s^2 inside the stellar envelope. But the accretion rates needed to explain the prompt emission from long-duration bursts far exceed this, suggesting that no such regulation occurs. A starlike accretion flow, producing orders of magnitude more energy than can be wicked away by the accretion flow, would suffer a similar energy crisis to that in the jetted TDEs.[59] Like the TDEs, GRBs are evidently able to dispose of the excess energy through powerful jets punching through a quasi-spherical envelope.

These jets are remarkable for their enormous bulk Lorentz factors ($\Gamma \sim 100 -$ 1000) which are inferred from variability considerations and the requirement that the gamma-rays be able to escape. These Lorentz factors are 1–2 orders of magnitude higher than those found in other jets produced by black hole accretion, such as the jets from AGN and XRBs. I will comment on the possible significance of this below.

5. Unsolved Problems

Here is a selection of problems that are currently of interest to workers in the field of black hole accretion. These are problems that particularly interest me, and I make no claim that they are widely agreed upon as among the most pressing issues.

– *What triggers state transitions*?

While there is fair agreement about the nature of each XRB state, the factors that cause transitions between states are much less clear. In particular, the transition from the thin disk state to the hot, radiatively inefficient (and presumably two-temperature) state, is poorly understood. Both states are possible for a range of accretion rates spanning many decades, and there is a large "potential barrier" against converting a thin disk to a hot torus, because the high density of the thin disk keeps electrons and ions extremely tightly coupled. Some kind of bootstrap evaporation process may be necessary, as has been proposed for cataclysmic variables.[63] Formation of coronae above thin disks has been observed in magnetohydrodynamic simulations with radiative cooling,[64] but the effects of microphysical plasma processes (such as viscosity and electrical resistivity[65]), which may regulate the level of magnetic activity, have yet to be assessed.

The question of what triggers state transitions is tied up with the question of what causes the hysteresis (Figure 2), in which the transition from a hard (radiatively inefficient) to a soft (radiatively efficient) state occurs at a higher luminosity than the reverse transition. Apparently there is a second parameter (in addition to $\dot{M}/\dot{M}_E$) affecting the transition, and this parameter is correlated with location in the cycle. The magnetic flux threading the disk, which can depend on accretion history, is an attractive candidate,[12,66] as is the magnetic Prandtl number, the ratio of viscosity to resistivity.[67] Other, less orthodox mechanisms, such as the history of disk warping, may also play a role.[68]

– *What stabilizes radiation-dominated accretion disks?*

The simple α-parameterization of angular momentum transport in thin accretion disks[18] predicts that disks dominated by radiation pressure should be thermally and viscously unstable,[69] i.e. they should heat up and thicken while clumping into rings. Yet the luminous, thin-disk states of XRBs show surprisingly little variability,[70] suggesting that this prediction is not borne out. This would not be surprising, given the oversimplification inherent in the α-model, were it not the case that simulations appear to show at least the kind of thermal instability predicted[71] (the models have not been run long enough to check viscous instability). Models to address this problem rely on providing an extra channel for release of energy, e.g. through disk turbulence[72] or winds,[73] and/or diluting radiative pressure support for the disk, e.g. through magnetic fields.[39] Until we understand the stability properties of luminous disks, it will be hard to develop a compelling explanation for state transitions.

– *How limiting is the Eddington limit?*

The Eddington limit is often regarded as an upper limit to the luminosity of an accreting black hole, and therefore as setting an upper limit to the mass accretion rate. In disklike flows with a super-Eddington mass supply, such as the XRB SS 433, there are indeed reasons to believe that most of the mass flux in excess of $\dot{M}_{\rm E}$ is reversed and flung away before getting anywhere near the black hole. The most luminous quasars also seem to respect the Eddington limit, although it is possible that this could result from a selection effect (super-Eddington quasars might be sufficiently obscured to have escaped identification) or a coincidence due to a distribution in mass supply rates that decreases rapidly with $\dot{M}$. But we have seen that in starlike flows, such as the super-Eddington phases of TDEs, such self-regulation may be impossible. And GRBs clearly violate the Eddington limit by many orders of magnitude.

Exceeding the Eddington accretion rate should be distinguished from violating Eddington's limiting luminosity. Black holes can grow at a rate that exceeds $L_{\rm E}/c^2$ for a number of reasons. For one, radiation produced in an accretion flow need not escape; it can be trapped and swept through the horizon.[29] This is one aspect of accretion that truly distinguishes black holes from other compact objects such as neutron stars. For another, the matter responsible for liberating most of the accretion energy is not necessarily the same matter that radiates it away. In a flow with large density contrasts on small scales, radiation could "go around" opaque clumps, escaping mainly through the low-density interstices. But if the low- and high-density regions are tied together by magnetic tension, a super-Eddington escaping flux will not exert enough force to stop accretion.[74,75] For a third, some gas in a radiatively inefficient accretion flow could be swallowed with low binding energy, allowing a large amount of accreted matter to release relatively little radiation.

From the point of view of escaping luminosity, the Eddington limit is a spherical idealization, sensitive to geometric modifications. To give a simple example, a geometrically thickened disk, its surface a cone about the rotation axis, can exceed the Eddington luminosity by a factor proportional to the logarithm of the ratio between the outer and inner radii.[76] The apparent luminosity is enhanced by an additional factor because this luminosity is focused into the solid angle subtended by the disk surface. Inhomogeneities in the disk density, such as those caused by "photon bubbles," can also create intrinsically super-Eddington luminosities.[75]

More subtle are global effects that make accretion flows unable to regulate their power outputs, such as the steep density profiles expected to develop in starlike accretion flows. Here we have no obvious theoretical reason to demand that the flow rid itself of the excess energy in an orderly way; it would be quite plausible if the flow blew itself apart. But the evidence from super-Eddington TDEs, and possibly from GRBs, indicates that somehow a pair of jets is able to carry off the excess energy. How efficiently this kinetic energy is converted into radiation is another challenging unsolved problem.

– *What causes quasi-periodic oscillations?*

Power density spectra of XRB variability display an array of rather narrow peaks, called *quasi-periodic oscillations* (QPOs), which often contain a significant amount of power (a percent or more) in all but the thermal, thin disk states. High-frequency QPOs typically have frequencies of hundreds of Hz, and are presumably associated with dynamical processes (orbital motions, p-mode oscillations, etc.) in the inner portions of the accretion flow. A 3:2 resonance which is often seen can be interpreted in terms of epicyclic motions in a relativistic gravitational field.[77] Low-frequency QPOs, with frequencies of less than 0.1 to a few Hz, are more mysterious because they are unlikely to be associated with disk dynamics very close to the black hole, where most of the energy is liberated, yet they can carry a significant fraction (up to a few percent) of the total accretion luminosity. Both kinds of QPOs are primarily a feature of the hard X-ray spectral bands, which is believed to be produced by a hot, radiatively inefficient (two-temperature) accretion flow in low luminosity states, and a corona surmounting a thin, magnetically active disk in high-luminosity states. For the low-frequency QPOs, this suggests that energy is being spread through the inner hot region by nonradiative processes, to be modulated by some kind of oscillation or rotational motion at a rather sharp outer boundary. Candidates for the modulation mechanism include coherent Lense-Thirring precession of a hot accretion flow with an outer radius of several tens of gravitational radii[78] and rigid rotation of an extended ($\sim 10^3 R_g$) magnetosphere,[79] which requires an enormous amount of trapped magnetic flux. Other suggestions appeal to thermal or viscous timescales, but no model has been particularly satisfactory to date.

– *Are jets always propelled by coherent magnetic fields?*

According to standard theories for jet formation, the power of a jet depends on the total magnetic flux threading the engine, $L_{\rm J} \sim \Phi^2 \Omega^2 / c$, where Φ is the net magnetic flux and Ω is the angular velocity of the crank. While the rms magnetic field *strength* in the inner region of a TDE accretion flow can be substantial, most of this is associated with turbulent field resulting from the MRI. The net poloidal *flux* is limited to the magnetic flux contained in the disrupted star, which is too small to power the observed jets by about five orders of magnitude.[80] Whether a starlike, super-Eddington accretion flow can generate large-scale fluctuating fields with enough coherence to power the observed jets is an open question. GRB jets face a similar shortfall, although the likely discrepancy is only one or two orders of magnitude.

This suggests that jets in TDEs and perhaps GRBs are propelled by the energy in chaotic magnetic fields,[81] which would quickly decay by turbulent reconnection into radiation pressure. Radiative acceleration of optically thin gas to relativistic velocities is severely limited by radiation drag effects, which are made worse by relativistic aberration,[82] but these effects can easily be ameliorated if the flow entrains a large enough optical depth in ambient matter to shield itself. A marginally self-shielded jet could be accelerated to a Lorentz factor that is some fractional power ($\sim 1/4$) of the Eddington ratio L_J/L_E, which could explain why GRB jets are so fast.

It is curious that the most highly relativistic jets known — those associated with GRBs — should emerge from the most optically thick, radiation-dominated regions. Perhaps this reflects the difficulty that large-scale coherent magnetic fields face in converting most of their energy into motion. Once they reach moderately relativistic speeds, corresponding to rough equipartition between kinetic energy and Poynting flux, the electric field cancels out nearly all of the accelerating force and the energy conversion process stalls. Some form of magnetic dissipation may be necessary to catalyze further conversion of Poynting flux into kinetic energy.[83,84]

These arguments also apply to the jets which are more likely to be propelled by coherent magnetic flux, i.e. those produced in AGN and XRBs. It is commonly believed that jets cannot be propelled to extremely relativistic speeds by thermal pressure, but this is true only if most of the energy is quickly transferred to electrons, which cool rapidly. At the low densities and optical depths present in AGN jets, there is no reason to exclude a sizable contribution from relativistic ion thermal pressure. On the other hand, the fact that these jets seem to be limited to fairly low Lorentz factors, in the range of a few to a few tens, might indicate that such dissipative processes are not very effective.

6. Looking Forward

Our understanding of black hole accretion seems to be in pretty good shape. Magnetohydrodynamic simulations have shown that MRI really can drive accretion, and many other observable phenomena besides, and gives rough scaling relations that are closer to the prescient α-model of Shakura and Sunyaev[18] than we had any right to expect. On the other hand, exquisite observations of time-dependence and spectral transformations in XRBs, and completely new regimes of black hole accretion in TDEs and GRBs, show that the phenomenology of black hole accretion is even richer than we thought.

MHD simulations have come into their own as mature laboratory tools. Shearing-box experiments have been crucial but are probably approaching the end of their useful life. Global simulations — which at the very least are essential for understanding the coupling of disks to winds — have begun to take their place, but do not yet have adequate dynamic range. This is presumably just a matter of computer speed and available CPU time. We will also need to incorporate microphysical effects, such as viscosity, resistivity and heat conduction, which appear to be important for regulating the level of MRI-driven turbulence and perhaps its coupling to large-scale dynamos. On small turbulent scales, the relevant microphysical effects are likely to be collisional, but we will also need codes that can identify current sheets on large scales and compute the collisionless reconnection that is likely to occur.[85] Finally, the increasing importance of radiation-dominated flows, with luminosities approaching or even greatly exceeding the Eddington limit, means that radiation magnetohydrodynamic codes will be essential. Given the likely role of radiation pressure in driving some of the fastest jets, these codes will have to do radiative transfer in regimes where simple closure schemes are likely to fail.

Not all of these tools are in place yet, but there is every likelihood that these advances will happen soon.

Acknowledgments

I thank Prof. Phil Armitage for numerous conversations, both over beers and above treeline, and for preparing the figures. My research on black hole accretion and its manifestations is supported in part by National Science Foundation grant AST 1411879, NASA Astrophysics Theory Program grants NNX11AE12G and NNX14AB375, and Department of Energy grant DE-SC008409.

References

1. J. B. Hartle, *Gravity: An Introduction to Einstein's General Relativity* (Addison-Wesley, San Francisco, California, 2003).
2. M. Kozłowski, M. Jaroszyński, and M. A. Abramowicz, *A&A*, **63**, 209 (1978).
3. R. Narayan, I. V. Igumenshchev, and M. A. Abramowicz, *PASJ*, **55**, L69 (2003).
4. S. A. Balbus and J. F. Hawley, *RMP*, **70**, 1 (1998).
5. R. D. Blandford and D. G. Payne, *MNRAS*, **199**, 883 (1982).
6. R. D. Blandford and M. C. Begelman, *MNRAS*, **303**, L1 (1999).
7. M. C. Begelman, *MNRAS*, **420**, 2912 (2012).
8. J. F. Hawley, C. F. Gammie, and S. A. Balbus, *ApJ*, **440**, 742 (1995).
9. X.-N. Bai and J. M. Stone, *ApJ*, **767**, 30 (2013).
10. K. Beckwith, P. J. Armitage, and J. B. Simon, *MNRAS*, **416**, 361 (2011).
11. S. H. Lubow, J. C. B. Papaloizou, and J. E. Pringle, *MNRAS*, **267**, 235 (1994).
12. M. C. Begelman and P. J. Armitage, *ApJ*, **782**, L18 (2014).

13. J. Guilet and G. I. Ogilvie, *MNRAS*, **424**, 2097 (2012).
14. R. D. Blandford and R. L. Znajek, *MNRAS*, **179**, 433 (1977).
15. C. S. Reynolds, D. Garofalo, and M. C. Begelman, *ApJ*, **651**, 1023 (2006).
16. A. Tchekhovskoy, R. Narayan, and J. C. McKinney, *ApJ*, **711**, 50 (2010).
17. H. Bondi, *MNRAS*, **112**, 195 (1952).
18. N. I. Shakura and R. A. Sunyaev, *A&A*, **24**, 337 (1973).
19. J. E. McClintock, R. Narayan, and J. F. Steiner, *SSRv*, **183**, 295 (2014).
20. M. A. Abramowicz, B. Czerny, J.-P. Lasota, and E. Szuszkiewicz, *ApJ*, **332**, 646 (1988).
21. R. Narayan and I. Yi, *ApJ*, **428**, L13 (1994).
22. R. Narayan and I. Yi, *ApJ*, **444**, 231 (1995).
23. R. Narayan, I. V. Igumenshchev, and M. A. Abramowicz, *ApJ*, **539**, 798 (2000).
24. E. Quataert and A. Gruzinov, *ApJ*, **539**, 809 (2000).
25. J. M. Stone, J. E. Pringle, and M. C. Begelman, *MNRAS*, **310**, 1002 (1999).
26. J. M. Stone and J. E. Pringle, *MNRAS*, **322**, 461 (2001).
27. J. F. Hawley and S. A. Balbus, *ApJ*, **573**, 738 (2002).
28. R. D. Blandford and M. C. Begelman, *MNRAS*, **349**, 68 (2004).
29. M. C. Begelman, *MNRAS*, **187**, 237 (1979).
30. M. J. Rees, M. C. Begelman, R. D. Blandford, and E. S. Phinney, *Nat*, **295**, 17 (1982).
31. F. Meyer and E. Meyer-Hofmeister, *A&A*, **104**, L10 (1981).
32. R. P. Fender, T. M. Belloni, and E. Gallo, *MNRAS*, **355**, 1105 (2004).
33. R. A. Remillard and J. E. McClintock, *ARA&A*, **44**, 49 (2006).
34. C. Done, M. Gierliński, and A. Kubota, *A&ARev*, **15**, 1 (2007).
35. J. Dexter, J. C. McKinney, S. Markoff, and A. Tchekhovskoy, *MNRAS*, **440**, 2185 (2014).
36. T. Muñoz-Darias, R. P. Fender, S. E. Motta, and T. M. Belloni, *MNRAS*, **443**, 3270 (2014).
37. I. Shlosman and M. C. Begelman, *ApJ*, **341**, 685 (1989).
38. J. Goodman, *MNRAS*, **339**, 937 (2003).
39. M. C. Begelman and J. E. Pringle, *MNRAS*, **375**, 1070 (2007).
40. E. Gaburov, A. Johansen, and Y. Levin, *ApJ*, **758**, 103 (2012).
41. A. R. King and J. E. Pringle, *MNRAS*, **377**, L25 (2007).
42. J. H. Krolik, *Active Galactic Nuclei: From the Central Black Hole to the Galactic Environment* (Princeton University Press, Princeton, New Jersey, 1999).
43. M. C. Begelman, R. D. Blandford, and M. J. Rees, *RMP*, **56**, 255 (1984).
44. J. Cuadra, S. Nayakshin, V. Springel, and T. Di Matteo, *MNRAS*, **366**, 35 (2006).
45. M. Sikora and M. C. Begelman, *ApJ*, **764**, L24 (2013).
46. J. G. Hills, *Nat*, **254**, 295 (1975).
47. J. H. Lacy, C. H. Townes, and D. J. Hollenbach, *ApJ*, **262**, 120 (1982).
48. M. J. Rees, *Nat*, **333**, 523 (1988).
49. S. Komossa and N. Bade, *A&A*, **343**, 775 (1999).
50. S. Gezari, in *Tidal Disruption Events and AGN Outbursts, Madrid, Spain*, Ed. R. Saxton and S. Komossa, *EPJ Web of Conferences*, **39**, 03001 (2012).
51. C. R. Evans and C. S. Kochanek, *ApJ*, **346**, L13 (1989).
52. J. Guillochon and E. Ramirez-Ruiz, *ApJ*, **767**, 25 (2013).
53. D. N. Burrows, J. A. Kennea, G. Ghisellini, et al., *Nat*, **476**, 421 (2011).
54. J. S. Bloom, D. Giannios, B. D. Metzger, et al., *Sci*, **333**, 203 (2011).
55. A. J. Levan, N. R. Tanvir, S. B. Cenko, et al., *Sci*, **333**, 199 (2011).
56. S. B. Cenko, H. A. Krimm, A. Horesh, et al., *ApJ*, **753**, 77 (2012).
57. A. Loeb and A. Ulmer, *ApJ*, **489**, 573 (1997).

58. L. E. Strubbe and E. Quataert, *MNRAS*, **415**, 168 (2011).
59. E. R. Coughlin and M. C. Begelman, *ApJ*, **781**, 82 (2014).
60. S. E. Woosley, *ApJ*, **405**, 273 (1993).
61. T. Piran, *RMP*, **76**, 1143 (2005).
62. C. C. Lindner, M. Milosavljevic, S. M. Couch, and P. Kumar, *ApJ*, **713**, 800 (2010).
63. F. Meyer and E. Meyer-Hofmeister, *A&A*, **288**, 175 (1994).
64. Y.-F. Jiang, J. M. Stone, and S. W. Davis, *ApJ*, **784**, 169, (2014).
65. S. A. Balbus and P. Henri, *ApJ*, **674**, 408 (2008).
66. P.-O. Petrucci, J. Ferreira, G. Henri, and G. Pelletier, *MNRAS*, **385**, L88 (2008).
67. W. J. Potter and S. A. Balbus, *MNRAS*, **441**, 681 (2014).
68. C. Nixon and G. Salvesen, *MNRAS*, **437**, 3994 (2014).
69. N. I. Shakura and R. A. Sunyaev, *MNRAS*, **175**, 613 (1976).
70. M. Gierliński and C. Done, *MNRAS*, **347**, 885 (2004).
71. Y.-F. Jiang, J. M. Stone, and S. W. Davis, *ApJ*, **778**, 65 (2013).
72. Y. Zhu and R. Narayan, *MNRAS*, **434**, 2262 (2013).
73. S.-L. Li and M. C. Begelman, *ApJ*, **786**, 6 (2014).
74. N. J. Shaviv, *ApJ*, **494**, L193 (1998).
75. M. C. Begelman, *ApJ*, **643**, 1065 (2006).
76. A. R. King, *MNRAS*, **385**, L113 (2008).
77. M. A. Abramowicz and W. Kluzńiak, *A&A*, **374**, L19 (2001).
78. A. Ingram, C. Done, and P. C. Fragile, *MNRAS*, **397**, L101 (2009).
79. D. L. Meier, *Black Hole Astrophysics: The Engine Paradigm* (Springer-Verlag, Berlin, Heidelberg, 2012).
80. A. Tchekhovskoy, B. D. Metzger, D. Giannios, and L. Z. Kelley, *MNRAS*, **437**, 2744 (2014).
81. S. Heinz and M. C. Begelman, *ApJ*, **535**, 104 (2000).
82. E. S. Phinney, *MNRAS*, **198**, 1109 (1982).
83. Y. Lyubarsky, *ApJ*, **698**, 1570 (2009).
84. A. Tchekhovskoy, J. C. McKinney, and R. Narayan, *ApJ*, **699**, 1789 (2009).
85. D. A. Uzdensky, *ApJ*, **775**, 103 (2013).

Discussion

S. Tremaine I would like to defer comments on tidal disruption events since Martin will be giving us a prepared comment on that. Are there any immediate comments or questions on other issues?

J. Ostriker I would like to just add one phenomenon to the list, and this is for AGN. The radiation that you see coming out of them, however it's been generated, will interact with the matter that's flowing into them and then if you ask: that matter that's flowing into them, at the distance away that it is, when it has more or less zero energy, when the gravitational energy equals the thermal energy, will absorb that energy, it will stop it and blow it away. You didn't mention that phenomenon. It's very dramatic in our calculations, and it occurs at about a tenth of the Eddington limit. It may be what regulates the AGNs; it's basically heating of the radiation near the bondi or the equivalent radius is very important in regulating things.

M. Begelman I agree, I actually have a slide discussing some of the complications with the AGN that I didn't have the time to go over. So, I think that's right. One of the challenges in understanding AGNs is we do not actually know very much about the gas supply in the outer parts of the accretion disk. What we can say pretty confidently is that a standard thin accretion disk cannot extend very far out and feed an AGN because it becomes self-gravitating and you just do not get a self-consistent model. So there has to be some kind of injection of mass from molecular clouds or hot gas condensing or something that we do not understand very well. We know a lot more about the boundary conditions for X-ray binaries, so it is probably easier to develop a full picture for them.

R. Blandford Mitch, can I turn to your comment about the flux problem in extracting spin-energy magnetically from the black hole. I completely agree with you that this is a problem. One way of circumventing it perhaps is, instead of requiring the magnetic field in the vicinity of the black hole to be essentially dipolar, to allow it to be quadrupolar. The current is then dipolar. That then circumvents the need to have that flux *ab initio*, but it brings the problem that numerical simulations do not produce stable jets for reasons that are relatively clear. What may still be true is that the spin-energy may be extracted rather efficiently, and you rely on some different mechanisms to collimate the jet.

M. Begelman I agree, yes.

R. Wijers I have a question of clarification for you Mitch. One of the questions you state is whether AGNs also go through state changes. We of course have plenty of evidence either statistically or from observing clusters that they go through very large variations in at least the amount of power that they put out in jets. Isn't that already evidence for state changes, or could you maybe be a bit more specific about what precisely you meant by that problem?

M. Begelman There's actually two things you can refer to when you're talking about state changes. The average state of an X-ray binary typically has a very low accretion rate, and so what triggers one of these cycles is apparently dumping of a much larger amount of gas into the central regions of the accretion disk. But once you do that it's not clear that the system will respond in a steady-state fashion, determined completely by the rate at which you're dumping matter in. What probably happens is that once you reach a certain threshold, maybe a few percent of the Eddington limit, the conditions of the accretion flow become very sensitive to exactly how the matter is supplied, and so small changes in the gas flowing in create large changes in the thermal and dynamical states of the accretion flow. What I meant was that we do not know if those events, the histories and other unique properties also apply to AGNs.

S. Tremaine Let me defer further discussion. I wanted to ask Sterl Phinney to make a contribution to describe the possibility of detecting gravitational radiation from merging black holes. So that nobody gets jealous, he's going to talk about two distinct experiments, so he gets two slides.

Prepared comment

S. Phinney: Black Holes and Gravitational Waves

Black holes accessible to astronomers come in two sizes: stellar mass $(1 - 10^2 M_\odot)$, and supermassive $(10^5 - 10^{10} M_\odot)$. The former can be thought of as overweight neutron stars, and the latter as the cesspools of galaxies, into which falls everything in a galaxy that loses its angular momentum. Black holes of intermediate mass may exist, but have so far proven hard to find and confirm. As of 2014, the strong evidence for black holes, of both stellar mass and the supermassive sort, has come from the orbital motions of visible stars orbiting the invisible black holes, and from the electromagnetic radiation liberated by gas spiralling into the black holes.

In the next few years, it is very likely that we will detect both these types of black holes in a new way, through their gravitational waves.[a]

Gravitational waves from supermassive black holes can be detected by timing the pulses from radio pulsars, as first proposed by Mikhail Sazhin (1978) and Steven Detweiler (1979). This became astrophysically interesting with the discovery of millisecond pulsars in 1982. The spins of millisecond pulsars are as regular as atomic clocks on earth, and many have narrow pulses whose times of arrival can be determined to better than microsecond accuracy. Passing gravitational waves of dimensionless strain amplitude h and periods $T_{gw} = 1/f_{gw}$ of years wiggle both the earth and the pulsar (thousands of light-years distant from earth), producing shifts in the pulse arrival times of order $\delta t \sim hT_{gw}$. Timing with microsecond

---

[a]Providing, of course, Einstein was correct in asserting that "Subtle is the Lord, but malicious he is not." The nightmare of both LIGO experimentalists and pulsar timers is that astrophysical processes are malicious, and *only* make binary black holes at separations so large that the lifetime to gravitational wave inspiral is longer than the current age of the Universe.

precision for several years can thus detect gravitational waves with amplitude $h \sim 10^{-14}(\delta t/\mu s)(3y/T_{gw})$. This could be produced by a pair of $10^9 M_\odot$ black holes merging in the nearby Virgo cluster. There are several such black holes in the nuclei of galaxies there, and cosmological models predict they ought each to have experienced a few mergers in their lifetime. But when the black hole orbits have periods as short as $T_{gw} \sim 3y$, gravitational radiation will drive the black holes together in only $\sim 10^5 y$, 10^{-5} times the age of the galaxies. So mankind would have to be very lucky to have such an event now in progress so nearby.

Much better prospects arise from the superposition of all the (much more distant) merging supermassive black holes on our past lightcone. The gravitational binding energy of black holes of masses M_1, and $M_2 \ll M_1$ orbiting each other with separation a is $E = GM_1M_2/(2a)$. Using Kepler's third law, and that the gravitational wave frequency is twice the orbital frequency, one easily finds that a fraction

$$\frac{1}{M_1 c^2}\frac{dE}{d\ln f_{gw}} = \frac{1}{3}\frac{M_2}{M_1}\left(\frac{GM_1}{c^3}\pi f_{gw}\right)^{2/3} = 0.002\frac{M_2}{M_1}\left(\frac{M_1}{10^9 M_\odot}\frac{f_{gw}}{10^{-8}\text{Hz}}\right)^{2/3}$$

of the big black hole's mass must be radiated in gravitational waves as the orbital frequency sweeps through the e-fold around f_{gw}, if they are the main energy loss at that orbital frequency. Surveys of nearby galaxies show that virtually all big galaxies have supermassive black holes in their nuclei, and adding up their masses, one finds that black holes in ellipticals have a cosmic mass density relative to the closure density $\rho_c = 3H_0^2/(8\pi G) = 1.3 \times 10^{11} M_\odot \text{Mpc}^{-3}$ of $\Omega_{\bullet E} = 2.5 \times 10^{-6}$, (mainly $10^{9\pm1} M_\odot$) and those in spirals have $\Omega_{\bullet S} = 7 \times 10^{-7}$ (mainly $10^{7\pm0.5} M_\odot$). If these black holes experienced mergers at redshift z when a fraction η_m of their mass had grown by gas accretion (which does not produce gravitational waves!), i.e. if $\sum M_2 = \eta_m M_1$ on average for final mass M_1, then the above values of Ω, times the fraction of energy radiated at $f_{gw} = 10^{-8}f_{-8}$Hz from the displayed equation, give

$$\Omega_{gw}(f_{gw}) \equiv \frac{2\pi^2}{3H_0^2}f_{gw}^2 h_c^2(f_{gw}) \sim \frac{(5 \times 10^{-9}\eta_{mE} + 7 \times 10^{-11}\eta_{mS})}{\langle 1+z\rangle^{1/3}}f_{-8}^{2/3},$$

where the two terms are respectively from elliptical and spiral galaxies. This gives the rms strain $h_c \sim 5 \times 10^{-15}(\eta_{mE} + 0.014\eta_{mS})^{1/2}f_{-8}^{-2/3}$. Australian, European, and North American groups have been discovering and improving timing of millisecond pulsars for decades, and in the next several years should be able to either detect the background if $\eta_{mE} > 0.2$, or constrain $\eta_{mE} < 0.1$.

Reasons why one might expect $\eta_{mE} < 0.1$ include the possibility that most $10^9 M_\odot$ black hole binaries in giant ellipticals, which dominate the mass density, might not have enough stars and gas on orbits getting close enough to remove the binary's angular momentum, and could therefore get stuck at $a > 0.1 - 1$ parsec (gravitational wave periods $> 300y$), where the inspiral time is longer than the age of the universe. This is much less likely to be true of the $\sim 10^7 M_\odot$ black holes in the denser, more gas rich spiral galaxies (i.e. $\eta_{mS} \sim 1$), but for these, stars and gas drag can rapidly bring their smaller black holes to radii where $f_{gw} > 10^{-8}$Hz, so less gravitational radiation would be produced at low frequencies (where the timing measurements get most sensitive as the time baseline increases). Most robustly conservative seems a prediction that from mergers of spirals alone $h_c > 2 \times 10^{-16} f_{-7.5}^{-2/3}$ for $f > 10^{-7.5}$Hz. The impending limits and frequency dependence will tell us much about the dynamics of black hole binaries in galactic nuclei.

Gravitational waves from stellar mass black holes can be detected by the current generation of gravitational wave detectors, the most sensitive of which is expected to be Advanced LIGO, starting science operations in late 2015. From their electromagnetic emissions, astronomers have found binary neutron stars (pulsars), and neutron stars and black holes orbiting ordinary stars (X-ray binaries; astronomers could have discovered a pulsar-black hole binary, but haven't yet). But we remain in the dark about binary black holes: they are not expected to emit electromagnetic radiation. Uncertainties in massive star evolution and the gas dynamics of the common envelope needed to get orbital periods short enough for gravitational wave emission to merge the black hole and neutron star remnants to mean that we cannot reliably predict the merger rate of black hole binaries even in our own galaxy. But Advanced LIGO will be sensitive to merging $50 M_\odot$ black hole binaries at redshift $z = 1$ at rates as low as 10^2 per 10^{10}y in Milky-Way-like galaxies, 10^{-3} of the rate of merging neutron stars inferred from the observed binary neutron stars, so prospects are good. Thanks to advances in numerical relativity, we now have accurate gravitational waveforms even for rapidly spinning, precessing black holes with spins at arbitrary angles to the orbital axis. If and when these are detected, they will constrain formation models, and most important, will provide our first measurements of dynamical strong-field gravity.

Discussion

M. Rees Just a comment that in the final coalescence there's an asymmetry in the radiation which leads to recoil. Even a one percent asymmetry leads to recoil which can kick the merged black hole out of its parent galaxy, and this is a phenomenon that can be important in studying the way in which mergers build up black holes, because those that escape will hurtle through intergalactic space and not participate in further mergers.

M. Begelman One thing I wanted to say and we didn't have time to is that the occupation fraction in the centers of galaxies is actually an interesting test of that. So far as we know it seems to be a hundred percent, but if there are these special cases it should be less than one, or eventually we should find some missing ones.

S. White I would like to just come back to a comment I made to Reinhard. It's an interesting question particularly for this meeting. Reinhard very properly emphasized possible alternatives, and the one he had did not seem particularly plausible.

There are other plausible possibilities, for example like a boson star of some strange field theory.

But it's quite striking that people talking about stellar mass black holes seem to assume that the case for a black hole is proven. Is that really true that we consider now on the stellar mass scales that the existence of black holes is proven? That this couldn't be anything else, that there really is an event horizon, and that general relativity is proven all the way down to the event horizon? And if not, what are the other possibilities?

J. Ostriker This is on the second talk and the possibility of merging. There are many indications that the merger rate of galaxies has been underestimated in the dark matter-only simulations and that in fact there is a very high rate of minor mergers. That would give you locally a large number of mergers of black holes from these minor mergers between galaxies. And so if you put that in you'll just find higher rates, but smaller energy per event. And I think it's an interesting phenomenon.

S. Phinney That doesn't have any effect on the background. There is a simple theorem analogous to Soltan's theorem which is that the background just depends on the total mass of merged black holes in the local universe, so there's an upper limit set by the measured mass-density of black holes locally at redshift zero. So it doesn't matter if you double their mass by merging equal masses or you have a hundred mergers, each with a hundredth of the mass, it does not affect the gravitational-wave background.

S. Phinney Let me say something about Simon's question. I think if Ramesh Narayan were here he would say yes, the existence of an event horizon has been proven by the fact there is not a solid surface and therefore you can have higher accretion rates with lower efficiency. He's not here but he would say what I think.

E. van den Heuvel I think the evidence is strong, but it's circumstantial. The masses, of course, of these objects in these X-ray binaries, are going up to ten, twenty solar masses, in what we call black hole binaries. We have of course this theorem that causality does not allow neutron stars more massive than something like 3.5 solar masses. On the other hand there is no evidence at all in these systems for a solid surface like you have in the neutron star binaries where you have all kinds of indications that there is a solid surface. And then of course there is what Mitch showed, this whole behaviour in the color-color diagram which the neutron star systems do not show but the black holes do show. So there is another type of beast in these systems, and it looks likely that these are the characteristics of a black hole, but it's all circumstantial evidence.

M. Begelman There was a paper a few weeks ago claiming that neutron stars show exactly the same hysteresis with slightly different normalization and slightly different tilts in the trajectories. This is from a fairly large sample analyzed by Fender's group.

S. Tremaine I would like to start with a couple of short contributions on a different topic that was alluded to by earlier speakers: the topic of feedback. The basic point of feedback is that, as Reinhard said, the ratio of black hole mass to the typical galaxy solar mass is a few tenths of one percent. The ratio of the typical speed in the galaxy to the speed of light is 10^{-3}. So there is much more energy released in the process of forming the black hole than there is in the process of forming the rest of the galaxy. Even if a tiny fraction of that is fed back in the galaxy formation process it can radically alter and even suppress the formation process.

Prepared comment

A. Fabian: X-ray Evidence for Feedback in Local Galaxies and Clusters

The X-ray image in Figure 1 is centred on NGC1275 in the core of the Perseus cluster of galaxies. It is about 180 kpc from North to South and clearly shows the impact of the central massive black hole on a large region of surrounding intracluster gas, a billion times larger that the event horizon of the central black hole. The luminous accreting black hole is the white spot in the centre, above and below which lie cavities or bubbles in the X-ray emitting gas, blown by jets of relativistic plasma emanating from close to the black hole. Further away are seen another pair of bubbles to the NW and SE which have buoyantly detached from the centre over 10^8 yr ago. The sharp edge to the pale arc of emission to the NE of the Northern inner bubble is a weak shock where the gas density jumps inward by a factor of 1.3. This arc is seen in a pressure map to completely encircle the bubble and another high pressure ring encircles the lower one. They are presumably caused by the inflation of the bubbles. Faint ripples are seen further out.

Sharp blue patches occur running East to West across the Northern inner bubble. These are due to photoelectric absorption by cold gas in an intervening galaxy which is falling into the cluster. The depth of the absorption is very high with little sign of any foreground emission between us and the cluster, implying that the galaxy lies at least 150 kpc in front of NGC1275 and thus there is no direct interaction between it and the bubble system. Faint red streaks of soft X-ray emission coincide with a web of H-α emitting filaments which surround NGC1275. The filaments are basically molecular and have a total mass of at least 40 billion M_{sun}. The horsehoe-shaped filament just inside the outer NW bubble is shaped by gas flow as the bubble rises outward in the gravitational potential of the cluster. This graphically demonstrates that the filaments are dragged out by rising bubbles. The filaments extend to the North by over 60 kpc, presumably due to past bubbles no longer seen here. The filament sytem appears to be powered by inter penetration by the hot gas, which means that it is growing at about 100 M_{sun}/yr.

The interpretation of the image implies that the bubbling activity is long-lived and probably continuous, rather like bubbles rising from a fish-tank aerator. The process generates pressure waves and turbulence which propagate outward, dissipating energy over a wide region. The power of the process, estimated from

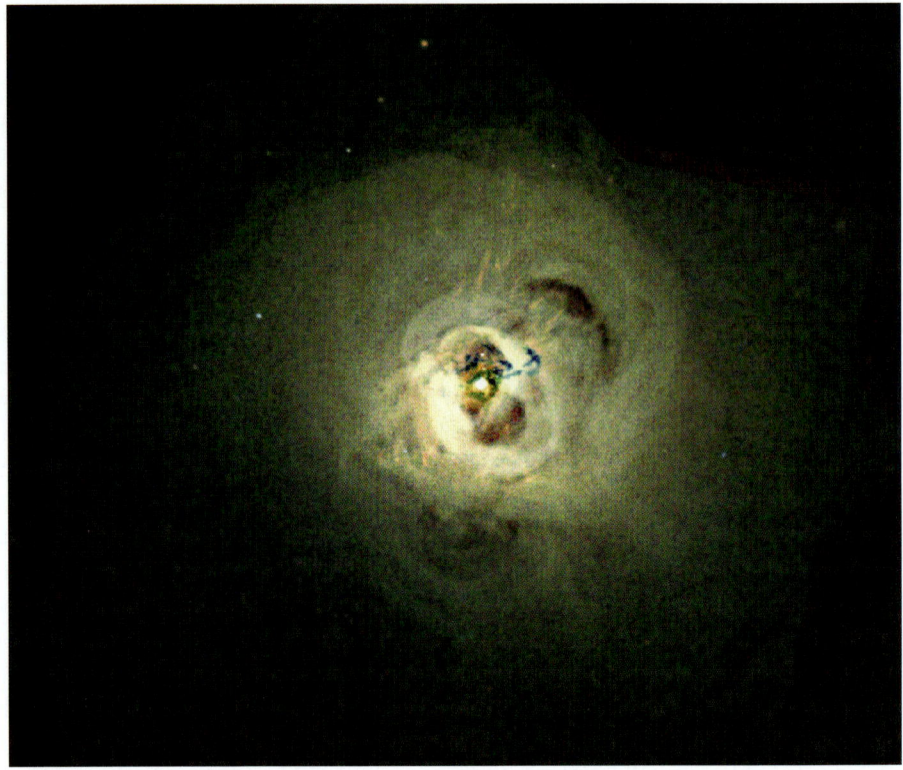

Fig. 1.   Slide shown by A. Fabian at the conference.

the PV energy content of the bubbles and their risetime, is approaching 10^{45} erg/s which is sufficient to balance radiative cooling of the intracluster gas, which would otherwise form a cooling inflow onto NGC1275 of several 100 M_{sun}/yr. AGN feedback acting in this way thereby stems the growth of massive galaxies such as NGC1275: the accreting black hole has a controlling influence over its much more massive host. This process is seen operating in the centres of most cool core clusters and groups of galaxies and also occurs in many elliptical galaxies.[a]

Discussion

S. Tremaine One short question. You have chosen a posterchild for this kind of feedback. Could you briefly say how common similar less well-resolved structures are in other clusters and galaxies.

---

[a]For further reading: Fabian AC: 2012 ARAA 50 455; Fabian AC et al.: 2003 MNRAS 344 L48 and 2008 Mat 454 968; McNamara, BR Nulsen: PEJ 2007 ARAA 45 117; Salome P et al.: 2001 A&A 531 85.

A. Fabian When we look at clusters in which the cooling time is short at the centers, we find that most of them have bubbles. I've done a recent paper with a student indicating basically the ones where we can't yet see bubbles tend to be ones where we do not have good enough data. And essentially the key is you need more than 20 thousand counts from the central 20 kiloparsecs before you can see bubbles. But generally, the bubbling duty cycle is close to 100% percent. And it would be very interesting to find one that didn't have bubbles.

S. Tremaine I wish you many more megaseconds.

A. Fabian Thank you. I wish you we are on the TAC.

M. Begelman Do you have any constraints on the cosmic ray distribution. Either from gamma rays or radio, or something else.

A. Fabian It hasn't been detected in gamma rays. The central AGN has been detected but not the intracluster medium. And as far as I know, no one has yet detected gamma rays from intracluster galaxies. There is a mini radio halo coincident with the region you could see there but we looked very hard but found no correlations between the radio flux and the X-ray structure or anything else.

S. Tremaine Before taking more questions, Jerry also has a prepared comment on feedback and we can then have questions on both.

Prepared comment

J. Ostriker: AGN (black hole) "feedback"

- It has been known for some time that energy, mass and momentum input from central black holes (BH) in galaxies ("Active Galactic Nucleus" – "AGN" – feedback) plays a central role in regulating the evolution of massive galaxies.
- The ratio of the mass accumulating in the central BH to the stellar mass is known to be relatively constant at roughly 1/1000. Straightforward "Bondi" accretion of gas from the central regions of the galaxy can account for this mass accumulation as determined by many investigators.
- What is less well understood are the exact physical mechanisms by which the accretion of this mass leads to AGN outputs that affect the whole galaxy sufficiently to reduce or terminate star formation in the systems.
- The observed output of AGN consists of

  (1) Relativistic jets.
  (2) Electromagnetic radiation in IR, UV and X-ray bands.
  (3) Conical winds emanating from the central region with velocities of thousands of kilometers/s.

- In modeling the output of AGN most investigations have adopted the heuristic algorithm of "thermal feedback" – for every mass element DM accreted by the BH, an energy DE is added to the surrounding gas, where DE =

epsilon c^2 * DM. This energy has been assumed to be added to the thermal content of gas near the BH. No physical mechanism is employed to transfer the energy and the results depend strongly on the mass to which it is transferred.

– The work described and performed by JPO with E. Choi and T. Naab consists of allowing for the wind output and electromagnetic output of the AGN in ways that mimic observed AGN and then computing in detail the effect on the surrounding galaxy based on detailed hydrodynamic numerical simulations. Published results so far are very promising and indicate that we may have a good handle on understanding the BH-Galaxy interaction.

– In particular, we can examine the X-ray output of the simulated galaxies and it agrees well with observations – far better than is the case with "thermal feedback".

– The ongoing work has not yet included the radio jets, which tend to drill through the galaxy and are most effective in heating the intergalactic or circum-galactic media, but that will be an additional undertaking.

Discussion

S. Tremaine Thank you. Comments on the feedback?

R. Romani Effective suppression of the inflow seems to require that you have relatively isotropic deposition of energy and momentum. Yet we know that we have this fantastic fly wheel in the middle. We do see of course in the fossil bubbles, as you showed for Perseus A, that there are several axes. How was this affected? And do you have a comment on how we can isotropize something that is essentially a jet?

J. Ostriker As Andy likes to say: the heating of a medium with a jet is like heating your house with a laser beam. It just burns a hole in it, and the roof. Through M87 you can see it burning the hole. This is not a jet. The BAL lines are seen in about a third of the quasars. And so, we assumed that the two cones of the wind occupy about a third of the 4π steradians which is again what Daniel Proga got in his acceleration mechanism. That is not very finely calculated but it is not 4π. And that seems to do the trick quite quantitatively.

A. Fabian So Jerry has provided a theoretical explanation. Just in terms of the observations it is well-known that the FRI jets which are common in these objects just puff out. And it is thought that basically there is a lot of pick up of surrounding material which mass slows them down, makes the bubbles and then the bubbles rise possibly oscillating as they do so. And if it is sound waves that take the energy outwards then they go out pretty isotropically.

Could I make just one comment again to Jerry's presentation about feedback and the quasars? I really believe in feedback but I should be truthful. The only object locally where we see evidence for a cooling flow taking place in a cluster is an object H1821+643 which has a very powerful quasar in the middle of it. So it seems where you have a very powerful quasar you at the same time get a cooling flow. And the possibility there is that the quasar actually Compton cools the intracluster medium, Jerry, and facilitates the cooling flow.

J. Ostriker Could you tell for sure which is the cause and the effect? Could it be that the cooling flow caused the quasar to turn on?

A. Fabian Could be. But somehow there is a positive feedback which is keeping it turned on.

S. White Yes, I agree that it is very plausible. I also have published papers claiming that this is the mechanism which suppresses the formation of massive galaxies. But I think to be fair that the objects where we see this, like the one Andy showed us or M87, are really massive systems whereas the plot that Reinhard showed us where this process kicks in is only for galaxies only slightly more massive than our own Milky Way. So I think where we need the process more effectively we have actually very little direct evidence that it is working. So we can run simulations such as the one Jerry showed and they look plausible but I do not think we have direct observational proof that this is actually the mechanism in this regime for things like the Milky Way or Andromeda.

J. Ostriker I agree and I was urging Andy to make other observational searches to look. We do not have direct observational evidence.

S. White The problem is that things like Andromeda and the Milky Way do not have X-ray halos. You can't see it so it is unclear whether this is a consequence or not of the same process, but it makes it very hard to do the observations.

R. Sunyaev The absolute majority of clusters in the vicinity shows that there is very strong heating by local AGN. I was very glad when Andy mentioned that one object in the vicinity where there is cooling flow. Another object which is an extremely interesting object is Phoenix cluster of galaxies discovered by South Pole Telescope led by the group of John Carlstrom. This is the object which then was observed in ultraviolet by other radio telescopes and Hubble. And I can tell you this is a tremendous object. There is cooling flow on the level of 3000 solar masses a year. Today it is observed in the cluster. Star formation rate exceeds 400 stars a year. It is on cosmological distance of this cluster of galaxies. I think that in the new sample which eROSITA and also SPT and ACT will discover in the next year, we will find a lot of distant clusters where there is strong cooling flow and which permit these giant galaxies to form at redshifts higher than unity.

S. Tremaine I had one question. Many people here have simulated feedback and other aspects of galaxy formation. And given our inability to resolve all the scales involved in feedback either the scale of the black hole or the scale of star formation these simulations normally rely on some complicated interplay in which you put in subgrid physics with multiple free parameters and then adjust the free parameters to match some sets of observations, carry on the simulation and then claim that the simulation matches another set of observational parameters. This makes me a little uncomfortable and I wonder if there is anybody else here who either feels uncomfortable or can tell me how we should be thinking about this as a scientific process.

S. White I think that you are right to be uncomfortable but if we want to study these objects I do not think that we have other options because we can see the scales involved differ by these large factors. We can't make the event horizon of the black hole larger with respect to the size of the galaxy than it is and we can't carry out a simulation over this range of scales. We are forced always to carry out partial studies and if you carry out partial studies, in the parts you can't study you have to put in some simplified model, which is usually in parametrized form. So you are right the way it is usually done is to put in a model in parametrized form and adjust its parameters according to some set of data. Just recently this has become to be called calibration because it sounds more acceptable if you call it calibration instead of tuning, which is how it used to be called. In any case when you've done this the data that you use to carry out your calibration, your tuning, you fit it by construction. So you can only see if this is enhancing your understanding of the system by going to other things which you did not use. For example the evolution in time or the behavior in the system at completely different scales and see if it also fits there. Typically it doesn't work. What that means is that your model that you put in was too simple. And you put in another model with even more parameters which you then calibrate to fit the new observations. But I think in a sense it's OK if the things you're adjusting are physically interesting values. For example the efficiency of the accretion into the black hole or the efficiency with which radiation is converted into a wind then in a sense rather than looking at it as calibration you could see it as measurement of physical parameters. But I think that it is a complicated question. I do not think that we have many other options.

E. S. Phinney Let me put it another way. Phenomenology is a means not an end.

S. White Absolutely.

J. Ostriker There are two kinds of calibration that one can do. One kind: adjust the parameters until the result fits the observations. That's one thing and Simon gave a good explanation for why that's plausible. Another approach which we tried to follow is to make the input numbers fit observations. We use the spectral output of the quasars to match what Sunyaev and Sazonov said quasars actually put out. We used the velocity and the mass outflow as determined by BAL quasars so we are matching observational numbers to begin with and then we make no adjustments.

M. Begelman There is another boundary condition that you have some flexibility with, which is the density of the gas that you are irradiating because if the ionization parameter which is the ratio of the flux to density is low enough, then the gas will just re-radiate all the X-rays it receives. It will not heat up. It will not evaporate the gas.

J. Ostriker That just comes out of the cosmological simulation. It is the infall of gas given ΛCDM.

M. Begelman But you have to assume something about the state of that gas, that it doesn't clump into molecular clouds or something like that.

J. Ostriker Well, in so far the code calculates that correctly. We are just using cosmological initial conditions.

R. Genzel Very briefly, I would say: isn't there another way of doing all of this, which is talking between the observational and the theoretical communities. So that you can give each other guidelines and then you can check it out and you can see if it works out and so forth?

C. Frenk I think you and Simon are both right in feeling uncomfortable about this. Me too even though I make a living from this. But I think that the question really is what do you think the simulations are for? Of course because we cannot resolve the entire range of scales that are relevant you have to make approximations. But making approximations is a perfectly respectable part of physics. The question is what is it that you are trying to do with the simulations? If you are trying to predict certain things which are completely sensitive on your parameters, well of course you will not advance the subject. But let me give you two examples of where you take a simulation, you have parameters as you often do in many other branches of physics. You adjust these parameters the best you can and then you have a model and then you can ask other questions from the model of which are on the scales that are resolved.

So let me give you two examples. One is what we have been talking about today: the AGN feedback. So the idea is that AGN feedback was needed came from various sources but one of them was a form of simulation if you want to call them that. It is an analytic modeling that showed that if gas was allowed to cool in the halos that we know formed, then you will end up with extremely massive galaxies that are not observed. It became clear then that some form of energy injection was needed. And the whole idea of AGN feedback emerges from this sort of consideration. One example of where you have models with approximations and parameters. You fix them in this particular case to match the stellar mass function. And then you use it to ask other interesting questions. And the other one is one that has been solved recently. It is the question of what does it take to make a galactic disk? For many years gas dynamics simulations could never make a disk. And only recently it has become clear what are the conditions that one needs in order to make a thin galactic disk and one of them for example is to make sure that the gas gets kept out of small halos that are subsequently merged and allow to fall in later. Of course the details are model-dependent and depend on how you think the feedback from supernovae couples to the gas and so on. So those are model-dependent but the general statement that if you want to have a small disk it is essential to keep the gas out of small halos and wait until they merge and let the gas cool after that, I think that is an important result that comes from simulations. So I think that you have to be careful when you do a simulation. What is it exactly that you are trying to learn from it?

T. Abel Yes, we should all be nervous of course with anything smaller than the grid scale because as gas collapses it can pick up an enormous amount of gravitational energy and so it takes care to properly interpret simulation results. On the other hand we can do many experiments that happen on faster time scales: the first initial collapse for some of the supermassive black hole scenarios, the formation of the first proto-stars, are such problems. And I ask then in this context both speakers who brought up the angular momentum problem in black holes. I would summarize the last 10 years or so as: people who actually tried extremely high dynamic range calculations of turbulent clouds find that the angular momentum is in fact not a particularly severe problem as long as you are only trying to collapse a few per cent of the available gas mass that you have. As long as there is this large reservoir that the collapsing core can share its angular momentum with, that envelope enables a very fast central collapse.

S. Tremaine I think that I am going to cut this off because I now feel much more comfortable.

S. Tremaine We did have some discussion on firewalls this morning and I asked Eva Silverstein to give us a brief tutorial on firewalls.

Prepared comment

E. Silverstein: Brief Overview of Firewalls

The Hawking calculation of black hole evaporation produces a thermal result (carrying no information), starting from the assumption of vacuum effective quantum field theory (EFT) near the horizon. Turning this around, if black hole evaporation is unitary, then the system is not described by EFT in its vacuum near the horizon.

This can be described in the language of quantum information, with the monogamy of entanglement violated by the combined hypothesis of near-horizon vacuum (where the late Hawking radiation modes and their interior partners are entangled) and unitarity (which requires entanglement between the late and early Hawking radiation).

'Complementarity' was the idea that this problem cannot be seen by a single observer. But a team from UCSB (AMPS '12) showed that there exist thought-experimental observers who can see too much for this idea to be consistent.

Even without astrophysical observational constraints, these thought experimental constraints are difficult to satisfy, with no complete model yet satisfying them. Something has to give:

– Unitarity: but the AdS/CFT correspondence provides compelling evidence in favor of unitarity (a point that Hawking has conceded).

– EFT outside the horizon and the smooth infall of a late observer.

If the latter is what gives, we need to understand the dynamics behind it. I will now share my approach (including work in collaboration with my student Matthew Dodelson, with useful input from others including J. Polchinski and S. Giddings). This is simply to check the breakdown of EFT in string theory in the context of the above thought experiments. String-theoretic effects going beyond EFT are not automatically negligible despite the low curvature, because of the large relative boost that develops between the late infaller and early matter. Our calculations yield some 'drama' due to string-theoretic spreading and non-adiabatic effects, and it will be interesting to see if this is enough. Once the dynamics and its time and mass scales are understood, it will be interesting to see if the theory is conceivably testable using astrophysical observations.

Discussion

T. Piran To what extent does this drama result from the assumption that the evaporation really ends with a flat spacetime and the black hole evaporates completely? Isn't it the case that if you drop the assumption that the black hole completely evaporates and allow for some way that for some information to stay in the black hole or for some reason that Bardeen suggested recently that the evaporation slows down as the black hole evaporates and it doesn't go to complete evaporation this problem is solved in a very different way.

E. Silverstein I think that what you are describing is known as remnants, where you just assume indeed as you said that the black hole doesn't fully evaporate and the information is stored in remnants. The argument against that basically as I understand it is that you end up with an enormous number of these remnants cause the number of states of the black holes is the exponential of the entropy of the original black hole which is enormous. And then you have to fold that into your calculations and even in just particle physics where you have this enormous number of species that can contribute to quantum processes. So there are arguments like this against remnants. Maybe other people want to comment but that's my understanding. As far as what Bardeen wrote recently I shouldn't comment because I haven't studied it in detail.

D. Gross Again just to answer Tsvi. What has convinced everybody is this example of black holes in AdS space where we have a duality which has passed so many tests that everyone believes in it, that's looked at carefully. And in that case there is no question that if you have infalling gravitons or whatever, they form a black hole, it evaporates. We have a duality which is strongly supported by much evidence that relates that to an absolutely unitary process in which you start out with anti-de Sitter space which can be pretty close to Minkowski space and end up with the same and with a unitary time evolution. So I think that's really what has convinced everybody. We do not have such an adequate description for say black holes in asymptotically Minkowski space but I think that that's why people are convinced, including Hawking.

U. Pen If we are to not trust things like the equivalence principle for infalling observers of which textbooks say that in any introductory GR we all know that for an infalling observer there is nothing unusual happening at all at the horizon, that if there is no sign saying you're at the horizon it is only when you go back in the long term you have a hard time doing that. If we now say that these intuitive things that we have developed over a century are not necessarily true because of subtleties on how boosting works and how information holds, how do we know that even more fundamental premises that lead to this problem are to be trusted. For example, the whole problem really comes because of a no hair premise of black holes. Hawking's assumption assumes that you have this vacuum background that doesn't have hair. I mean if we are now asking could it be a connection to observables. For example the event horizon telescope or other observables that kind of rely on being no hair. But the geometry is it trustworthy at all? If one thing fails, couldn't everything fail?

E. Silverstein I would really like to respond to that and I will do so in the context of the approach that I mentioned at the end because I can understand that approach. One can't really speak for anybody else in this field I have to say. The point is that the drama is catalyzed by the late observer. It will not be there as a property of the horizon by itself. The relative boost I talked about is again between an early infaller which could be part of the matter and a late infaller. And it is only then that you get an effect. So it is consistent with what you were saying about the equivalence principle in that sense.

Another question you might ask which we will ask is what about other types of horizons, cosmological horizons, ... The same information theoretic problem doesn't arise there but you could ask about galaxies falling through some observer's horizon relatively boosted and so on. So that's on the table. But it is really premature to make any connection to the observations but I think that it is a great question for the future. In any case, I'm saying in the drama that I at least understand how it plays out the answer to your question is that it is catalysed by the late observer, not a property of the horizon without it.

M. Zaldarriaga I just would like to ask you if you could expand a bit if there is a connection with other possible horizon or any kind of other drama that can happen in cosmological horizons or things like that.

E. Silverstein I think it's a great question. Here is my understanding at the moment but it could change. Let's talk about inflation and then later about acceleration. And again I will do it in the context of this approach that I have mentioned at the end. Again it is not a property of the horizon itself so it requires these relatively boosted probes and I think that in the late universe that may happen if you take a distribution of galaxies and you slice it much what happens is as those galaxies cross some observer's horizon you can get the relative boost but the late universe is not that late. We're in this phase where it is not old in a sense that I described. The black hole is. It is kind of marginal. So we want to use that as a constraint if it works to do so but it might just be that we're not quite late enough. In the inflationary case it is not at all clear that the same effect happens just because we start in the Bunch-Davies vacuum I do not know who is falling into the horizon. It could be there is other junk around or enough period but we really have no evidence for that so it is hard to use that as a constraint. That's my answer for that particular approach but it is a great question in general.

F. Englert If I remember correctly, the first paper I do not know if you were in it or not...

Eva Silverstein No.

F. Englert ... about this and there was some idea that there was a firewall instead of the horizon, the usual thing, in order to solve the same problem. Now, I think the problem one way or another when it is treated this way or I think the way you do it is based essentially on entropic considerations. The entropy at the end has to be essentially zero back. Right?

E. Silverstein Right.

F. Englert That's the way to state the unitarity.

E. Silverstein Yes.

F. Englert But in the first paper I remember this argument was used in detail. Now what bothers me is the following and is related to the remark that was just made that any entropy consideration has to be subtantiated by a microscopic argument. And in this case the only microscopic argument that I can see has to do with the horizon, or the would-be horizon if there is a problem there. Now therefore and in the argument neither comes the idea that there is a scale of the horizon nor that the horizon is the same for everyone like in de Sitter but I think that the problem is general. If the argument that way is correct, it should apply to the black hole it should apply to the de Sitter space and actually also apply to Minkowski space I think where you can also have horizons for the accelerated observer. So the problem should be there and as far as I see for the present at least nobody is going to say that there is a firewall in Minkowski space. Or maybe I do not know. Anyway I think therefore that the problem is not solved and the problem should be related to a deeper understanding of what we mean by horizon in quantum mechanics in general, whether it is a black hole or not and the question of the singularity which has been put is in my opinion secondary. The problem fundamentally is the black hole and how do we treat black hole quantum mechanics. Complementarity or not, or whatever but something there and maybe the classical limit, general relativity, is not what we think it is. That's just my remark and we haven't solved the problem and one should focus on an interesting way of understanding horizons.

V. Mukhanov Here is in fact a question because here people are practical. Let's specify the meaning of firewall. Let's take our big black hole at the center of the galaxy and make experiments. We will go together through the horizon. So what does it mean firewall, it means that when we will be crossing the horizon we will smash on it or we will not even see it. So what is the answer from the point of view of firewalls? Or what is going on with the accretion gas?

E. Silverstein Ok, short answer is that nobody knows the precise answer to that. But I actually tried to address it. First of all, the age of the black hole in the center of the galaxy is nowhere near the timescale at which this firewall if it is there would need to set in. However there are shorter time scales in the problem that might be relevant. Again if I answer in the approach that I am taking, there are shorter scales involving just the black hole radius itself as opposed to the radius cubed in Planck units which is the decay time related to the Page time. So first of all it is not at all clear that you should see anything. The general arguments do not say that you should see something in the practical sense. And then what form the drama would take is a great question. I have been not convinced by the arguments of the original paper even though I think they are completely right about the paradox. I'm not convinced by the arguments that it should be some sharply localized wall of radiation. That's a UV sensitive statement. It depends on what the completion of gravity is. That relative boosting has been keeping on emphasizing in some ways a known thing in the field, it is called the trans-Planckian problem of black hole physics which I do not think is a problem but it might be part of the solution. For that we just need the dynamics to answer your question and once we understand the dynamics we will understand the timescales involved. And then we will understand whether or not there is any hope of seeing it in these wonderful telescopes.

V. Mukhanov You see, normally people write equations and then put words to it. If we put words then there should be some equations.

E. Silverstein We do. I will send you our paper on string scattering in the linear dilaton background and you can enjoy the equations.

V. Mukhanov Ok. What is the conclusion from your equations? What is the answer to the question? Because there is also the other thing which was raised, this time. A black hole decreases by a factor 2, and then you have some remnant and it starts to emit some information. But the other says that you can get actually mass by just contracting black holes. It means that you can form one mass of black hole for instance of radius of one solar mass black hole. Or you can get the same black hole just evaporating two masses by a factor 2. It means that black holes also acquire new hairs or what? Because in general relativity as we know there is only one parameter, right?

S. Tremaine I am going to ask to cut this off because I did have one more contribution and it is almost time to break. Martin Rees has volunteered to say something brief about tidal disruption events. So I would like to get our feet back on the ground...

Prepared comment

M. Rees: Tidal Disruption of Stars

I want to comment on the tidal disruption of stars by massive holes. This process makes only a minor contribution to the overall fuelling of powerful AGNs, but could give rise to characteristic 'flares' in the nuclei of galaxies that are otherwise quiescent.

Tidal capture and disruption of stars attracted interest back in the 1970s, when theorists started to address the dynamics of stars concentrated in a high-density 'cusp' surrounding black holes in the centres of galaxies (and perhaps in some globular star clusters as well). It was recognized that stars could be captured and swallowed by the central hole if they were in a 'loss cone' of near-radial orbits.

If the central hole is sufficiently massive, tidal forces at the horizon may be too gentle to disrupt the star while it is still in view, in which case it is captured without any conspicuous display. For a solar-type star, this requires a hole mass exceeding $10^8 \, M_\odot$; for white dwarfs the corresponding mass is $10^4 \, M_\odot$. And neutron stars are swallowed whole by black holes with masses above about $10 \, M_\odot$ (this is important for the gravitational wave signal in coalescing binary stars). For a spinning hole, the cross-section for capture, and the tidal radius for disruption, depend on the relative orientation of the star's orbital angular momentum and the hole's spin. Stars on orbits counter-rotating with respect to the hole are preferentially captured: this is a process that would reduce the spin of a hole in a galactic nucleus.

The physics is much messier when the star is tidally disrupted rather than swallowed whole. This phenomenon has been studied since the 1970s, first via analytic models and subsequently by progressively more powerful numerical simulations. There are several key parameters: the type of star; the pericentre of the star's orbit relative to the tidal (or Roche) radius, and the orientation of the orbit relative to the hole's spin axis. In most astrophysical contexts, the captured stars would be on highly eccentric orbits (i.e. their orbital binding energy would be small compared to that of a circular orbit at the tidal radius). If the pericentre were equal to the tidal radius, a main sequence star would be disrupted, and the debris would continue on eccentric orbits, but with a spread of energies of order the binding energy of the original star. Indeed nearly half the debris would escape from the hole's gravitational field completely; the rest would be on more tightly bound (but still eccentric) orbits, and would be fated to dissipate further, forming a disc much of which would then be accreted into the hole.

A pericentre passage at (say) 2 or 3 tidal radii would not disrupt a star completely, but would remove its envelope, and induce internal oscillations, thereby extracting orbital energy and leaving the star vulnerable on further passages. On the other hand, a star that penetrates far inside the tidal radius (but not so close to the hole that it spirals in) will be drastically distorted and compressed by the tidal forces, perhaps to the extent that a nuclear explosion occurs, leading to a greater spread in the energy of the debris than would result from straight gas dynamics.

(These comments apply to main sequence stars. For giants it is a bit more complicated as there is a large range of pericentre distances for which the envelope is removed but the core survives).

There have in recent years been detailed computations of these processes, and also of the complicated and dissipative gas dynamics that leads to the accretion of the debris, and the decline of the associated luminosity as the dregs eventually drain away. There are two generic predictions: after debris returns to the hole and dissipates into an axisymmetric configuration, it should initially have a thermal emission with a power comparable to the Eddington luminosity of the hole; and at late times, when the emission comes from the infall of debris from orbits with large apocentre, the bolometric luminosity falls as $L \propto^{-5/3}$.

Some flares in otherwise quiescent galactic nuclei, where the X-ray luminosity surges by a factor exceeding 100, have been attributed to tidal disruptions. But perhaps the best candidate for an event triggered by tidal capture of a star is a remarkable gamma-ray burst, Swift J1644+57, located at the centre of its host galaxy, and which was exceptionally prolonged in its emission compared to other bursts. The high energy radiation, were this model correct, would come from a jet generated near the hole. Modelling is still tentative, and is difficult because there is no reason to expect alignment between the angular momentum vectors of the hole and of the infalling material. Be that as it may, this exceptional burst offers model-builders an instructive 'missing link' between the typical long ('Type 1') gamma rays burst, involving a massive star, and the jets in AGNs which are generated by processes around supermassive holes.

Discussion

S. Tremaine Thank you. I would like to have further discussion on this but because of the time I think that we're going to have to stop. I just would like to take the liberty of closing with another quote from an essay by Wigner on the unreasonable effectiveness of mathematics in the natural sciences. The relevant quote is that "The miracle of the appropriateness of the language of mathematics for the formulation of the laws of physics is a wonderful gift which we neither understand nor deserve. We should be grateful for it and hope that it will remain valid in future research". And I think that we are very grateful that this gift enabled us to find astrophysical black holes after they were predicted by an arcane theory. And we very much hope that it will remain valid in continuing efforts to understand firewalls around them.

Session 3

Cosmic Dawn

Chair: *Matias Zaldarriaga*, IAS, Princeton, USA
Rapporteurs: *Richard Ellis*, Caltech, USA and *Steven Furlanetto*, UCLA, USA
Scientific secretaries: *Thomas Hambye* and *Michel Tytgat*, Université Libre de Bruxelles, Belgium

Rapporteur Talk by R. Ellis: Massive Black Holes: Evidence, Demographics and Cosmic Evolution

Abstract

I review recent progress and challenges in studies of the earliest galaxies, seen when the Universe was less than 1 billion years old. Can they be used as reliable tracers of the physics of cosmic reionization thereby complementing other, more direct, probes of the evolving neutrality of the intergalactic medium? Were star-forming galaxies the primary agent in the reionization process and what are the future prospects for identifying the earliest systems devoid of chemical enrichment? Ambitious future facilities are under construction for exploring galaxies and the intergalactic medium in the redshift range 6 to 20, corresponding to what we now consider the heart of the reionization era. I review what we can infer about this period from current observations and in the near-future with existing facilities, and conclude with a list of key issues where future work is required.

Keywords: Galaxy evolution; cosmology.

1. Introduction

Most would agree that the final frontier in piecing together a coherent picture of cosmic history concerns studies of the era corresponding to a redshift interval from 25 down to about 6; this corresponds to the period 200 million to 1 billion years after the Big Bang. During this time the Universe apparently underwent two vitally important changes. Firstly, the earliest stellar systems began to shine, bathing the Universe in ultraviolet radiation from their hot, metal-free stars. Although isolated massive stars may have collapsed and briefly shone earlier, the term *cosmic dawn* usually refers to the later arrival of dark matter halos capable of hosting star clusters or low mass galaxies. Secondly, the intergalactic medium transitioned from a neutral and molecular gas into one that is now fully ionized — a process termed *cosmic reionization.*

It is tempting to connect these two changes via a cause and effect as illustrated in Figure 1. Young stellar systems forming at a redshift of 25, corresponding to 200 Myr after the Big Bang, emit copious amounts of ultraviolet radiation capable of ionizing their surroundings. These ionized spherical bubbles expand with time and, as more stellar systems develop, they overlap and the transition to a fully ionized intergalactic medium is completed.

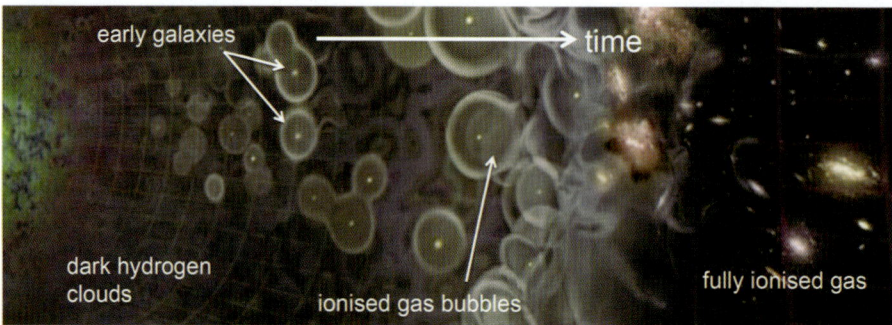

Fig. 1.   An illustration of how early populations of star-forming galaxies reionized the Universe. Baryonic gas is attracted into assembling dark matter halos and it cools and collapses to form the first stellar systems. Ultraviolet radiation from their hot young stars photoionizes the surrounding neutral hydrogen creating ionized bubbles. As more systems collapse and the ionized bubbles increase in size, their volumes overlap and cosmic reionization is complete.[1]

In addition to determining when this transition occurred and whether this simple picture is correct, studies of galaxies and the nature of the intergalactic medium during this period are valuable in further ways. The relevant physical processes governing star formation at this time determine which primitive systems survive and which form the basic material for the subsequent evolution of galaxies. Indeed, relics of this period may be present in local low mass dwarf galaxies devoid of star formation. The abundance of the earliest low mass systems depends sensitively on the assembly history of the dark matter halos which, in turn, depends on its streaming velocity. Although the cold dark matter picture is favored by large scale structure observations, early galaxy formation would be delayed if the dark matter was somewhat warmer and so direct observations of very early galaxies could verify or otherwise the standard picture.[2]

Ambitious facilities are now under construction, motivated in part by studies of the reionization era. These include the James Webb Space Telescope (JWST) which has the unique capability to undertake spectroscopy longward of 2μm, thereby accessing familiar rest-frame optical nebular lines as measures of the ionizing radiation field and the evolution of the gas phase metallicity. Three next-generation 25-40 meter aperture ground-based telescopes (the European Extremely Large Telescope, the Thirty Meter Telescope and the Giant Magellan Telescope) are also under development which will improve the spectroscopic capabilities. High order adaptive optics will give these facilities impressive imaging capabilities, a highly relevant advantage as the faintest sources at early epochs are otherwise unresolved. Deep near-infrared imaging over large areas of sky by survey facilities such as the European Space Agency's *Euclid* and NASA's WFIRST-AFTA missions will significantly improve information on the demographics of early galaxies which is currently limited by cosmic variance uncertainties associated with the small fields of view of the Hubble and Spitzer Space Telescopes.

These impressive upcoming facilities will be complemented by independent probes of the distribution of cold and ionized gas charted tomographically using the redshifted 21cm line. Initial *pathfinder* projects such as the Low Frequency Radio Array (LOFAR) will address the statistical distribution over a limited redshift range, whereas the Square Kilometer Array (SKA) will have the power to directly image the evolving distribution of neutral gas. The combination of clustering statistics for the early galaxy distribution and equivalent data for the neutral gas will delineate the evolution of ionized regions in the context of the radiation from observed sources. This will revolutionize our understanding of the reionization era.

In this brief review I take stock of what we currently know about the two principal questions that address the picture illustrated above: when did reionization occur and were galaxies the primary reionizing agents? Although we can address these questions using a variety of approaches, I will focus primarily on what we are learning from studies of early star-forming galaxies. This naturally leads to a discussion of the prospects for the next few years, including those possible with the future facilities listed above. Finally, I list some of the fundamental challenges faced in interpreting the growing amount of data on early galaxies. My review is to be read in conjunction with a complementary discussion presented by Steve Furlanetto in this volume which focuses more on the theoretical aspects of reionization and the future prospects with 21cm tomography.

2. When Did Reionization Occur?

The earliest constraints on the reionization history arose from the Gunn-Peterson test[3] applied to the absorption line spectra of $z > 5.5$ QSOs (see [4]). The decreasing transmission due to thickening of the Lyman alpha forest was initially used to argue that the reionization process ended at a redshift close to 6. However, only a very small change in the volume-averaged fraction of neutral hydrogen, $x(HI) \simeq 10^{-3}$, is required to completely suppress the spectroscopic signal shortward of Lyman alpha in the spectrum of a QSO, above which saturation rapidly occurs. Accordingly, this method is only useful for detecting a subtle change at the end of the reionization process. Since the bulk of the high redshift QSOs were analyzed some 8-10 years ago,[4,5] progress in locating higher redshift QSOs has been slow. Fortunately, some additional constraints have been provided through equivalent spectroscopy of a handful of $z > 6$ long duration gamma ray burst (GRB) afterglows.[6] Unfortunately, none of the more distant GRBs discovered beyond $z \simeq 7$ was followed up in detail. Indeed, only one source above a redshift of 7 - a QSO - has a relevant absorption line spectrum above a redshift of 7.[7] The initial analysis of this spectrum suggested that the IGM may indeed be significantly neutral ($x(HI) \simeq 10^{-1}$) at this redshift ([7,8] but see Boseman & Becker, in prep.), although confirmation from additional lines of sight is clearly desired.

A second constraint on the reionization history arises from the optical depth τ to electron scattering to cosmic microwave background (CMB) photons and the cross-correlation of the polarization signal induced by these electrons and the temperature fluctuations. τ therefore acts as a integral constraint on the line of sight distribution of ionized gas. The angular correlation can be interpreted in structure formation theory as providing an approximate redshift of the reionization era. Usually the quoted result corresponds to that assuming an (unrealistic) instantaneous reionization. Over the past few years WMAP has provided a series of improved constraints[9] corresponding to instantaneous reionization at $z \simeq 10.6 \pm 1.1$. No polarization results are yet available from Planck mission but early constraints based on temperature fluctuations alone[10] are consistent. It will be very important to secure independent confirmation of τ from the Planck mission. The prospects of using higher order CMB data to improve our understanding of reionization in the future is discussed by Calabrese et al.[11]

The most recent development in tracing reionization history follows studies of the rate of occurrence of Lyman alpha (Lyα) emission in star-forming galaxies. Miralda-Escudé[12] and Santos[13] discussed the prospect of using Lyα as a resonant transition, one which is readily absorbed if a line emitting galaxy lies in a neutral IGM. Early results based on the luminosity functions of narrow-band selected Lyα emitting galaxies over the redshift range $5.7 < z < 6.5$ supported the notion of a rapidly-changing IGM via a marked decline in the abundance of emitters over a short period of cosmic history (corresponding to an interval of less than 200 Myr).[14,15] However, although a striking result, it is hard to separate the effect of an increasingly neutral IGM at high redshift from the declining abundance of star-forming galaxies deduced from the overall population observed beyond $z \simeq 4$.[16]

An improved test that removes this ambiguity involves measuring the *fraction* of line emission in well-controlled, color-selected Lyman break galaxies. First introduced as a practical proposition by Stark[17] this method has been variously applied in the last 3 years[18–20] and most recently, by Schenker et al.[21] The availability of large numbers of $z > 7$ candidates from deep HST imaging and new multi-object near-infrared spectrographs has enabled considerable progress of late. These observations confirm a marked decline in the visibility of Lyα beyond a redshift $z \simeq 6.5$, consistent with the Gunn-Peterson constraints discussed above (Figure 2). Although Schenker et al report spectroscopic data for 102 $z > 6.5$ Lyman break galaxies, only a handful beyond $z \simeq 7$ show Lyα emission, the current record-holder being at $z = 7.62$.

The challenge lies in interpreting the fairly robust decline in the visibility of Lyα emission in the context of an increasing neutral fraction $x(HI)$ at earlier times. Radiative transfer calculations have suggested the fast decline in Figure 2 could imply a 50% neutral fraction by volume as late as $z \simeq 7.5$.[22,23] The uncertainties in this interpretation include (i) cosmic variance given the limited volumes so far probed with ground-based spectrographs,[24] (ii) the assumed velocity offset of Lyα with respect to the systemic velocity of the galaxy which is critical in understanding whether the line resonates with any neutral gas[25,26] and (iii) the possible presence of optically-thick absorbing clouds within the ionized regions.[8] A final variable is the escape fraction of ionizing photons from the galaxy, f_{esc}. If this were much higher at earlier times as a result of less neutral gas in the galaxies, the production of Lyα in the intrinsic spectrum would be reduced.[27]

A complementary and promising method for tracing reionization is to statistically chart the evolving distribution of neutral gas directly via redshifted 21cm emission using radio interferometers such as LOFAR[28] and the Murchison Wide Field Array.[29] No direct detections are yet available but the prospects are discussed by Steve Furlanetto elsewhere in this volume.

Figure 3 represents a recent summary of the various constraints on reionization and includes several methods not described in this brief review.[30] As can be seen, the redshift range 6 to 20, corresponding to a period of 800 Myr is considered to be the window of interest.

3. Were Galaxies Responsible for Cosmic Reionization?

Potential contributors to the reionizing photons include star-forming galaxies, non-thermal sources such as quasars and low luminosity active galactic nuclei, primordial black holes and decaying particles. Luminous QSOs decline rapidly in their abundance beyond $z \simeq 6$ so the only prospect for non-thermal sources contributing significantly to reionization might be if the faint end of their luminosity function is unusually steep. Current estimates of the high redshift AGN luminosity func-

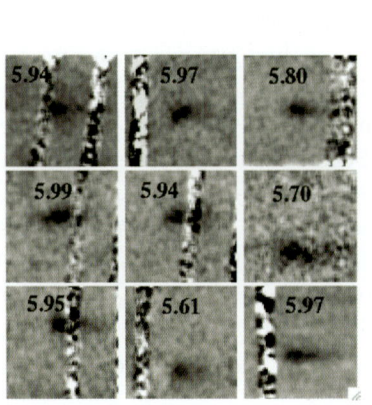
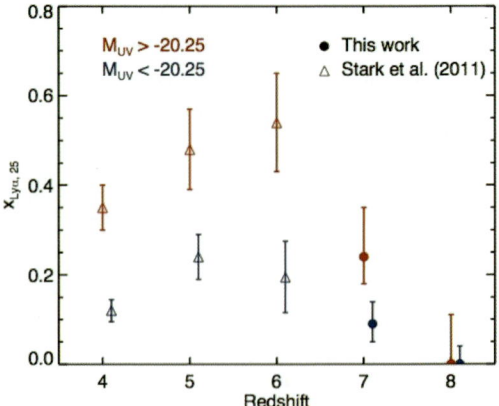

Fig. 2.   Left: Keck spectra for $z \simeq 6$ galaxies showing the high rate of occurrence of Lyα emission. Each sub-panel represents a 2-D spectrum of a Lyman break galaxy and the black regions represent line emission at the redshift marked.[17] Right: The evolving fraction of Lyman break galaxies in various luminosity bins that present a detectable Lyα emission line from the recent survey of Schenker et al.[21] The rising fraction over $4 < z < 6$ is interpreted via a reduced dust extinction at early times, whereas the sudden reversal beyond $z \simeq 6$ is attributed to an increasingly neutral intergalactic medium.

tion suggest this is not the case although the observational uncertainties are still large.[31,32]

Star-forming galaxies represent the most promising reionizing source given they are now observed in abundance in the relevant redshift range from deep surveys such as the Hubble Ultra Deep Field (UDF).[33] These and other data reveal a steep luminosity function at the faint end,[34–36] such that it is reasonable to assume we are only observing the luminous fraction of a much larger population. However, a quantitative calculation of the photon budget requirements for maintaining reionization involves additional parameters, some of which are largely unconstrained (see recent review by [30]).

In this case, the reionization process is a balance between the recombination of free electrons with protons to form neutral hydrogen and the ionization of hydrogen by Lyman continuum photons. The dimensionless volume filling factor of ionized hydrogen Q_{HII} can be expressed as a time-dependent differential equation:

$$\dot{Q}_{HII} = \frac{\dot{n}_{ion}}{<n_H>} - \frac{Q_{HII}}{t_{rec}}$$

The recombination time t_{rec} depends on the baryon density, the primordial mass fraction of hydrogen, the case B recombination coefficient and the clumping factor $C_{HII} \equiv <n_H^2> / <n_H>^2$ which takes into account the effects of IGM inhomogeneity through the quadratic dependence of recombination on density. Simulations

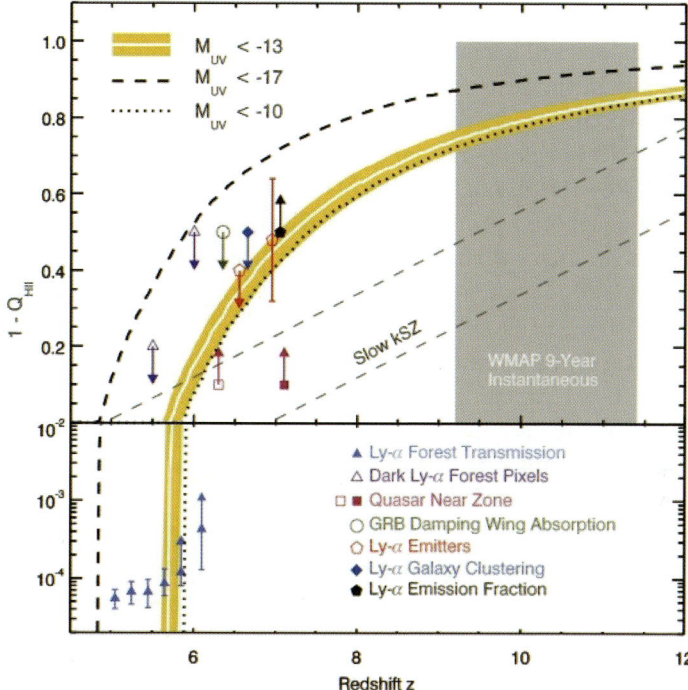

Fig. 3. Reionization histories for models that include galaxies to various luminosity limits from the UDF12 survey ($M_{UV} < 13$ white line; 68% credibility region: orange area; < 17, dashed line; < 10, dotted line) plus other claimed constraints on the neutral fraction $1 - Q_{HII}$ (see lower panel for legend). Methods not discussed in the text include the fraction of dark pixels in the Lyα forest (purple open triangles), QSO near-zone measurements (open and solid magenta squares), damping wing absorption in a GRB (open green circle), the clustering of Lyα emitters (filled dark blue diamond). The gray dashed lines labeled 'Slow kSZ' illustrate the slowest evolution permitted by small-scale CMB temperature data and the shaded gray region shows the redshift of instantaneous reionization according to WMAP.[30]

suggest $C_{HII} \simeq 1\text{-}6$ at the relevant redshifts,[37] although there has been much discussion of its redshift dependence depending on the epoch when the ultraviolet (UV) background becomes uniform. If the clumping factor C_{HII} is time invariant, t_{rec} declines with increasing redshift. For the expected values above, at redshifts $z < 10$, t_{rec} exceeds 100-200 Myr[38,39] ensuring recombination is unlikely. However, if the source of ionizing photons is not steady in the redshift range $10 < z < 25$, there remains the possibility of an intermediate recombination era, perhaps inbetween reionization from the first isolated massive stars and that subsequently from early galaxies.

The main uncertainty in understanding the contribution of galaxies can be understood via the relative contributions to the ionizing photon rate $\dot{n}_{ion}$:

$$\dot{n}_{ion} = f_{esc}\xi_{ion}\rho_{SFR}$$

where ρ_{SFR} represents the most direct observable, the integrated volume density of star-forming galaxies. This involves measuring the redshift-dependent luminosity function, typically in the rest-frame UV continuum ($\simeq 1500$ Å) which is accessible at $z \simeq 7 - 10$ with HST's near-infrared camera WFC3/IR, and above $z \simeq 10$ with NIRCam on JWST. The faint end slope of the luminosity function is a critical factor given it contributes the major portion of the integrated luminosity density.[34-36] ξ_{ion} is the ionizing photon production rate which encodes the number of photons more energetic than 13.6 eV that are produced per unit star formation rate. This requires knowledge of the stellar population which can currently only be estimated by modeling the average galaxy color. Finally, f_{esc} represents the fraction of ionizing photons below the Lyman limit which escape to the IGM. This is the least well-understood parameter. It can only be directly evaluated through rest-frame UV imaging or spectroscopy at $z \simeq 2 - 3$([40,41]) where values as low as 5% are typical. At higher redshift, any photons below the Lyman limit are obscured along the line of sight by the lower redshift Lyman alpha forest.

There are several ways to address the question of whether galaxies can meet the ionization budget and these depend critically on the assumed value of the currently unobserved quantities, e.g. f_{esc}. A fundamental requirement is that the integrated electron path length to the start of reionization should match the optical depth of Thomson scattering, τ, in the CMB. When this requirement is imposed, in the context of the results from the Hubble UDF, three conditions are necessary for galaxies to be the main reionization agents.[30] Firstly, the escape fraction f_{esc} has to rise with redshift or be sufficiently luminosity-dependent so that at least 20% on average of the photons escape a typical low luminosity $z \simeq 7 - 10$ galaxy. Secondly, galaxies must populate the luminosity function to absolute magnitudes below the limits of the deepest current HST images at $z \simeq 7 - 8$ ($M_{UV} = -17$). Finally, the galaxy population must extend beyond a redshift $z \simeq 10$ to provide a sustained source of ionizing radiation. Various combinations of these three requirements have been discussed in the literature and presented alternatively as reasonable assumptions or as critical shortfalls in the ionizing budget!

A further constraint on the above is the requirement that the sum of the star formation during the reionization era cannot exceed the stellar mass density observed using the Spitzer satellite at the end of reionization, say $z \simeq 5 - 6$.[43] This mid-infrared satellite is uniquely effective in this regard given its infrared camera, IRAC, surveys high redshift galaxies at rest-frame optical wavelengths where longer-lived stars can be accounted for. Formally, this can be expressed:

$$\rho_*(z = 6) = C \int_{z=6}^{\infty} \int \Phi(L_{UV}, z) L_{UV} dL dz$$

where ρ_* is the required stellar mass density per comoving volume at the end of reionization, and C represents the necessary factor to convert the observed redshift-dependent UV luminosity function $\Phi(L_{UV}, z)$ and its associated luminosity density, into a star formation rate density. Stellar masses for individual galaxies are usually

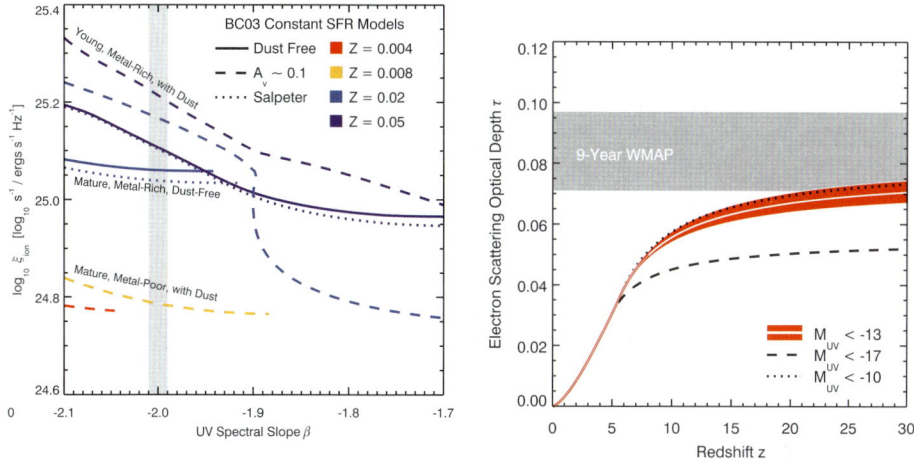

Fig. 4.   Left: Degeneracies in inferring the ionizing photon production factor ξ_{ion} in terms of the observed slope β of the ultraviolet continuum, the gray shaded area being that observed for $z \simeq 7-8$ galaxies.[42] Time tracks are shown for stellar population synthesis models of varying dust content, metallicity and the initial mass function.[30] Right: One aspect of the UV 'photon shortfall' for galaxies as agents of reionization given the abundance of galaxies in the UDF. Assuming a 20% escape fraction and continuity in the declining star formation rate density beyond $z \simeq 10$, the figure shows the need to extend the UV luminosity function lower than the current $M_{UV} = -17$ detection limit to reproduce the optical depth of electron scattering in the WMAP data.[30]

determined by deriving a mass/light ratio from fitting the spectral energy distribution and multiplying by the luminosity. To secure the integrated mass density is challenging given only a more massive subset of the $z \simeq 6$ population is currently detectable with Spitzer. Additionally the Spitzer photometric bands are likely contaminated by nebular line emission at $z \simeq 6$ and significant, but uncertain, downward corrections are required to estimate the true mass density.[25,44] When reasonable estimates are made of the unseen stellar mass at $z \simeq 5\text{-}6$ and corrections applied for nebular emission based on spectroscopic evidence at lower redshift, the stellar mass densities ρ_* can be reconciled with the earlier star formation history.[30]

4. The Near Future

Fortunately we observers have not yet reached a threshold in exploring the early galaxy population pending the arrival of new facilities such as JWST and the next generation of large ground-based telescopes. There are several interesting and immediate initiatives available for making further progress.

In addition to probing the reionization history with the fractional rate of occurrence of Lyα emission, the spatial distribution of line emitters in principle contains data on the topology of ionized regions where emission can be transmitted. Narrow-band filters are being used with panoramic cameras to locate Lyα emitters at discrete redshifts where the line is favorably placed with respect to the night sky emission, for example at redshifts z=5.7, 6.6 and 7.1 with the HyperSuprime-Cam 1.5 degree field imager on the Subaru 8.2m telescope (see an example of earlier work of this nature in Figure 5). The correlation of such line emission with redshifted 21cm emission would be a particularly fruitful program.

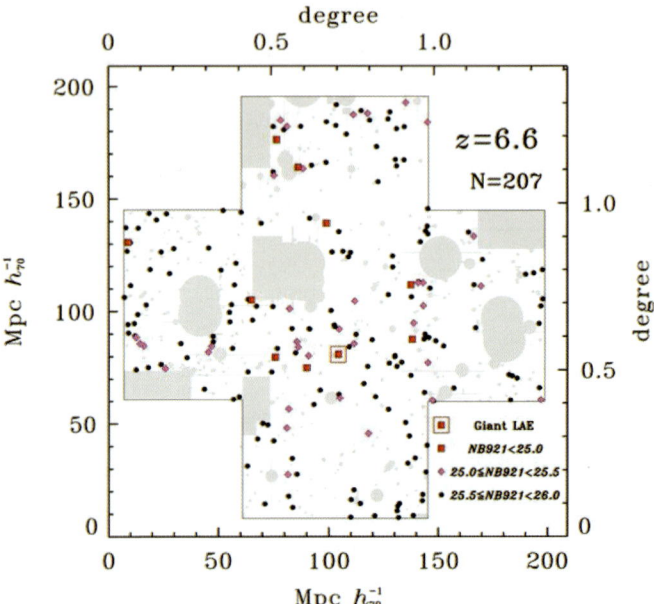

Fig. 5.   Angular distribution of 207 Lyα emitters at a redshift of z=6.565 $\pm$ 0.054 selected from a mosaic of narrow band images taken with the Suprime Camera on the 8.2m Subaru telescope, color coded according to their luminosity (decreasing from red squares, through magenta diamonds to black circles.[15] The open red square denotes the extended and luminous emitter 'Himiko' (see Section 5).

Strong gravitational lensing by foreground clusters offers a valuable tool for exploring the redshift range $7 < z < 10$ population. HST and Spitzer are investing significant resources in deep imaging of selected clusters via the CLASH[45] and Frontier Fields[a] programs. Lensing facilitates two broad applications depending on the source magnification involved. Bradley et al[46] discuss the magnification distribution for the CLASH survey and Richard et al[47] for the upcoming Frontier Field clusters. Most of the lensed sources have magnifications of $\times$1.5-3 with less than 5% greater than $\times$10 (Figure 6a).

The first regime involves very highly-magnified and usually multiply-imaged sources observed close to the critical line of the cluster. With magnifications of $\times 10 - 30$[48-51] such systems offer the prospect of valuable detailed studies. A good example is the $z \simeq 6.02$ galaxy in the rich cluster Abell 383 which has a magnification

---

[a]http://frontierfields.org

of ×11.4±1.6 corresponding to a 0.4 L^* galaxy.[52] The significant boost in brightness enables a much more precise spectral energy distribution for a representative sub-luminous system than would otherwise be the case providing a fairly robust stellar age of 640-940 Myr, corresponding to a formation redshift of $z > 15$. However, such configurations are rare and do not represent a straightforward route to large samples.

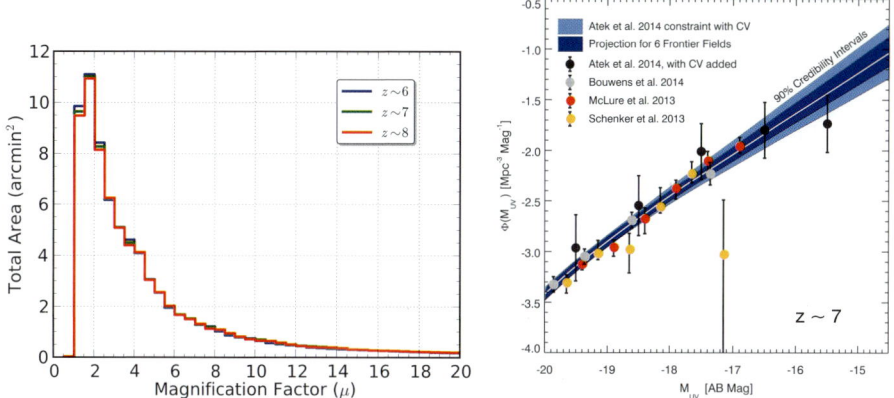

Fig. 6. Left: The distribution of lensing magnifications deduced for high redshift galaxies in the HST CLASH survey of foreground clusters.[46] Most sources are only modestly magnified but there are rare sources magnified by factor as large as ×10-20. Right: Using gravitational lensing in the HST Frontier Field program to extend the $z \simeq 7$ luminosity function fainter than was possible in the UDF. Using the available data for one cluster,[53] a projection is made for all 6 clusters.[54]

The second regime involves more modest magnifications of larger numbers of background sources. The benefits here are not in detailed studies of individual sources but rather for statistical purposes, e.g. in extending the $z \simeq 7$-8 luminosity function fainter than was possible in the deepest blank field studies[53] (Figure 6b). Robertson et al[54] recently projected the likely gain in depth over all 6 Frontier Field clusters incorporating the increased cosmic variance in lensed surveys. They claim the uncertainty in the faint end slope α of the luminosity function would be significantly reduced compared to the value in the UDF ($\Delta\alpha = \pm0.05$ c.f. ± 0.18).

Detailed spectroscopy of $z \simeq 7 - 8$ galaxies can also provide further information on the ionization state and metallicity of the gas. Stark et al[26] illustrate how, even when Lyα is suppressed by neutral gas, other nebular lines such as [CIII] 1909 and CIV 1550 Å are within reach of current near-infrared spectrographs, although this is highly challenging work even for lensed sources.

This leads naturally to the longer term goal of gathering *gas-phase metallicities* for early galaxies thereby adding *chemical enrichment* as the next logical tracer of earlier activity. Metallicity measurements will very much be the province of JWST given all the familiar rest-frame optical lines ([O II], [O III], Hα, [N II]) used locally and at intermediate redshifts as well-calibrated metallicity indicators, are shifted beyond $2\mu m$ where ground-based spectroscopy of faint objects is impractical. However, there are valuable sub-mm lines accessible with ALMA at high redshift which may give information on both the metallicity and dust content of early galaxies. Although the currently-held view is that the blue UV colors of most of the $z > 7$ galaxies imply little or no dust, strong ALMA upper limits on far-infrared continua would provide a more convincing argument.

The [CII] 158 μm line has traditionally been one of the most valuable tracers of star formation in energetic sources and a correlation is often claimed between the [C II] luminosity and the star formation rate estimated from the far infrared flux although its interpretation remains unclear.[55] Early ALMA studies of luminous $z \simeq 5 - 7$ dusty starbursts recovered prominent [CII] emission[56,57] consistent with this correlation. However, an intense Lyα emitter, dubbed 'Himiko' at $z=6.595$ (see Figure 5) with a high star formation rate ($\simeq 100 M_{\odot}$ yr^{-1}) reveals no far infrared or [CII] emission,[58] and thus deviates significantly from the normal relation. As the Lyα emission is particularly extended and the source is unusually luminous compared to its cohorts, conceivably it is being observed during a special moment in its history e.g. an energetic burst of early activity in a very low metallicity system. Such studies with ALMA may shed light on metal formation in the most luminous early systems ahead of the launch of JWST.

Ultimately one might hope to identify systems with minimal pollution from metals. Such 'Population III' sources initially represented something of a 'Holy Grail' for the next generation facilities - specifically, the charge to find a star-forming galaxy or stellar system devoid of metals. More recent numerical simulations[59] indicate the self-enrichment of halos from early supernovae is surprisingly rapid (<100 Myr) and so such primordial 'first generation' stellar systems may be very rare.

5. Outstanding Issues

Although there are gaps in our quantitative knowledge of the reionization history and the role of galaxies, it has perhaps become commonplace to regard sketched histories such as Figures 3 and 4b as the correct framework within which future facilities can fill in the details. In this concluding section I want to highlight some outstanding issues and puzzles that will serve to focus our collective research in the near future.

The extent of star formation beyond $z \simeq 10$: The Ultra Deep Field 2012 campaign argued for a near-continuous decline in the cosmic star formation rate density over $4 < z < 10$ (Ref [60]) and Robertson et al[30] used this continuity plus the mature ages of the $z \simeq 7$-8 galaxies,[42] as indirect evidence that the star formation history was beyond $z \simeq 10$. However, recent work exploiting the wider, but shallower CANDELS data[61] together with several analyses exploiting early Frontier Field lens data[62] point to a discontinuity in this decline at $z \simeq 8$. Such a downturn would be hard to reconcile with the stellar mass density evolution[52,63] and, if correct, would seriously increase the UV photon budget shortfall. A key issue here, given the paucity of data beyond $z \simeq 8$, is uncertainties arising from cosmic variance.[54] Hopefully with further data from the Frontier Fields and more Spitzer age measures of individual galaxies at $z \simeq 7\text{-}8$, the situation will be clarified ahead of the launch of JWST.

Missing star-forming galaxies: The high redshift galaxies discussed in this review have almost exclusively been located by their ultraviolet emission, either via continuum colors or through Lyα emission. In addition to assuming there are yet fainter galaxies further down the luminosity function beyond HST's limits, is it conceivable there are additional sources perhaps dusty or those not selected via the current methods? An unresolved puzzle is the anomalously high rate of long duration gamma ray bursts seen beyond $z \simeq 5$ compared to that expected using a GRB rate normalized to the star formation rate observed at lower redshift[64] (Figure 7a). This discrepancy may be telling us more about the evolving production rate of GRBs in low metallicity environment rather than something fundamental about the cosmic star formation history. Nonetheless, it acts as a warning that some aspects of early massive star formation may not be understood.

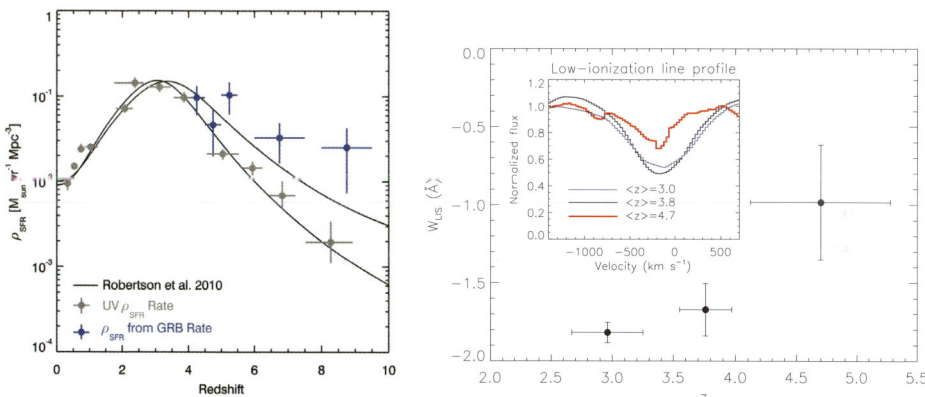

Fig. 7. Left: The star formation history beyond a redshift 4 as inferred from the rate of gamma ray bursts (GRBs).[64] The number of GRBs is converted into a volume-averaged star formation by matching their cumulative redshift distribution over $0 < z < 4$ with the cosmic star formation history. There is a worrying excess in the number of high z GRBs compared to expectations based on the rate at lower redshift. Right: The equivalent width of low ionization gaseous absorption lines from Keck spectra of Lyman break galaxies stacked at various redshifts. The inset shows the stacked absorption line profiles whose depth becomes shallower as the redshift increases. Such data suggests that the covering fraction of neutral gas is less at high redshift and hence the escape fraction increases.[65]

The escape fraction of ionizing photons: The largest uncertainty in addressing the role of galaxies in completing the reionization process is the average fraction of ionizing photons that can escape a typical low-luminosity galaxy. Even with a fraction $f_{esc} \simeq$, 20% there is significant tension in the ionizing budget and in reproducing the optical depth τ of electron scattering by the CMB (Figure 4b).

Most likely the escape fraction varies significantly from galaxy to galaxy according to the geometric viewing angle, kinematic state, star formation rate and physical size of each galaxy. Even at redshifts $z \simeq 2$-3, determining f_{esc} has been a challenging endeavor although the consensus points to the range 0-5%.[40,41] At high redshift, the only practical route is to examine the *covering fraction*, f_{cov}, of neutral or low ionization gas on the assumption that, typically, $f_{esc} = 1 - f_{cov}$. Even so, measuring f_{cov} requires high signal/noise absorption line spectroscopy which is only practical for stacks of galaxies[65] or strongly-lensed examples.[66] Such data to $z \simeq 4$-5 shows some evidence for a rising escape fraction with increased redshift (Figure 7b) but the method needs to be extended to larger samples at yet higher redshifts.

When did the Universe produce dust? To these more immediate issues of observational interpretation should be added the question of whether dust is present beyond $z \simeq 7$. Its presence would seriously confuse interpretations of the UV colors (e.g. Figure 4a) as well as raise the question of obscured star formation. An example has recently been found of a convincing ALMA continuum detection for a star-forming galaxy at $z=7.58$ (Watson et al in prep) which raises very interesting consequences. This early result highlights the key role that ALMA can play in complementing studies of high redshift galaxies with HST and Spitzer.

6. Summary

Although many puzzles remain as indicated above, the pace of observational discovery is truly impressive and will continue as we see the first convincing results from 21cm interferometry in the next 1-2 years, launch JWST in 2018 and commission the next generations telescopes in the early 2020's. The observational promise is evident and I encourage our theoretical colleagues to get ready for the next revolution in observational data at the redshift frontier!

Acknowledgments

I acknowledge valuable discussions with my co-Rapporteur, Steve Furnaletto, a colleague on the Hubble Ultra Deep Field (UDF) campaign. I likewise acknowledge the support and scientific input from my other UDF colleagues, Brant Robertson, Jim Dunlop, Ross McLure and Anton Koekemoer, as well as my Keck spectroscopic co-workers Dan Stark, Matt Schenker, Tucker Jones and Adi Zitrin. I thank the organizers of this memorable meeting for their organizational efforts and hospitality in Brussels.

References

1. A. Loeb, *Scientific American* **295**, 46(November 2006).
2. F. Pacucci, A. Mesinger and Z. Haiman, *Mon. Not. R. Astron. Soc.* **435**, L53(August 2013).
3. J. E. Gunn and B. A. Peterson, *Astrophys. J.* **142**, 1633(November 1965).

4. X. Fan, C. L. Carilli and B. Keating, *Ann. Rev Astron. Astrophys.* **44**, 415(September 2006).
5. X. Fan, V. K. Narayanan, M. A. Strauss, R. L. White, R. H. Becker, L. Pentericci and H.-W. Rix, *Astron. J.* **123**, 1247(March 2002).
6. R. Chornock, E. Berger, D. B. Fox, W. Fong, T. Laskar and K. C. Roth, *ArXiv e-prints* (May 2014).
7. D. J. Mortlock, S. J. Warren, B. P. Venemans, M. Patel, P. C. Hewett, R. G. McMahon, C. Simpson, T. Theuns, E. A. Gonzáles-Solares, A. Adamson, S. Dye, N. C. Hambly, P. Hirst, M. J. Irwin, E. Kuiper, A. Lawrence and H. J. A. Röttgering, *Nature* **474**, 616(June 2011).
8. J. S. Bolton, M. G. Haehnelt, S. J. Warren, P. C. Hewett, D. J. Mortlock, B. P. Venemans, R. G. McMahon and C. Simpson, *Mon. Not. R. Astron. Soc.* **416**, L70(September 2011).
9. G. Hinshaw, D. Larson, E. Komatsu, D. N. Spergel, C. L. Bennett, J. Dunkley, M. R. Nolta, M. Halpern, R. S. Hill, N. Odegard, L. Page, K. M. Smith, J. L. Weiland, B. Gold, N. Jarosik, A. Kogut, M. Limon, S. S. Meyer, G. S. Tucker, E. Wollack and E. L. Wright, *Astrophys. J. Supp.* **208**, p. 19(October 2013).
10. Planck Collaboration, P. A. R. Ade, N. Aghanim, C. Armitage-Caplan, M. Arnaud, M. Ashdown, F. Atrio-Barandela, J. Aumont, C. Baccigalupi, A. J. Banday and et al., *Astron. Astrophys.* **571**, p. A16(November 2014).
11. E. Calabrese, R. Hložek, N. Battaglia, J. R. Bond, F. de Bernardis, M. J. Devlin, A. Hajian, S. Henderson, J. C. Hil, A. Kosowsky, T. Louis, J. McMahon, K. Moodley, L. Newburgh, M. D. Niemack, L. A. Page, B. Partridge, N. Sehgal, J. L. Sievers, D. N. Spergel, S. T. Staggs, E. R. Switzer, H. Trac and E. J. Wollack, *J. Cos. Astroparticle Phys.* **8**, p. 10(August 2014).
12. J. Miralda-Escudé, *Astrophys. J.* **501**, 15(July 1998).
13. M. R. Santos, *Mon. Not. R. Astron. Soc.* **349**, 1137(April 2004).
14. N. Kashikawa, K. Shimasaku, M. A. Malkan, M. Doi, Y. Matsuda, M. Ouchi, Y. Taniguchi, C. Ly, T. Nagao, M. Iye, K. Motohara, T. Murayama, K. Murozono, K. Nariai, K. Ohta, S. Okamura, T. Sasaki, Y. Shioya and M. Umemura, *Astrophys. J.* **648**, 7(September 2006).
15. M. Ouchi, K. Shimasaku, H. Furusawa, T. Saito, M. Yoshida, M. Akiyama, Y. Ono, T. Yamada, K. Ota, N. Kashikawa, M. Iye, T. Kodama, S. Okamura, C. Simpson and M. Yoshida, *Astrophys. J.* **723**, 869(November 2010).
16. R. J. Bouwens, G. D. Illingworth, P. A. Oesch, M. Stiavelli, P. van Dokkum, P. Trenti, D. Magee, I. Labbé, M. Franx, C. M. Carollo and V. Gonzalez, *Astrophys. J.* **709**, L133(February 2010).
17. D. P. Stark, R. S. Ellis, K. Chiu, M. Ouchi and A. Bunker, *Mon. Not. R. Astron. Soc.* **408**, 1628(November 2010).
18. L. Pentericci, A. Fontana, E. Vanzella, M. Castellano, A. Grazian, M. Dijkstra, K. Boutsia, S. Cristiani, M. Dickinson, E. Giallongo, M. Giavalisco, R. Maiolino, A. Moorwood, D. Paris and P. Santini, *Astrophys. J.* **743**, p. 132(December 2011).
19. M. A. Schenker, D. P. Stark, R. S. Ellis, B. E. Robertson, J. S. Dunlop, R. J. McLure, J.-P. Kneib and J. Richard, *Astrophys. J.* **744**, p. 179(January 2012).
20. T. Treu, K. B. Schmidt, M. Trenti, L. D. Bradley and M. Stiavelli, *Astrophys. J.* **775**, p. L29(September 2013).
21. M. A. Schenker, R. S. Ellis, N. P. Konidaris and D. P. Stark, *Astrophys. J.* **795**, p. 20(November 2014).
22. A. Mesinger, A. Aykutalp, E. Vanzella, L. Pentericci, A. Ferrara and M. Dijkstra, *ArXiv e-prints* (June 2014).

23. A. Smith, C. Safranek-Shrader, V. Bromm and M. Milosavljević, *ArXiv e-prints* (September 2014).

24. J. Taylor and A. Lidz, *Mon. Not. R. Astron. Soc.* **437**, 2542(January 2014).

25. M. A. Schenker, R. S. Ellis, N. P. Konidaris and D. P. Stark, *Astrophys. J.* **777**, p. 67(November 2013).

26. D. P. Stark, J. Richard, S. Charlot, B. Clement, R. Ellis, B. Siana, B. Robertson, M. Schenker, J. Gutkin and A. Wofford, *ArXiv e-prints* (August 2014).

27. M. Dijkstra, S. Wyithe, Z. Haiman, A. Mesinger and L. Pentericci, *Mon. Not. R. Astron. Soc.* **440**, 3309(June 2014).

28. G. Mellema, L. V. E. Koopmans, F. A. Abdalla, G. Bernardi, B. Ciardi, S. Daiboo, A. G. de Bruyn, K. K. Datta, H. Falcke, A. Ferrara, I. T. Iliev, F. Iocco, V. Jelić, H. Jensen, R. Joseph, P. Labroupoulos, A. Meiksin, A. Mesinger, A. R. Offringa, V. N. Pandey, J. R. Pritchard, M. G. Santos, D. J. Schwarz, B. Semelin, H. Vedantham, S. Yatawatta and S. Zaroubi, *Experimental Astronomy* **36**, 235(August 2013).

29. J. D. Bowman, I. Cairns, D. L. Kaplan, T. Murphy, D. Oberoi, L. Staveley-Smith, W. Arcus, D. G. Barnes, G. Bernardi, F. H. Briggs, S. Brown, J. D. Bunton, A. J. Burgasser, R. J. Cappallo, S. Chatterjee, B. E. Corey, A. Coster, A. Deshpande, L. deSouza, D. Emrich, P. Erickson, R. F. Goeke, B. M. Gaensler, L. J. Greenhill, L. Harvey-Smith, B. J. Hazelton, D. Herne, J. N. Hewitt, M. Johnston-Hollitt, J. C. Kasper, B. B. Kincaid, R. Koenig, E. Kratzenberg, C. J. Lonsdale, M. J. Lynch, L. D. Matthews, S. R. McWhirter, D. A. Mitchell, M. F. Morales, E. H. Morgan, S. M. Ord, J. Pathikulangara, T. Prabu, R. A. Remillard, T. Robishaw, A. E. E. Rogers, A. A. Roshi, J. E. Salah, R. J. Sault, N. U. Shankar, K. S. Srivani, J. B. Stevens, R. Subrahmanyan, S. J. Tingay, R. B. Wayth, M. Waterson, R. L. Webster, A. R. Whitney, A. J. Williams, C. L. Williams and J. S. B. Wyithe, *Proc. Astron. Soc. Aust.* **30**, p. 31(April 2013).

30. B. E. Robertson, S. R. Furlanetto, E. Schneider, S. Charlot, R. S. Ellis, D. P. Stark, R. J. McLure, J. S. Dunlop, A. Koekemoer, M. A. Schenker, M. Ouchi, Y. Ono, E. Curtis-Lake, A. B. Rogers, R. A. A. Bowler and M. Cirasuolo, *Astrophys. J.* **768**, p. 71(May 2013).

31. E. Glikman, S. G. Djorgovski, D. Stern, A. Dey, B. T. Jannuzi and K.-S. Lee, *Astrophys. J.* **728**, p. L26(February 2011).

32. I. D. McGreer, L. Jiang, X. Fan, G. T. Richards, M. A. Strauss, N. P. Ross, M. White, Y. Shen, D. P. Schneider, A. D. Myers, W. N. Brandt, C. DeGraf, E. Glikman, J. Ge and A. Streblyanska, *Astrophys. J.* **768**, p. 105(May 2013).

33. A. M. Koekemoer, R. S. Ellis, R. J. McLure, J. S. Dunlop, B. E. Robertson, Y. Ono, M. A. Schenker, M. Ouchi, R. A. A. Bowler, A. B. Rogers, E. Curtis-Lake, E. Schneider, S. Charlot, D. P. Stark, S. R. Furlanetto, M. Cirasuolo, V. Wild and T. Targett, *Astrophys. J. Supp.* **209**, p. 3(November 2013).

34. M. A. Schenker, B. E. Robertson, R. S. Ellis, Y. Ono, R. J. McLure, J. S. Dunlop, A. Koekemoer, R. A. A. Bowler, M. Ouchi, E. Curtis-Lake, A. B. Rogers, E. Schneider, S. Charlot, D. P. Stark, S. R. Furlanetto and M. Cirasuolo, *Astrophys. J.* **768**, p. 196(May 2013).

35. R. J. McLure, J. S. Dunlop, R. A. A. Bowler, E. Curtis-Lake, M. Schenker, R. S. Ellis, B. E. Robertson, A. M. Koekemoer, A. B. Rogers, Y. Ono, M. Ouchi, S. Charlot, V. Wild, D. P. Stark, S. R. Furlanetto, M. Cirasuolo and T. A. Targett, *Mon. Not. R. Astron. Soc.* **432**, 2696(July 2013).

36. S. L. Finkelstein, R. E. Ryan, Jr., C. Papovich, M. Dickinson, M. Song, R. Somerville, H. C. Ferguson, B. Salmon, M. Giavalisco, A. M. Koekemoer, M. L. N. Ashby, P. Behroozi, M. Castellano, J. S. Dunlop, S. M. Faber, G. G. Fazio, A. Fontana, N. A.

Grogin, N. Hathi, J. Jaacks, D. D. Kocevski, R. Livermore, R. J. McLure, E. Merlin, B. Mobasher, J. A. Newman, M. Rafelski, V. Tilvi and S. P. Willner, *ArXiv e-prints* (October 2014).

37. K. Finlator, S. P. Oh, F. Özel and R. Davé, *Mon. Not. R. Astron. Soc.* **427**, 2464(December 2012).

38. A. H. Pawlik, J. Schaye and E. van Scherpenzeel, *Mon. Not. R. Astron. Soc.* **394**, 1812(April 2009).

39. B. E. Robertson, R. S. Ellis, J. S. Dunlop, R. J. McLure and D. P. Stark, *Nature* **468**, 49(November 2010).

40. B. Siana, H. I. Teplitz, H. C. Ferguson, T. M. Brown, M. Giavalisco, M. Dickinson, R.-R. Chary, D. F. de Mello, C. J. Conselice, C. R. Bridge, J. P. Gardner, J. W. Colbert and C. Scarlata, *Astrophys. J.* **723**, 241(November 2010).

41. D. B. Nestor, A. E. Shapley, C. C. Steidel and B. Siana, *Astrophys. J.* **736**, p. 18(July 2011).

42. J. S. Dunlop, A. B. Rogers, R. J. McLure, R. S. Ellis, B. E. Robertson, A. Koekemoer, P. Dayal, E. Curtis-Lake, V. Wild, S. Charlot, R. A. A. Bowler, M. A. Schenker, M. Ouchi, Y. Ono, M. Cirasuolo, S. R. Furlanetto, D. P. Stark, T. A. Targett and E. Schneider, *Mon. Not. R. Astron. Soc.* **432**, 3520(July 2013).

43. D. P. Stark, A. J. Bunker, R. S. Ellis, L. P. Eyles and M. Lacy, *Astrophys. J.* **659**, 84(April 2007).

44. D. P. Stark, M. A. Schenker, R. Ellis, B. Robertson, R. McLure and J. Dunlop, *Astrophys. J.* **763**, p. 129(February 2013).

45. M. Postman, D. Coe, N. Benítez, L. Bradley, T. Broadhurst, M. Donahue, H. Ford, O. Graur, G. Graves, S. Jouvel, A. Koekemoer, D. Lemze, E. Medezinski, A. Molino, L. Moustakas, S. Ogaz, A. Riess, S. Rodney, P. Rosati, K. Umetsu, W. Zheng, A. Zitrin, M. Bartelmann, R. Bouwens, N. Czakon, S. Golwala, O. Host, L. Infante, S. Jha, Y. Jimenez-Teja, D. Kelson, O. Lahav, R. Lazkoz, D. Maoz, C. McCully, P. Melchior, M. Meneghetti, J. Merten, J. Moustakas, M. Nonino, B. Patel, E. Regös, J. Sayers, S. Seitz and A. Van der Wel, *Astrophys. J. Supp.* **199**, p. 25(April 2012).

46. L. D. Bradley, A. Zitrin, D. Coe, R. Bouwens, M. Postman, I. Balestra, C. Grillo, A. Monna, P. Rosati, S. Seitz, O. Host, D. Lemze, J. Moustakas, L. A. Moustakas, X. Shu, W. Zheng, T. Broadhurst, M. Carrasco, S. Jouvel, A. Koekemoer, E. Medezinski, M. Meneghetti, M. Nonino, R. Smit, K. Umetsu, M. Bartelmann, N. Benítez, M. Donahue, H. Ford, L. Infante, Y. Jimenez-Teja, D. Kelson, O. Lahav, D. Maoz, P. Melchior, J. Merten and A. Molino, *Astrophys. J.* **792**, p. 76(September 2014).

47. J. Richard, M. Jauzac, M. Limousin, E. Jullo, B. Clément, H. Ebeling, J.-P. Kneib, H. Atek, P. Natarajan, E. Egami, R. Livermore and R. Bower, *Mon. Not. R. Astron. Soc.* **444**, 268(October 2014).

48. R. Ellis, M. R. Santos, J.-P. Kneib and K. Kuijken, *Astrophys. J.* **560**, L119(October 2001).

49. J.-P. Kneib, R. S. Ellis, M. R. Santos and J. Richard, *Astrophys. J.* **607**, 697(June 2004).

50. D. Coe, A. Zitrin, M. Carrasco, X. Shu, W. Zheng, M. Postman, L. Bradley, A. Koekemoer, R. Bouwens, T. Broadhurst, A. Monna, O. Host, L. A. Moustakas, H. Ford, J. Moustakas, A. van der Wel, M. Donahue, S. A. Rodney, N. Benítez, S. Jouvel, S. Seitz, D. D. Kelson and P. Rosati, *Astrophys. J.* **762**, p. 32(January 2013).

51. A. Zitrin, W. Zheng, T. Broadhurst, J. Moustakas, D. Lam, X. Shu, X. Huang, J. M. Diego, H. Ford, J. Lim, F. E. Bauer, L. Infante, D. D. Kelson and A. Molino, *Astrophys. J.* **793**, p. L12(September 2014).

52. J. Richard, J.-P. Kneib, H. Ebeling, D. P. Stark, E. Egami and A. K. Fiedler, *Mon.*

Not. R. Astron. Soc. **414**, L31(June 2011).

53. H. Atek, J. Richard, J.-P. Kneib, M. Jauzac, D. Schaerer, B. Clement, M. Limousin, E. Jullo, P. Natarajan, E. Egami and H. Ebeling, *ArXiv e-prints* (September 2014).

54. B. E. Robertson, R. S. Ellis, J. S. Dunlop, R. J. McLure, D. P. Stark and D. McLeod, *ArXiv e-prints* (October 2014).

55. I. de Looze, M. Baes, G. J. Bendo, L. Cortese and J. Fritz, *Mon. Not. R. Astron. Soc.* **416**, 2712(October 2011).

56. D. A. Riechers, C. M. Bradford, D. L. Clements, C. D. Dowell, I. Pérez-Fournon, R. J. Ivison, C. Bridge, A. Conley, H. Fu, J. D. Vieira, J. Wardlow, J. Calanog, A. Cooray, P. Hurley, R. Neri, J. Kamenetzky, J. E. Aguirre, B. Altieri, V. Arumugam, D. J. Benford, M. Béthermin, J. Bock, D. Burgarella, A. Cabrera-Lavers, S. C. Chapman, P. Cox, J. S. Dunlop, L. Earle, D. Farrah, P. Ferrero, A. Franceschini, R. Gavazzi, J. Glenn, E. A. G. Solares, M. A. Gurwell, M. Halpern, E. Hatziminaoglou, A. Hyde, E. Ibar, A. Kovács, M. Krips, R. E. Lupu, P. R. Maloney, P. Martinez-Navajas, H. Matsuhara, E. J. Murphy, B. J. Naylor, H. T. Nguyen, S. J. Oliver, A. Omont, M. J. Page, G. Petitpas, N. Rangwala, I. G. Roseboom, D. Scott, A. J. Smith, J. G. Staguhn, A. Streblyanska, A. P. Thomson, I. Valtchanov, M. Viero, L. Wang, M. Zemcov and J. Zmuidzinas, *Nature* **496**, 329(April 2013).

57. D. A. Riechers, C. L. Carilli, P. L. Capak, N. Z. Scoville, V. Smolcic, E. Schinnerer, M. Yun, P. Cox, F. Bertoldi, A. Karim and L. Yan, *ArXiv e-prints* (April 2014).

58. M. Ouchi, R. Ellis, Y. Ono, K. Nakanishi, K. Kohno, R. Momose, Y. Kurono, M. L. N. Ashby, K. Shimasaku, S. P. Willner, G. G. Fazio, Y. Tamura and D. Iono, *Astrophys. J.* **778**, p. 102(December 2013).

59. J. H. Wise, M. J. Turk, M. L. Norman and T. Abel, *Astrophys. J.* **745**, p. 50(January 2012).

60. R. S. Ellis, R. J. McLure, J. S. Dunlop, B. E. Robertson, Y. Ono, M. A. Schenker, A. Koekemoer, R. A. A. Bowler, M. Ouchi, A. B. Rogers, E. Curtis-Lake, E. Schneider, S. Charlot, D. P. Stark, S. R. Furlanetto and M. Cirasuolo, *Astrophys. J.* **763**, p. L7(January 2013).

61. P. A. Oesch, R. J. Bouwens, G. D. Illingworth, I. Labbé, R. Smit, M. Franx, P. G. van Dokkum, I. Momcheva, M. L. N. Ashby, G. G. Fazio, J.-S. Huang, S. P. Willner, V. Gonzalez, D. Magee, M. Trenti, G. B. Brammer, R. E. Skelton and L. R. Spitler, *Astrophys. J.* **786**, p. 108(May 2014).

62. M. Ishigaki, R. Kawamata, M. Ouchi, M. Oguri, K. Shimasaku and Y. Ono, *ArXiv e-prints* (August 2014).

63. I. Labbé, P. A. Oesch, R. J. Bouwens, G. D. Illingworth, D. Magee, V. González, C. M. Carollo, M. Franx, M. Trenti, P. G. van Dokkum and M. Stiavelli, *Astrophys. J.* **777**, p. L19(November 2013).

64. B. E. Robertson and R. S. Ellis, *Astrophys. J.* **744**, p. 95(January 2012).

65. T. Jones, D. P. Stark and R. S. Ellis, *Astrophys. J.* **751**, p. 51(May 2012).

66. T. A. Jones, R. S. Ellis, M. A. Schenker and D. P. Stark, *Astrophys. J.* **779**, p. 52(December 2013).

Discussion

J. Ostriker This was a very good summary of the UV. I wonder whether you thought about reionization from the X-rays as a component. If you take our galaxy, LMC and SMC, the ratio of X-rays from high mass binaries to star formation increases by a factor 10 when you go from low metallicity systems, so you might assume that at early times you had more X-rays binaries and more X-rays and that could give you low level of reionization for long period and help to explain the WMAP results. Have you included that?

R. Ellis No, I have not. I focused here on a very simple question, which is how close can we get to the reionizing budget from galaxies alone. As you see there is plenty of scope for a deficit.

J. Ostriker X-rays cannot dominate because the most they can ever reionize things to, is by 10 percents, and after that it just goes onto heat, but it can give you an extensive period (for metal poor systems).

M. Begelman I think the X-ray comment also applies to AGN because the fraction of AGN that are heavily obscured goes up with both luminosity and redshift, so can you start putting some constraints on how many obscured AGNs there are at these red shifts?

R. Ellis The difficulty with AGN is the statistic at high redshifts is still very poor but in the last calculation I have seen about AGN, even if the luminosity function evolves in such a way that at redshift 4 there were many obscured low luminosity AGN, here and at earlier time there were luminous, if this luminosity function evolves to become very steep at redshift 7-8, and if you took this luminosity function and you gave a 100% escape fraction, then it still gives only 25% of the ionizing budget. So it may be a contributor but it is not the dominant one.

S. Zaroubi I have a comment about what we learn from the Lyman-α forest. There are a number of other things you did not mention. One thing that I think is solid, is the number of ionizing photons per baryon that you can deduce from these systems. It put a very strong constraint, it says that roughly there are about 2 or 3 ionizing photons per baryon at redshift 6, and this is I think a very strong thing that we have to cope with. Any of these models have to explain that. It is called the photon star reionization and it basically says that this is a very extended and very low level type of reionization process that we have, if it is correct.

T. Piran I just want to make a warning about using gamma ray bursts as a tracer of star formation. The problem comes with that we do not really know what are the progenitors that produce GRB and it is clear that it is a subclass of stars and possibly everybody is guessing this is the class of most massive stars, but this is just a guess. The point is that if you just look at the positions of GRB within galaxies, you see that they are really concentrated in the region of highest star formation, unlike SN that are everywhere, and they are of course more numerous in regions of large star formation and less numerous in regions of less star formation but this is just proportional to the star formation. GRB are proportional to something like the star formation squared. So, while they are tracing something, we do not exactly know what they are tracing and this is why until we understand this, it is a difficult tool and can lead to completely different conclusions.

R. Ellis Let me address that. I think you are right, this correlation here on the left is purely empirical and in fact if you read the paper I wrote, there are various ways in which we fit this including dependance on metallicity. What strikes me is that there is a tight correlation over a significant fraction of the age of the Universe at redshift around 4 to 0, so I think what is puzzling is why we find so many GRB in a period that in cosmic history occurred not that much earlier, so if there is an explanation I would love to hear it because when we wrote this paper we were unable to think of an explanation that would explain this discrepancy.

T. Piran I can only remark that when we did the same exercise, the results also in low redshift are incompatible, what we find is that the star formation rate and the GRB, even at low redshift, are incompatible with each other and at low redshift in fact the GRB sort of follow the star formation rate in the lowest mass lowest metallicity galaxy. If you take just this sub-population they follow the star formation but this sub-population is producing much higher star formation rate now than the average one.

Rapporteur Talk by S. Furlanetto: The Cosmic Dawn: Theoretical Models and the Future

Abstract

I review several important open questions in the study of the Cosmic Dawn, the earliest epoch of luminous structure formation. I describe some of the key uncertainties in galaxy formation and evolution during this early times, highlighting the importance of feedback on both local and large scales. I then describe the relatively advanced state of models of the early stages of reionization but emphasize that its timing remains uncertain, as does a quantitative treatment of the final stages of the process. Finally, I describe how observations of the spin-flip transition of neutral hydrogen can allow us to observe most of the Cosmic Dawn and settle some of these questions.

1. Introduction

The Cosmic Dawn is one of the most compelling frontiers of modern astrophysics. Richard Ellis has discussed many of the current observations and their implications for the timing of reionization. Here, I will expand the discussion to consider the current state of theoretical models of high-z galaxies and the reionization process. I will then turn to two important questions about the Cosmic Dawn: understanding the end of reionization (for which there are good prospects in the near-term) and mapping the process of reionization in the intergalactic medium (or IGM, for which there are excellent prospects in the future).

2. How Do Galaxies Evolve During the Cosmic Dawn?

Galaxy formation is an enormous topic, even in well-observed eras like the peak of star formation. During the Cosmic Dawn, when observations leave the physics almost wide open, it can be difficult to even know where to begin. I will therefore focus on a couple of particularly interesting questions, one the "big picture" idea of understanding how galaxies first form (and how that process differs qualitatively from normal galaxy evolution models) and one much closer to home, on the state-of-the-art in understanding the galaxies we have seen at $z < 8$.

2.1. *When Did the First Galaxies Form?*

Over the past 15 years, a great deal of attention has been focused on the physics of primordial star formation. Although there are many unanswered questions, the problem of first *star* formation appears to be a tractable one: the initial conditions are well-posed and the physics (dark matter and baryonic collapse, chemistry of the primordial gas, accretion disk formation, and radiative feedback) is straightforward enough that one can at least imagine solving the problem in full.[1,2]

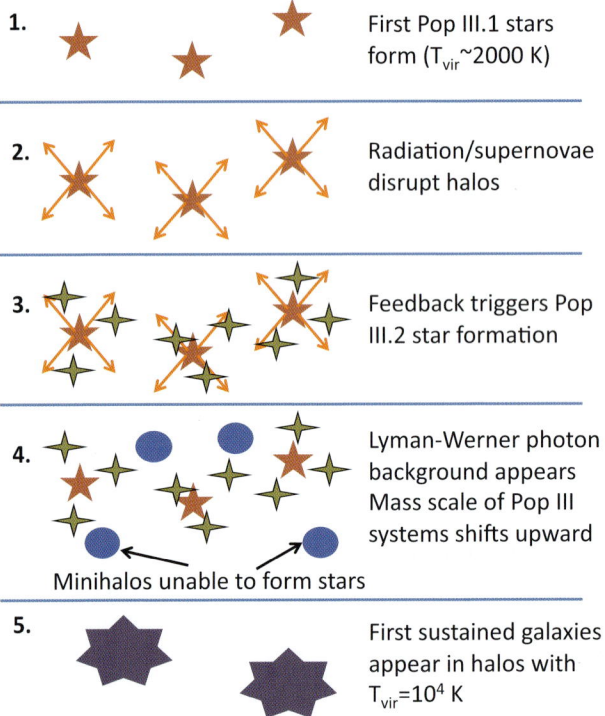

1. First Pop III.1 stars form ($T_{vir} \sim 2000$ K)

2. Radiation/supernovae disrupt halos

3. Feedback triggers Pop III.2 star formation

4. Lyman-Werner photon background appears Mass scale of Pop III systems shifts upward

Minihalos unable to form stars

5. First sustained galaxies appear in halos with $T_{vir} = 10^4$ K

Fig. 1.   Stages in a plausible scenario for the birth of the first stars and galaxies. See text for a detailed description of the stages.

However, the problem becomes *qualitatively* more difficult as soon as this first star forms, thanks to the myriad feedback mechanisms generated by these stars and their descendants. These processes are particularly important in the context of understanding the transformation of the first stars to the first *galaxies*,[3] which one can think of as gravitationally-bound systems of stars embedded in dark matter halos that exhibit *sustained* star formation over cosmological time periods (i.e. a substantial fraction of the Hubble time). Such objects must be able to form stars *and* be stable against feedback both from those stars and their surroundings.

Figure 1 shows a "best guess" (driven entirely by theoretical prejudices rather than actual data) for how this transformation might proceed.[4] It is most certainly wrong, but it provides a useful baseline paradigm for understanding the key questions in early galaxy formation.

(1) The first stars form inside halos cooled by molecular hydrogen, roughly 10^6 times smaller than the Milky Way.[5] Massive "Population III.1" stars form at

the center of these halos after cooling to low temperatures thanks to H_2 lines.[1,2] At present, the key uncertainty is whether the gas cloud fragments before the material accretes onto the protostar.[6,7] If not, the final mass is likely regulated by radiative feedback (with $M_\star > 100\ M_\odot$);[8,9] otherwise, the characteristic mass may be several times smaller.

(2) The first star (or star cluster) likely clears out its host halo's gas supply through radiation pressure[10,11] or a supernova.[12] The latter would also enrich the entire halo with heavy elements. Dense clumps already well on their way to star formation may survive this feedback, but nothing else will. Thus Population III.1 star formation in any individual halo may only occur in a single rapid burst.

(3) These same feedback mechanisms also operate on somewhat larger scales, as the H II region and supernova blastwave are able to penetrate to $\sim$ kpc scales. Any nearby halos will therefore be subject to the same effects: many will be disrupted, but others will likely have their star formation accelerated by shocks. These will experience more efficient cooling and hence (probably) a smaller characteristic mass; this mechanism creates so-called "Population III.2 stars."[13,14] Overall, close neighbors will experience "bursts" of Population III stars followed by long breaks after gas evacuation.[15]

(4) Feedback also operates on even larger scales. All Population III stars produce photons in the UV *Lyman-Werner* bands that photodissociate H_2.[16] As more stars form, the Lyman-Werner background increases, gradually raising the minimum halo mass required for cold gas formation inside minihalos. Because more massive halos are also more rare, this will tend to self-regulate the global rate of star formation.

(5) Eventually, the Lyman-Werner background will become intense enough to choke off Population III star formation in pristine minihalos entirely. Then star formation will shift to halos large enough to cool through atomic line radiation. Most of these halos will have already been chemically enriched, so they will form Population II stars.

(6) Systems somewhat above this threshold can likely maintain reasonable (though still small) star formation rates without completely disrupting their own gas supplies through supernovae or radiation pressure. *It is therefore this "second-generation" of star-forming halos that host the first sustained galaxies.*

(7) Nevertheless, feedback continues to be important in regulating galaxy formation at later times. For example, photoheating from reionization will gradually increase the Jeans mass and thus the minimum mass scale for galaxy formation (potentially completely shutting off star formation in small halos).[17]

Although this is a very plausible picture consistent with detailed theoretical work, there are a number of points at which seemingly minor differences may dramatically alter the results. For example, if the mass scale of the first stars is small enough, or if they do not produce efficient supernovae, feedback would be less efficient at evacuating halo gas, lowering the mass scale of the first sustained galaxies. Obviously there is a great deal of uncertainty in how the first stars will grow into the first galaxies – most likely, observations will be necessary to settle the question. However, theoretical work on these issues has progressed far enough that the individual underlying processes are well-understood in isolation: it is their complex interplay that makes the problem challenging and exciting to explore observationally.

Qualitatively, these phases of galaxy formation differ in two important ways from "normal" galaxy evolution at later times. First, the halos' small size and fragility mean that, at least in some cases, a single feedback process can dominate the physics (for a time). This offers hope both in successful modeling and in extracting physical parameters from observations without significant degeneracy (though neither of these hopes has yet been demonstrated in the real world, of course). Second, the relevant feedback mechanisms operate both within individual galaxies (which is true for all galaxies) and also over cosmological distances (unique to the Cosmic Dawn). The presence of large-scale feedback processes highlights the importance of following evolving galaxy *populations* rather than individual galaxies, and it also suggests that evolution can proceed at different rates in different environments. **These elements require tools above and beyond those familiar to galaxy evolution at lower redshifts.**

2.2. *How Can We Model their Subsequent Evolution?*

Once star clusters settle down into long-lived galaxies, they must evolve onto the populations we can observe now at $z < 8$. In the crudest representation, galaxies are machines that accrete gas (through mergers or slow growth) and transform it into stars and black holes. The crucial complication is feedback, which can both prevent gas from accreting onto a halo in the first place and expel material that is already present (preventing it from forming stars, or providing potential fuel for later accretion episodes). Models of feedback can be arbitrarily complicated, but at the simplest level we can describe it with a single (mass-dependent) efficiency factor. Given the paucity of data at $z > 6$, simple models that relate dark matter halos to luminous galaxies with just a few parameters are the norm in the literature.[18–20]

This will likely change in the near future, as we get much more sophisticated data sets from JWST, TMT, and other instruments. For now, however, the high-level models, as well as the hope that these small, young, and often extreme objects may be more often dominated by single processes, have led the community to focus on isolating signatures of such aspects. For example, the galaxy evolution community is split between two approaches to understanding star formation on galactic scales. The first is to build on our understanding of the "microphysics" of star formation in local galaxies.[21,22] Such models relate the star formation rate as directly as possible to the fuel supply (molecular gas). The model thus requires computing the molecular fraction, which is very difficult but can be calibrated to various local measurements, and asking how efficiently molecular clouds can form stars, which – at least locally – appears to have a well-defined answer. The primary disadvantage of this "bottom-up" approach is that it relies heavily on local calibrations that may not hold in the very different environments of small high-z galaxies.

The second approach to understanding star formation begins with global assumptions about each galaxy rather than focusing on the details of star formation.[23,24] The basic idea is that star formation can only occur if some sort of large-scale gravitational instability allows fragmentation to higher densities (in the presence of pressure). Once fragmentation begins, feedback from star formation will heat the gas, slowing further fragmentation. On the other hand, if star formation does not occur, the gas will cool rapidly, exacerbating the instability. The expectation (which appears to be realized in nearby galaxies) is therefore that galaxies will form stars sufficiently fast to remain marginally stable in a sort of self-regulated flow. The key input to this "top-down" view is the feedback mechanism providing the pressure support (which could be radiation pressure or supernovae). These models typically connect more closely to the cosmological environment, but they assume that small-scale star formation processes can automatically adjust to achieve the marginal stability criterion.

With recent data on high-z galaxies (including both Lyman-break populations and individual extreme objects, like quasar hosts), we can now begin to ask which picture is most consistent with the observations in this regime. For example, the top-down approach can fit Lyman-break galaxy abundances and predict the black hole accretion rates in these objects.[25] The model makes predictions for the star formation rate that can then be interpreted with the tools used in the bottom-up approach, which in turn imply specific observables.[26,27] Interestingly, even the sparse data currently available (just a couple of lines in a handful of quasar hosts) appear to require new physics in the star formation process.

3. How Do We Model Reionization?

Although much of astronomy focuses on the luminous material inside galaxies, the majority of matter today – and the vast majority at $z > 6$ – actually lies outside of these structures, in the IGM. This material ultimately provides the fuel for galaxy and cluster formation, and offers a cleaner view of structure formation and fundamental cosmology. It is therefore of great interest to study the properties of the IGM, especially during the era of the first galaxies (when the IGM undergoes major changes). The most important of these changes is reionization itself, and we will now turn toward understanding how that process unfolds.

3.1. *When Does Reionization Occur?*

The simplest approach to reionization is to compute the evolution of the average neutral fraction across the entire Universe. We can obtain a first estimate for the requirements of reionization by demanding one stellar ionizing photon for each hydrogen atom in the Universe. To zeroth order the accounting is relatively simple: let us define the efficiency parameter ζ to be the number of ionizing photons produced per baryon inside galaxies. If we ignore the possibility of recombinations, the neutral fraction is

$$Q_{\mathrm{HII}} = \zeta f_{\mathrm{coll}}, \tag{1}$$

where Q_{HII} denotes the average *filling factor* of ionized bubbles (i.e. the fraction of the Universe's volume inside of H II regions) and the collapse fraction f_{coll} is the fraction of matter incorporated in galaxies (typically above some minimum mass threshold determined by cooling and/or feedback). The efficiency factor ζ depends on several uncertain parameters: the fraction of galactic gas that forms stars ($f_\star$), the fraction of ionizing photons that escape the ISM of their source galaxy without being absorbed (f_{esc}), and the stellar population itself (determining the number of ionizing photons produced per stellar baryon).

As simple as it is, this equation points to the difficulty of estimating the timing of reionization with either models or galaxy observations: if we define the event to complete when every hydrogen atom has been ionized once, we need to understand the *total* fraction of gas that has formed stars as well as the properties of those stars. This is no easy task, as it requires observing not only the galaxy population at, say, $z \sim 7$, but also all the galaxies produced at earlier times! At present, the best we can do is assume a single stellar population and try to count the density of stars (subject to uncertainties about objects below the detection threshold, of course). Assuming a Salpeter IMF and solar metallicity (most likely poor assumptions), this condition implies a minimum comoving density of stars after reionization of[28]

$$\rho_\star \sim 1.6 \times 10^6 f_{\mathrm{esc}}^{-1} \ M_\odot \ \mathrm{Mpc}^{-3}. \tag{2}$$

We can motivate the next level of sophistication by including recombinations. To do so, we treat each ionizing source as producing an isolated bubble and assume that their volumes add to give the total filling factor. Let us define the *clumping factor* $C = \langle n_e^2 \rangle / \langle n_e \rangle^2$, which represents the enhancement to the recombination rate in an inhomogeneous medium. Then we can statistically describe the transition from a neutral Universe to a fully ionized one via

$$\frac{dQ_{\mathrm{H\ II}}}{dt} = \zeta \frac{df_{\mathrm{coll}}}{dt} - \alpha(T) \frac{C}{a^3} \bar{n}_H^0 Q_{\mathrm{H\ II}}, \tag{3}$$

where $\alpha(T)$ is the temperature dependent recombination rate and $\bar{n}_H^0$ is the comoving number density of hydrogen atoms. In short, the net growth of ionized regions is the rate at which photons are produced minus the rate at which they are absorbed by recombinations inside the bubbles. Although this equation appears simple, even at this low level of sophistication it hides a number of uncertain parameters. Not only do each of the elements of ζ have large uncertainties, but they may also evolve in time; similarly, the clumping factor C depends on the IGM temperature and the pattern of ionization in the IGM. For concreteness, typical stellar populations have

$\zeta \sim 4000 f_\star f_{\rm esc}$, and $C < 5$, with an increase toward the end of reionization as the ionized bubbles try to penetrate dense regions where the recombination rate is high.

This second form suggests an alternate approach to constraining the timing of reionization, and one much more easily answered by current observations: can a particular set of ionizing sources *keep* the IGM ionized at a sufficiently high level? On a global scale, this requires balancing the recombination rate per unit volume with the emissivity (by number) of ionizing photons, The canonical requirement for the comoving star formation density in galaxies is[29]

$$\dot{\rho}_\star \sim 0.003 f_{\rm esc}^{-1} \left(\frac{C}{3}\right) \left(\frac{1+z}{7}\right)^3 \; M_\odot \; {\rm yr}^{-1} \; {\rm Mpc}^{-3}. \tag{4}$$

Here we have again assumed a Salpeter IMF and solar metallicity, both of which are likely conservative and so *overestimate* the required $\dot{\rho}_\star$. We emphasize, however, that without additional observational constraints on the source populations, equation (4) provides only a rough guide.

There is little hope that we can answer either of these questions from first principles: the uncertainties in the input parameters (both for the sources and IGM) are much too large. Theory is useful to interpret observations, but our models are not predictive so far as estimating *when* reionization should occur.

3.2. *How Does Reionization Unfold?*

Fortunately, theory fares much better at addressing the deeper questions of the spatial structure and local behavior of reionization. Conceptually, the process of hydrogen reionization follows three distinct stages:

(1) The initial "pre-overlap" phase consists of individual ionizing sources turning on and ionizing their surroundings. During this period, the IGM is nearly a two-phase medium characterized by highly-ionized regions separated from neutral regions by narrow ionization fronts, so that the ionizing intensity is highly inhomogeneous.

(2) Because these first sources are highly clustered, this early phase quickly enters the central, relatively rapid "overlap" phase when neighboring H II regions begin to merge. When this occurs, the ionizing intensity inside H II regions increases rapidly. By the end of this stage, most regions in the IGM are able to "see" many individual sources, making the ionizing intensity both larger and more homogeneous.

(3) Throughout these stages, dense neutral regions absorb any ionizing photons that strike them. During the early parts of reionization, such blobs are very rare and can largely be ignored. But eventually, the ionized bubbles become so large that most photons strike one of these LLSs before reaching the edge of a bubble. This final "post-overlap" phase thus has slower evolution in the ionizing background (at least in the simplest models), modulated by the evaporation of these LLSs, and that background becomes increasingly more uniform.

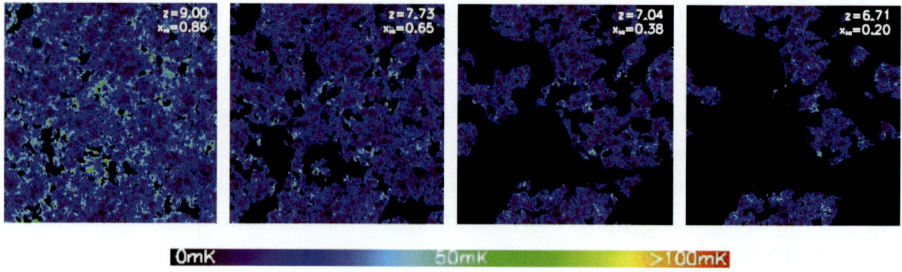

Fig. 2.   Simulations of the reionization process.[30] All slices are 143 Mpc on a side and 0.56 Mpc thick, and correspond to neutral fractions x_{HI} = 0.86, 0.65, 0.38, and 0.20, from left to right. (These go from early to late cosmic times, as labeled by the redshift z shown in each panel.) Black corresponds to ionized gas, while all other colors are neutral; the color scale shows the hydrogen spin-flip brightness (see §5).

Of course, this reionization process unfolds at different rates in different regions of the Universe; naturally, areas with an overabundance of sources undergo more rapid reionization, while those with relatively few sources require input of ionizing photons from external sources. Thus the three phases identified above are not clearly distinct from each other: overdense environments will rapidly reach the overlap (and even post-overlap) stages while void regions are still in the pre-overlap phase. This general march of reionization from high to low density is referred to as *inside-out* reionization. While most reionization models follow this behavior when averaged over large scales, on sufficiently small scales the process is actually *outside-in*, proceeding from low to high densities, since dense blobs remain partially neutral for a more extended period of time.

Figure 2 illustrates this patchiness.[30] Note the wide distribution of ionized bubble sizes, with the largest bubbles centered around the largest clusters of galaxies in the simulation. This pattern of ionized and neutral gas (often called the morphology of reionization) is an important observable that can reveal a great deal about the sources and IGM. For example, at a fixed neutral fraction, more massive sources tend to produce larger ionized regions, while more abundant absorbers will tend to shrink the largest bubbles.

The starting point for models of this patchiness is very simple: photon counting. Given a region of the Universe, we can check whether it has been reionized by comparing the number of ionizing photons produced in that region to the number of hydrogen atoms. If the former is larger, reionization is complete. In other words, we can take equation (1) and apply it *locally*. If ζf_{coll} exceeds unity, the region is reionized.

The challenge in this sort of model is defining "local." Suppose that the region we have chosen is largely empty of galaxies, but that a neighboring region has an overabundance of galaxies and has not only ionized itself but also sent the excess to our region. In that case our "local" criterion would fail. To attack the overlap problem successfully, we must allow ourselves freedom in determining the relevant scale.

An elegant way to accomplish this is with the *excursion set formalism*, which considers the density of a point in the Universe as the sum of contributions from all scales.[31] We can therefore work from large to small physical scales by watching how the smoothed density field evolves; if our photon-counting criterion is satisfied *at any scale*, we have found an ionized bubble (and its size).

This procedure can be implemented analytically[31] or in *semi-numeric simulations*.[30,32,33] The idea of the latter is to use the photon-counting trick to avoid radiative transfer inside a computer simulation (and possibly to use similar tricks to avoid following nonlinear gravitational dynamics as well). Figure 2 illustrates one of these semi-numeric techniques in a region 143 Mpc across, following it across four phases of reionization. These implementations show impressive agreement with full radiative transfer simulations (at least on large scales) but require only a fraction of the computational resources.[34]

The key insight from the analytic and semi-numeric calculations has been how large the ionized regions get throughout reionization: in many models, they reach ~ 10 Mpc well before reionization is halfway complete, and (if recombinations are ignored) they typically span the calculation volume not long after that point. This is crucial because the scales relevant to the ionizing sources are kiloparsecs or even smaller, requiring a huge dynamic range for reionization calculations. Moreover, such large scale features are relatively easy to observe, offering hope to observe the reionization process directly.

4. How Does Reionization End?

Although the models discussed in the previous section do very well in describing the early and middle phases of reionization, the final part is much more difficult to understand from first principles, as it depends on the interaction of the sources with the inhomogeneous structure of the IGM. In particular, recombinations become increasingly important as reionization progresses and the ionizing fronts try to penetrate denser and denser structures. The properties of these neutral blobs, with overdensities far into the nonlinear regime, depend sensitively on both structure formation and reionization itself (which affects the IGM temperature and hence pressure support). A self-consistent treatment of the end of reionization is therefore not yet possible.[35–38] However, this is also (at least so far) the easiest part of reionization to observe.

There is another important aspect to the end of reionization: whatever sources drive it, and however the IGM responds to the process, both those sources and IGM structures must match smoothly onto the (much better understood) later Universe. The past twenty years have seen an explosion in our understanding of both galaxies and the IGM at $z \sim 2$–5, and it is important to ask whether we can learn anything about reionization from those observations, or whether we can at least constrain models of the high-z universe by requiring such a matching.

4.1. *How Does the Lyman-α Forest Probe Reionization?*

The Lyman-α optical depth of the IGM is enormous. Letting $x_{\mathrm{H\,I}}$ being the IGM neutral fraction and δ be the fractional overdensity of a patch of the IGM,[39,40]

$$\tau_\alpha \approx 1.6 \times 10^5 x_{\mathrm{H\,I}}(1+\delta)\left(\frac{1+z}{4}\right)^{3/2}. \tag{5}$$

The enormity of this optical depth means that *any* transmission across these wavelengths is evidence that the diffuse IGM is highly ionized. At $z \sim 6$, quasars have very little transmission blueward of each quasar's rest Lyman-α wavelength. Some of the highest redshift quasars have long stretches of zero observable transmission, which are known as *Gunn-Peterson troughs*.

To interpret the optical depth distribution of the IGM, we need to know how the neutral fraction $x_{\mathrm{H\,I}}$ varies through space. Clearly the models of the previous section will not suffice: while they partitioned the Universe into highly-ionized and nearly neutral components, computing the Lyman-α transmission requires us to understand just how highly ionized those regions are. A simple model for the absorption pattern of the inhomogeneous IGM associates each gas element with its "local" optical depth from equation (5). With the assumption of ionization equilibrium, one can show that this local value is[41]

$$\tau_\alpha(\delta) \approx 13\frac{(1+\delta)^2}{\Gamma_{12}}\left(\frac{1+z}{7}\right)^{9/2}, \tag{6}$$

where we have assumed an IGM temperature of $\sim 10^4$ K and let the ionization rate (per atom) be Γ, which we measure in units of $\Gamma = \Gamma_{12} \times 10^{-12}$ s^{-1} (a typical value after reionization is complete). Equation (6) shows that at $z \sim 6$ only the most underdense regions will allow transmission (with $\tau_\alpha < 1$); gas at the mean density will be extremely opaque even if the ionizing background is comparable to its values at lower redshifts. This explains the deep absorption troughs toward many $z \sim 6$ quasars. (At lower redshifts, the density fluctuations δ cause strong variations in the local absorption, which causes the rich zoo of features we see in the Lyman-α forest.)

The key remaining requirements to interpret the observed transmission are the spatial distributions of δ and Γ, so that we can properly address the inhomogeneous IGM. (Typically, the transmission is averaged over large segments of each line of sight, spanning several tens of Mpc.) This is not trivial, as much of the transmission comes from structures well into the nonlinear regime, and the ionizing background's spatial fluctuations require high-dynamic range calculations.

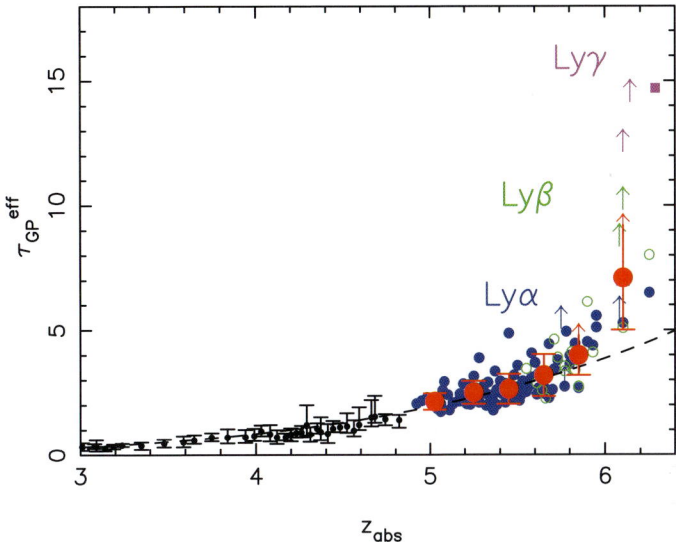

Fig. 3.   Measurements of the effective optical depth in the Lyman-α transition at high redshifts.[42] The small filled circles at $z > 5$ are direct measurements from the Lyman-α forest, the open circles are measurements from Lyman-β absorption translated into Lyman-α, and the filled squares are the same for Lyman-γ. The large filled circles with error bars show the average inferred from the Lyman-α and Lyman-β measurements. At $z > 5.5$, $\tau_{\rm eff} \propto (1 + z)^{11}$ or possibly even steeper.

Figure 3 shows how the "effective" optical depths, measured over large segments of the IGM, evolve at moderate and high redshfifts.[42] The absorption evolves slowly and smoothly at lower redshifts, but it is still quite strong by $z \sim 5.5$ when only $\sim 7\%$ of the light transmitted. Past that point, the forest thickens even more rapidly, so that very little light is transmitted. This is even more obvious if one examines the forest in higher Lyman-series transitions, which have smaller oscillator strengths and are thus more sensitive to low levels of transmission (at the cost of presenting a more serious modeling challenge), also shown in Figure 3. This rapid turnover has long been assumed to indicate the tail end of reionization, but recent models are now calling this into question, as we will discuss below.

One other aspect of the forest deserves mention in the context of reionization: the temperature distribution. Reionization is accompanied by a dramatic heating of the IGM, as the excess energy of each ionizing photon is deposited as heat (to $T > 10^4$ K). But after the gas is ionized, this heating channel switches off, and adiabatic cooling dominates the subsequent evolution. This implies that measurements of the IGM temperature evolution – even at $z < 6$ – may tell us about the timing of reionization.[43–45] In practice, this has proven very difficult to do, but the ever-increasing abundance of Lyman-α forest data suggests that it may soon be possible.

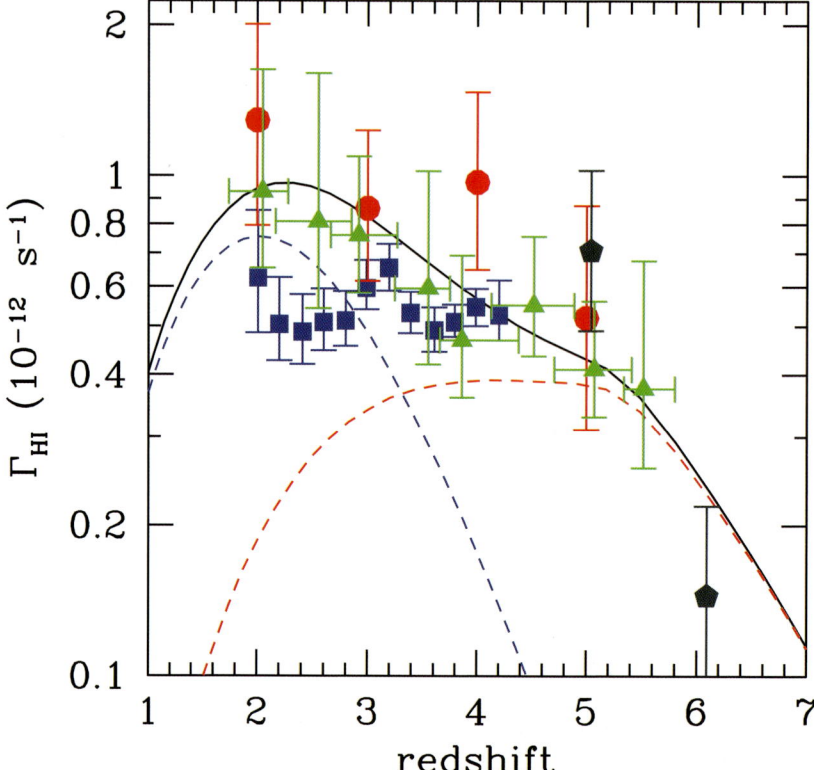

Fig. 4.   Measurements of the ionizing background at moderate redshifts (points with error bars), and models of it (curves).[46] The curves base the emissivity and mean free path inputs on measured values at $z < 6$ and use the method described in §4.2. The solid line shows the total Γ, while the dashed curves peaking at $z \sim 4$ and $z \sim 2$ show the separate contributions from star-forming galaxies and quasars, respectively.

4.2. *How Does the Ionizing Background Evolve?*

Given a model for the IGM density field, one can transform these optical depth measurements into constraints on the ionizing background (or, under the assumption of ionization equilibrium, the neutral fraction). The points with error bars in Figure 4 show several measurements of the ionizing background from $z \sim 2$–6, most using this method. To a reasonable approximation $\Gamma_{12} \sim 1$ over the range $z \sim 2$–5. The curves show a theoretical prediction of $\Gamma(z)$.[46] As expected, the data show a relatively flat Γ from $z \sim 2$–5 but rapid evolution beyond that – just when we believe reionization to be ending as well.

However, models now suggest that Γ has some rather subtle behavior at high redshifts. Let us consider a simple model where $\Gamma \approx \sigma_{H\,I} \varepsilon \lambda$, where σ_{HI} is the ionization cross section of hydrogen (suitably averaged over the metagalactic radiation spectrum), ε is the emissivity of ionizing photons, and λ is their mean free path. It is widely expected that ε does not evolve extremely rapidly, at least at $z < 8$. Thus rapid evolution is usually attributed to either changes in the mean free path or coupled changes in ε and λ. At first blush, the latter would be quite natural: an increased emissivity increases Γ, which decreases the neutral fraction in absorbers, increasing the mean free path, which increases Γ more, etc.

To use Γ as a probe of reionization, one would argue as follows. During the "bubble-dominated" phase of IGM evolution, the path over which photons travel is limited by the ionizing bubble in which they are born, effectively equating λ and the bubble size. If those bubbles are growing rapidly (as models predict), then rapid evolution in Γ could be a signature of bubble growth.

However, we have already argued that this mean free path will eventually be determined primarily by the abundance of (relatively rare) density peaks with high optical depths (as it certainly is in the post-reionization limit) rather than the ionized regions themselves. These "Lyman-limit systems" can be identified in quasar spectra relatively easily all the way to $z > 5$, but their physical nature remains obscure. Using the approximations described earlier, the typical density of these structures is[41]

$$1 + \delta_{\mathrm{LLS}} \sim 300 \Gamma_{12}^{2/3} \left(\frac{1+z}{4} \right)^{-3}. \tag{7}$$

At moderate redshifts, these objects thus have overdensities characteristic of virialized halos. But, crucially, this equation implies that the very nature of these LLSs may change qualitatively at $z \sim 6$, where $\delta_{\mathrm{LLS}} \sim 20$ – near the dividing line between halos and IGM. At higher redshifts, the densities become even more modest, characteristic of sheets and filaments in the IGM. One must therefore expect the nature of IGM absorption to change radically during reionization.

What might this imply for the evolution of Γ? If absorbers are primarily associated with virialized halos – and hence sources of the ionizing photons – we might expect relatively modest evolution in the ionizing background, as an increase in the emissivity (from having more sources) would be roughly balanced by a decrease in the mean free path (from having more absorbers in the same halos). On the other hand, once the diffuse IGM dominates the absorption, the emissivity and mean free path are no longer coupled. A modest change in emissivity can then lead to a rapid change in the ionizing background *without* any need for reionization (through the feedback cycle identified above). Instead, the observed change at $z \sim 5.5$ would imply something about the changing relation between absorbers and emitters, and very little about the neutral fraction.[47]

Two additional effects also complicate the interpretation of the Lyman-α forest data. First, reionization is highly inhomogeneous. Γ will evolve rapidly on a *local* scale as the bubbles grow around a particular point, sometimes increasing by orders of magnitude when neighboring ionized regions merge. But, averaged over the entire universe, the evolution could be much slower, growing (at best) with the characteristic size of the ionized regions (and possibly much slower once that size exceeds the mean free path from density inhomogeneities). In other words, reionization need not cause rapid evolution in Γ, and rapid evolution in Γ need not be a signature of reionization.[48,49]

Second, nearly all models (to date) have treated the post-reionization ionizing background as uniform. While this is an excellent approximation at $z < 4$, as we approach the reionization era the sources become more rare and the mean free path becomes smaller, suggesting that fluctuations may not be negligible anymore – as recent data may require.[50] Indeed, the same process that creates a feedback cycle on the average Γ operates on these fluctuations, potentially amplifying small emissivity fluctuations into much larger fluctuations in the ionizing background. These fluctuations, in turn, presumably map smoothly onto the fluctuations intrinsic to the reionization models, which generate the bubbles, and may contain very useful information on that process.

These subtleties cast doubt on a direct association of the Lyman-α forest data in Figure 3 with reionization. But of course they also suggest a deeper understanding of the process, which is already being turned into interesting constraints. For example, if one can properly model the IGM absorption, measurements of the optical depth evolution can be turned into measurements of the *total* emissivity ε, including that from otherwise unobserved sources. The latest results suggest that galaxies may be generating of order a few ionizing photons per hydrogen atom per Hubble time at $z \sim 6$, suggesting that reionization may be moderately rapid, at least near its conclusion.[51]

5. How Can We Observe the IGM Directly?

Although most observations of the Cosmic Dawn to date focus on understanding the luminous sources, reionization and other large-scale radiative feedback processes will likely not be understood in their entirety until we can observe them unfold directly in the IGM. There are a number of ways to do this:

(1) The cosmic microwave background is scattered by free electrons in the IGM. The total optical depth to electron scattering provides a measurement of the total column density of these electrons and hence an integrated measure of when reionization occurred. This has already been observed with the WMAP satellite,[52] which suggests that reionization occurred around $z \sim 10$. Moreover, the patchiness of reionization induces a secondary temperature anisotropy through the kinetic Sunyaev-Zeldovich effect that may have also been observed.[53,54] This can potentially constrain the duration of reionization.[55,56]

(2) Lyman-α emission lines from galaxies suffer severe scattering once the IGM becomes substantially neutral. The relative abundance of these lines therefore measures the progress of reionization, and the clustering of the line emitters may allow us to map out ionized regions in the future (see Richard Ellis' talk).

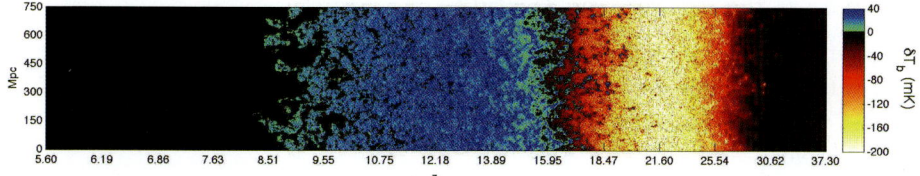

Fig. 5.   Time evolution of the expected 21 cm signal from a semi-numeric simulation 750 Mpc on a side, spanning the period before the first stars formed (at right) through the end of reionization (at left).[63] Galaxy parameters are similar to those of present-day galaxies. Coloration indicates the strength of the 21 cm brightness as it transitions from absorption (red) to emission (blue) and finally disappears (black) due to reionization.

(3) Similarly, Lyman-α transmission near luminous quasars depends sensitively on the abundance of neutral gas outside the "near-zone" of the quasar.[57] There are some indications of large absorption in the most distant quasar (at $z \sim 7.1$).[58,59] In principle, one can also observe the quasar's ionization front through its Lyman-α emission as it plows through neutral gas, but that is likely extremely faint.[60,61]

Although the Lyman-α line and the CMB are extremely powerful probes of reionization, they suffer from several shortcomings. Most importantly, the Gunn-Peterson optical depth is enormous, so that even a very small fraction of neutral hydrogen ($> 10^{-3}$) saturates the IGM absorption. The Lyman-α line is therefore difficult to interpret during the middle and early stages of reionization. On the other hand, the CMB probes are integrated measurements along the line of sight, offering no (direct) discriminatory power between events at different redshifts.

These problems can be avoided by observing the *spin-flip* or hyperfine line of neutral hydrogen, which is driven by the magnetic interactions of the proton and electron – though of course such a strategy introduces a new set of challenges. This transition is extremely weak, making the effective IGM optical depth only $\sim 1\%$. While the signal is therefore very faint, the neutral IGM is accessible over the entire epoch of reionization. Moreover, the transition energy is so low that it provides a sensitive calorimeter of the diffuse IGM, and – as a low-frequency radio transition – it can be observed across the entire sky and be used to "slice" the universe in the radial direction, thanks to the cosmological redshift (as shown in Figure 5). With such three-dimensional observations, the 21-cm line allows *tomography* of the neutral IGM, potentially providing a map of $> 90\%$ of the Universe's baryonic matter during the Dark Ages and cosmic dawn.[62] We shall focus on this as (potentially) the most revolutionary probe of the Cosmic Dawn. We shall briefly describe the physics that drives it, describe the signal qualitatively, and then describe some recent progress toward observing it.

5.1. *Why is the 21-cm Line Useful?*

The intensity of the spin-flip background is usually quantified through the brightness temperature T_b, which describes the intensity of the radiation relative to the CMB (against which it is observed) in the Rayleigh-Jeans limit. Straightforward radiative transfer yields

$$T_b(\nu) \approx \frac{T_S - T_\gamma(z)}{1+z} \, \tau_{\nu_0} \tag{8}$$

$$\approx \quad 9 \, x_{\mathrm{HI}}(1+\delta)(1+z)^{1/2} \left[1 - \frac{T_\gamma(z)}{T_S} \right] \left[\frac{H(z)/(1+z)}{dv_\parallel/dr_\parallel} \right] \mathrm{mK}, \tag{9}$$

where T_γ is the CMB temperature, T_S is the spin temperature (or excitation temperature of this transition), and $dv_\parallel/dr_\parallel$ is the velocity gradient along the line of sight. Here $T_b < 0$ if $T_S < T_\gamma$, yielding an absorption signal, while $T_b > 0$ otherwise, yielding emission. Both regimes are important for the high-z universe, with absorption in the earlier phases and emission later on (see Figure 5). In the latter case, δT_b saturates if $T_S \gg T_\gamma$ (though this is not true in the absorption regime).

Three processes compete to fix T_S:[64–66] (i) interactions with CMB photons; (ii) particle collisions; and (iii) scattering of UV photons. The CMB very rapidly drives the spin states toward thermal equilibrium with $T_S = T_\gamma$. The other two processes break this coupling, but collisions are weak during the Cosmic Dawn (at $z < 50$).

We therefore require a different process to break the coupling to the CMB during the era of galaxy formation. The *Wouthuysen-Field mechanism*[64,65,67] provides just such an effect. Suppose a hydrogen atom in the hyperfine singlet state absorbs a Lyman-α photon, reaches the excited state, and then spontaneously decays. The electric dipole selection rules allow the electron to decay to the triplet hyperfine state – thus making a hyperfine transition by absorbing and emitting a Lyman-α photon. If enough such photons fill the universe, this mechanism dominates in setting the spin temperature. The Wouthuysen-Field mechanism drives the spin temperature to the kinetic temperature of the gas,[64,68] $T_S \approx T_K$.

Next we consider the astrophysical processes that drive the 21-cm background. In general terms, three important radiation backgrounds affect the signal: (1) the metagalactic field near the Lyman-α resonance, which determines the strength of the Wouthuysen-Field effect; (2) the X-ray background, which determines the IGM temperature; and (3) the ionizing background, which eventually (nearly) eliminates the signal at the completion of reionization.

The Lyman-α background is generated by stars and most likely saturates fairly early in the Cosmic Dawn, as strong coupling requires only one photon per ~ 10 hydrogen atoms. The background is also fairly uniform through most of cosmic history: the effective "horizon" within which a given source is visible is ~ 250 comoving Mpc.[69,70] However, this horizon is comparable to the scales over which the relative baryon and dark matter velocities vary,[71] so those velocity fluctuations can induce much stronger variations in the Wouthuysen-Field coupling in some circum-

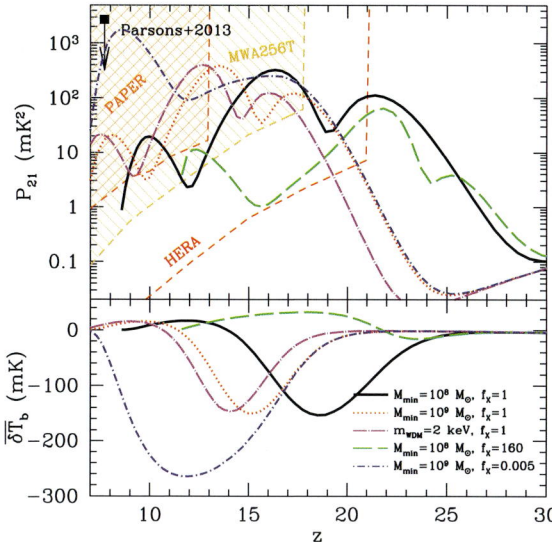

Fig. 6. Top panel: Amplitude of the 21-cm power spectrum at $k = 0.1$ Mpc^{-1} in several representative models (as labeled in the bottom panel).[83] We also plot the (1σ) sensitivities of 2000h observation with an expanded MWA with twice the current collecting area, PAPER, and the full HERA array. The recent upper limit from[84] is shown at $z = 7.7$. Bottom panel: The corresponding sky-averaged 21-cm brightness temperature (relative to the CMB) for these models. The input parameters are labeled in the bottom panel. $M_{\min}$ is the minimum halo mass allowed to form stars, f_X is the X-ray efficiency, and the magenta dot-dashed curve uses a cosmology with warm dark matter (which delays structure formation).

stances.[72,73] In §5.3 we will consider how fluctuations in this background translate into fluctuations in the 21-cm signal.

The X-ray background is generated by some combination of active galactic nuclei, supernova remnants,[74] stellar-mass black holes,[74–79] and hot ISM thermal emission.[80] We will consider stellar-mass black hole remnants of massive stars as a fiducial model, but any or all of these can be significant. Note that source models generally predict the overall intensity and spectrum of the background, but there is an additional step in computing the IGM heating, as X-rays only deposit a fraction of their energy as heat: they initially interact with the IGM by ionizing a neutral atom. The high-energy photoelectron then scatters through the IGM, ionizing more atoms, collisionally exciting others, and heating the gas through scattering off other electrons. The fraction of energy deposited in each of these processes varies with photon energy and the ambient conditions,[81,82] but as a rule of thumb each gets about 1/3 of the total for nearly-neutral gas. The important exception are very high energy X-rays, to which the IGM is largely transparent, so very hard sources have little direct influence on the IGM.

5.2. *How Does the Spin-flip Background Teach Us about Source Populations?*

We now turn to observable predictions of the 21-cm background. We begin in this section with the monopole, or sky-averaged brightness, as a function of frequency, which basically constrains how the three important radiation backgrounds (and hence source populations) evolve through the Cosmic Dawn. The bottom panel of Figure 6 shows the results (as a function of redshift) for several models of early star formation (the upper panel shows the corresponding fluctuations, which we will discuss next). The principal parameters varied here are $M_{\min}$, the minimum halo mass to host star formation, and f_X, the X-ray heating efficiency. We will take the solid black curve as our fiducial model: these choices are simplistic but representative of many models. The figure clearly shows that the signal can plausibly range by orders of magnitude over most of the epoch of interest. It also illustrates several important points about the 21-cm background. The most crucial is the presence of several critical points in the spin-flip background.[75,85] It is important to understand that all of these points depend sensitively on the properties of the first luminous sources, so their locations can vary, and they may not even occur in the order shown in our fiducial model (or indeed at all).

(1) The formation of the first stars (at $z \sim 25$) "activates" the 21-cm background through the Wouthuysen-Field mechanism. At this point, the IGM is cold (thanks to rapid adiabatic cooling as the Universe expands), so the first stars appear in *absorption* against the CMB.

(2) In most models, the next feature is the minimum in T_b, which occurs just before IGM heating becomes significant. The breadth and depth of this absorption feature is quite uncertain, as it depends on the relative efficiencies of X-ray heating and ionizing photon production. If the first is very large, this heating transition can precede strong coupling. If it is very small – as happens in some models of hard-spectra black holes – it can persist through the early stages of reionization.[79]

(3) The next turning point occurs at the maximum of T_b. As long as heating is relatively strong (at least comparable to the efficiency seen in local galaxies), this marks the point at which $T_K \gg T_\gamma$, so that the temperature portion of equation (9) saturates. The signal then starts to decrease rapidly once reionization begins in earnest. As just mentioned, it is not yet clear whether such a turning point appears before reionization.

(4) Finally, the monopole signal (nearly) vanishes when reionization completes.

Several efforts to observe this monopole signal are underway, including the Cosmological Reionization Experiment (CoRE), the Experiment to Detect the Global Epoch of Reionization Signal (EDGES),[a,86] the SCI-HI experiment,[87] the Large

---

[a]See http://www.haystack.mit.edu/ast/arrays/Edges/

Aperture Experiment to Detect the Dark Ages (LEDA),[b] and an ambitious program to launch a radio telescope to the moon in order to observe the high-redshift signal is also being planned (the Dark Ages Radio Explorer, or DARE).[c,88]

Because these experiments aim to detect an all-sky signal, small single dishes can easily reach the required mK sensitivity.[89] However, the much stronger synchrotron foregrounds from our Galaxy nevertheless make such observations extremely difficult: they have $T_{sky} > 200$–10^4 K over the relevant frequencies. Additionally, terrestrial interference and the ionosphere present very substantial challenges. The fundamental strategy for extracting the cosmological signal relies on the expected spectral smoothness of the foregrounds (which primarily have power law spectra), in contrast to the non-trivial structure of the 21-cm background. Nevertheless, isolating the high-redshift component will be a challenge that requires extremely accurate calibration over a wide frequency range and, most likely, sharp localized features in $T_b(z)$ that can be distinguished from smoother foreground features. Current estimates suggest that we can rule out rapid reionization histories that span a redshift range $\Delta z \lesssim 2$, provided that local foregrounds can be well modeled.[86] However, it may be necessary to perform such observations from space, in order to avoid systematics from terrestrial interference and the ionosphere, whose properties strongly vary spatially, temporally, and with frequency (in particular, the ionosphere crosses from absorption to emission in this range;[90]). In fact the best observing environment is the far side of the moon (though also the most expensive!), where the moon itself blocks any radio signals from Earth; this is the primary motivation for DARE.

5.3. *How Can We Learn More from the Spin-flip Background?*

While the 21 cm monopole contains a great deal of information about the mean evolution of the sources, every component in equation (9) can also fluctuate significantly. The evolving cosmic web imprints growing density fluctuations on the matter distribution. Ionized gas is organized into discrete H II regions (at least in the most plausible models), and the Lyman-α background and X-ray heating will also be concentrated around galaxies. The single greatest advantage of the 21-cm line is that it allows us to separate this fluctuating component both on the sky and in frequency (and hence cosmic time). Thus, we can study the sources and their effects on the IGM in detail. It is the promise of these "tomographic" observations that makes the 21 cm line such a singularly attractive probe.

Observing the 21-cm fluctuations has one practical advantage as well. The difficulty of extracting the global evolution from the enormously bright foregrounds lies in its relatively slow variation with frequency. On the small scales relevant to fluctuations in the signal, the gradients increase dramatically: for example, at the edge of an H II region T_b drops by ~ 20 mK essentially instantaneously. As a result,

---

[b]http://www.cfa.harvard.edu/LEDA/
[c]http://lunar.colorado.edu/dare/

separating them from the smoothly varying astronomical foregrounds may be much easier. Unfortunately, constructing detailed images will remain extremely difficult because of their extraordinary faintness; telescope noise is comparable to or exceeds the signal except on rather large scales. Thus, a great deal of attention has recently focused on using statistical quantities extractable from low signal-to-noise maps to constrain the IGM properties. This is motivated in part by the success of CMB measurements and galaxy surveys at constraining cosmological parameters through the power spectrum.[d]

We first define the fractional perturbation to the brightness temperature, $\delta_{21}(\mathbf{x}) \equiv [T_b(\mathbf{x}) - \bar{T}_b]/\bar{T}_b$, a zero-mean random field. We will be interested in its Fourier transform $\tilde{\delta}_{21}(\mathbf{k})$. Its power spectrum is defined to be

$$\left\langle \tilde{\delta}_{21}(\mathbf{k}_1)\, \tilde{\delta}_{21}(\mathbf{k}_2) \right\rangle \equiv (2\pi)^3 \delta_D(\mathbf{k}_1 - \mathbf{k}_2) P_{21}(\mathbf{k}_1), \tag{10}$$

where $\delta_D(x)$ is the Dirac delta function and the angular brackets denote an ensemble average. We will also consider $\Delta_{21}^2 = k^3 P_{12}/2\pi^2$, which quantifies the contribution of each physical scale to the variance.

Figure 7 shows several snapshots of a "semi-numeric" simulation of the spin-flip background and the corresponding power spectra (in the right column). The underlying model is very similar to the fiducial model in Figure 6, though the redshifts of the critical points differ slightly. We also illustrate this evolution in the top panel of Figure 6, which shows the evolution of the amplitude of the power spectrum at one particular wavenumber ($k = 0.1$ Mpc^{-1}, near the peak sensitivities of most arrays). Like the global signal, the fluctuations evolve through several phases during the Cosmic Dawn.

(1) The top row of Figure 7 shows the point where Lyman-α pumping begins to be significant. The hydrogen gas is cold ($T_K \ll T_\gamma$), and the spin temperature is just beginning to decouple from the CMB. In this case the fluctuations are driven by the discrete, clustered first galaxies: their radiation field drives $T_S \rightarrow T_K$ around those first sources, while leaving most of the IGM transparent. In Figure 6, the rightmost peak of the solid curve shows the effects of the Lyman-α fluctuations: they build up to a peak, with amplitude ~ 10 mK, before decreasing again once the Lyman-α background becomes strong throughout the universe (at which $T_S \approx T_K$ everywhere, and fluctuations in the radiation background become unimportant).

(2) The second row in Figure 7 shows a map shortly after X-ray heating commences. Near the first X-ray sources, the gas has $T_S \gg T_\gamma$, so these regions appear in emission, while more distant regions are still cold. The net effect is a very large fluctuation amplitude, with a strong contrast between emitting and absorbing regions, as we see in the middle peak of the solid curve in Figure 6.

---

[d]Other statistical measures, such as higher-order correlations, may also offer additional information.

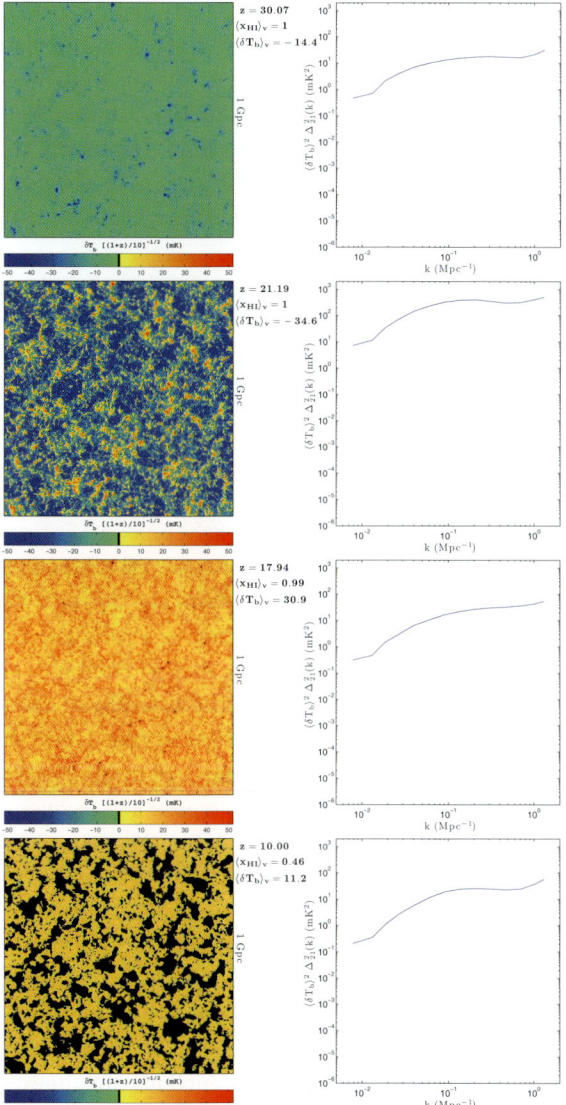

Fig. 7. Slices through a semi-numeric simulation (left), and the corresponding spherically-averaged power spectra (right), for a model of the spin-flip background at $z = 30.1, 21.2, 17.9, 10.0$ (top to bottom).[30] The slices were chosen to highlight various epochs in the cosmic 21-cm signal (from top to bottom): the onset of Lyman-α pumping (here the blue regions show the cold gas around the first galaxies), the onset of X-ray heating (here the blue regions are cold gas, while the compact red regions represent hot gas around the first black holes), the completion of X-ray heating (where all the gas is hot), and the mid-point of reionization (where black regions are ionized bubbles). All comoving slices are 1 Gpc on a side and 3.3 Mpc deep.

(3) The third row in Figure 7 shows the 21-cm signal after heating has saturated $(T_S \gg T_\gamma)$ throughout the IGM. At this point, spin temperature fluctuations

no longer contribute to T_b, and only the density field affects the overall signal. The fluctuations are thus relatively modest (also seen in Figure 6).

(4) Finally, the fluctuations increase again once reionization begins in earnest, as shown in the bottom row of Figure 7: here the contrast between the ionized bubbles and fully neutral gas in between them dominates the features. These bubbles are the key observable during reionization, as their pattern depends on the properties of the ionizing sources. This causes the leftmost peak in Figure 6.

The other curves in the top panel of Figure 6 show how the fluctuations on this scale can vary in a plausible range of models. Note that most provide the same overall structure, with three consecutive peaks, but their timing and amplitudes vary. Moreover, the blue short dashed-dotted curve, which assumes very weak X-ray heating, has only two peaks, as the IGM is not heated substantially until reionization is already underway. The broad range of possible signals makes the 21-cm line a powerful probe.

Currently, several experiments are in the early phases of attempting to observe this line: (1) the Giant Metrewave Radio Telescope (GMRT; in India), composed of thirty 45-m dishes, was the first to put limits on the spin-flip background in the summer of 2010;[91,92] (2) the Precision Array to Probe the Epoch of Reionization (PAPER, with its primary array in South Africa) combines signals from single dipoles into an interferometer and has placed the first physically relevant limits on the IGM at $z \sim 8$,[84] ruling out a cold, neutral IGM at that time (shown in Figure 6); (3) the Low Frequency Array (LOFAR; with the core in the Netherlands and outlying stations throughout Europe) is a general-purpose low-frequency radio telescope that began science operations in 2010 and is the largest of the current generation of instruments; and (4) the Murchison Widefield Array (MWA) in Western Australia is an interferometer built almost entirely to observe the 21-cm background.

In addition to this impressive suite of ongoing efforts, larger experiments are planned for the future, with their designs and strategies informed by this present generation. These include the Hydrogen Epoch of Reionization Array (HERA), which will eventually have hundreds of 14-m dishes optimized to use the strategies developed to analyze PAPER and MWA data (currently beginning the first stage of the array, with an eye toward completion by the end of the decade) and the Square Kilometer Array-Low, which will be even larger and should be capable of imaging large IGM structures.

Raw sensitivity to the very faint cosmic 21-cm fluctuations is hard enough to achieve, but an additional difficulty is separating that signal from the many (and much, much brighter) astrophysical foregrounds, especially synchrotron emission from our Galaxy. Conceptually, the way to proceed is straightforward:[93,94] most known foregrounds have very smooth spectra, while the cosmological signal varies rapidly as any given line of sight passes through density and temperature fluctuations and/or ionized bubbles. If one imagines transforming the data into an "image cube" (with observed frequency a proxy for radial distance), one ought to be able to fit a smooth function to each line of sight, subtract that smooth component, and be left a measurement of the rapidly-varying cosmological component (plus any rapidly varying foregrounds).

As a mathematical exercise applied directly to, e.g. simulation boxes, these strategies work extremely well: they remove both foregrounds and signal below some minimum wavenumber but do not contaminate the data in the remainder of the measured region.[95,96] However, the practical details of this foreground removal are quite challenging,[97] and there has not yet been a successful application of these removal strategies to real-world data. The simplest challenge to understand is the intrinsic chromaticity of the interferometer: each baseline in the interferometer measures a scale $\propto D/\lambda$, where D is the physical distance between the interferometer elements and λ is the observed wavelength. Thus the instrument response changes across the measurement band, introducing spurious frequency-dependent features from foregrounds.

Chromatic effects such as these manifest themselves along a "wedge" in Fourier space at large $k_\perp$ and small $k_\parallel$.[98–100] Crucially, at least to the limits of current MWA and PAPER data, the "extra" foreground contamination *only* appears in this wedge. This understanding (which can be described analytically[101,102]) can be used to minimize the impact of the foregrounds in new experiments (a key motivation in the design of HERA, for example), *even without attempting to model them in detail*. In the near-term, much of the community's focus has thus shifted from *foreground removal* to *foreground avoidance*: the simplest approach is to simply ignore data in that region and work inside the *EoR window* that remains uncontaminated. In practice, of course, there will be some residual contamination even here, from such factors as baseline gridding,[103] ionospheric refraction and reflection, and polarized foreground leakage (which has strong frequency dependence due to Faraday rotation). But the PAPER team has already demonstrated a four order-of-magnitude reduction in foreground contamination.[84] If the EoR window remains clean to another order of magnitude in depth, a detection (and eventually detailed characterization) of the spin-flip signal will simply require more collecting area and careful design to maximize the EoR window, which both HERA and SKA-Low will do.

6. Concluding Thoughts

The Cosmic Dawn is a tremendously exciting frontier that will be explored extensively over the next decade, using tried-and-true astrophysical techniques (like galaxy observations) and using novel probes (like the 21-cm line). In this chapter, I have reviewed some of the important questions and techniques in studying this field, with an emphasis on theoretical investigations.

In the near term, we are struggling to understand the nature of galaxies at the end of the cosmic dawn (at $z > 6$) and how they evolve into the well-studied galaxies at later epochs. High-level models drawing on the lessons of galaxy evolution studies at lower redshifts fare reasonably well, but more detailed investigations are starting to reveal hints that internal processes, like star formation, may be changing.

Equally interesting is the effort to understand the end of reionization by studying the Lyman-α forest. While models for the early and middle stages of the reionization process are fairly mature (at least until they are confronted by data!), the end stages require a detailed understanding of the IGM that is at the bleeding edge of our modeling abilities. While the naive interpretation that the forest observations suggest the end of reionization at $z \sim 6$ has been challenged, these investigations have revealed the rich information that is available with better modeling and more data. These efforts are our best hope for understanding the transition from the "bubble-dominated" morphology characteristic of the reionization process to the "web-dominated" morphology characteristic of the later universe.

Farther in the future, we will begin exploring deeper questions of how the first sources formed and influenced the universe around them. For galaxy formation studies, the next frontier in both theory and observation is likely to be the transition from the first (short-lived) star clusters to the first true galaxies. The physical processes regulating this transition are complex individually and even more difficult to understand when they interact with each other, but observations may help us disentangle the most important drivers of the transition. Meanwhile, instrument-builders and observers are making tremendous strides toward observing the spin-flip background from these early epochs, and – if the current understanding holds true – the next generation of instruments are poised to make detailed observations of both the reionization process and earlier epochs, when the first stars and black holes flooded the universe with UV and X-ray photons.

The next decade will likely see many of our expectations borne out and many more overturned. It is indeed an auspicious time to study the Cosmic Dawn!

References

1. T. Abel, G. L. Bryan and M. L. Norman, *Science* **295**, 93(January 2002).
2. V. Bromm, P. S. Coppi and R. B. Larson, *Astrophys. J.* **564**, 23(January 2002).
3. V. Bromm and N. Yoshida, *Ann. Rev Astron. Astrophys.* **49**, 373(September 2011).
4. A. Loeb and S. R. Furlanetto, *The First Galaxies in the Universe* (Princeton, NJ: Princeton University Press, 2013).
5. M. Tegmark, J. Silk, M. J. Rees, A. Blanchard, T. Abel and F. Palla, *Astrophys. J.* **474**, p. 1(January 1997).
6. M. J. Turk, T. Abel and B. O'Shea, *Science* **325**, 601(July 2009).
7. P. C. Clark, S. C. O. Glover, R. J. Smith, T. H. Greif, R. S. Klessen and V. Bromm, *Science* **331**, 1040(February 2011).
8. J. C. Tan and C. F. McKee, *Astrophys. J.* **603**, 383(March 2004).
9. A. Stacy, T. H. Greif and V. Bromm, *Mon. Not. R. Astron. Soc.* **422**, 290(May 2012).
10. D. Whalen, T. Abel and M. L. Norman, *Astrophys. J.* **610**, 14(July 2004).
11. D. Whalen, B. W. O'Shea, J. Smidt and M. L. Norman, *Astrophys. J.* **679**, 925(June 2008).
12. T. H. Greif, J. L. Johnson, V. Bromm and R. S. Klessen, *Astrophys. J.* **670**,

1(November 2007).

13. S. P. Oh and Z. Haiman, *Astrophys. J.* **569**, 558(April 2002).

14. J. L. Johnson and V. Bromm, *Mon. Not. R. Astron. Soc.* **366**, 247(February 2006).

15. N. Yoshida, S. P. Oh, T. Kitayama and L. Hernquist, *Astrophys. J.* **663**, 687(July 2007).

16. Z. Haiman, M. J. Rees and A. Loeb, *Astrophys. J.* **476**, p. 458(February 1997).

17. A. A. Thoul and D. H. Weinberg, *Astrophys. J.* **465**, p. 608(July 1996).

18. M. Trenti, M. Stiavelli, R. J. Bouwens, P. Oesch, J. M. Shull, G. D. Illingworth, L. D. Bradley and C. M. Carollo, *Astrophys. J.* **714**, L202(May 2010).

19. J. A. Muñoz, *J. Cos. Astroparticle Phys.* **4**, p. 15(April 2012).

20. P. S. Behroozi and J. Silk, *ArXiv e-prints* (April 2014).

21. A. Burkert, R. Genzel, N. Bouché, G. Cresci, S. Khochfar, J. Sommer-Larsen, A. Sternberg, T. Naab, N. Förster Schreiber, L. Tacconi, K. Shapiro, E. Hicks, D. Lutz, R. Davies, P. Buschkamp and S. Genel, *Astrophys. J.* **725**, 2324(December 2010).

22. M. R. Krumholz, A. Dekel and C. F. McKee, *Astrophys. J.* **745**, p. 69(January 2012).

23. T. A. Thompson, E. Quataert and N. Murray, *Astrophys. J.* **630**, 167(September 2005).

24. R. Davé, K. Finlator and B. D. Oppenheimer, *Mon. Not. R. Astron. Soc.* **421**, 98(March 2012).

25. J. A. Muñoz and S. Furlanetto, *Mon. Not. R. Astron. Soc.* **426**, 3477(November 2012).

26. J. A. Muñoz and S. R. Furlanetto, *Mon. Not. R. Astron. Soc.* **435**, 2676(November 2013).

27. J. A. Muñoz and S. R. Furlanetto, *Mon. Not. R. Astron. Soc.* **438**, 2483(March 2014).

28. B. E. Robertson, S. R. Furlanetto, E. Schneider, S. Charlot, R. S. Ellis, D. P. Stark, R. J. McLure, J. S. Dunlop, A. Koekemoer, M. A. Schenker, M. Ouchi, Y. Ono, E. Curtis-Lake, A. B. Rogers, R. A. A. Bowler and M. Cirasuolo, *Astrophys. J.* **768**, p. 71(May 2013).

29. P. G, H. C. Ferguson, M. E. Dickinson, M. Giavalisco, C. C. Steidel and A. Fruchter, *Mon. Not. R. Astron. Soc.* **283**, 1388(December 1996).

30. A. Mesinger, S. Furlanetto and R. Cen, *Mon. Not. R. Astron. Soc.* **411**, 955(February 2011).

31. S. R. Furlanetto, M. Zaldarriaga and L. Hernquist, *Astrophys. J.* **613**, 1(September 2004).

32. O. Zahn, M. Zaldarriaga, L. Hernquist and M. McQuinn, *Astrophys. J.* **630**, 657(September 2005).

33. A. Mesinger and S. Furlanetto, *Astrophys. J.* **669**, 663(November 2007).

34. O. Zahn, A. Mesinger, M. McQuinn, H. Trac, R. Cen and L. E. Hernquist, *Mon. Not. R. Astron. Soc.* **414**, 727(June 2011).

35. S. R. Furlanetto and S. P. Oh, *Mon. Not. R. Astron. Soc.* **363**, 1031(November 2005).

36. T. R. Choudhury, M. G. Haehnelt and J. Regan, *Mon. Not. R. Astron. Soc.* **394**, 960(April 2009).

37. D. Crociani, A. Mesinger, L. Moscardini and S. Furlanetto, *Mon. Not. R. Astron. Soc.* **411**, 289(February 2011).

38. E. Sobacchi and A. Mesinger, *Mon. Not. R. Astron. Soc.* **440**, 1662(May 2014).

39. J. E. Gunn and B. A. Peterson, *Astrophys. J.* **142**, 1633(November 1965).

40. P. A. G. Scheuer, *Nature* **207**, p. 963(August 1965).

41. J. Schaye, *Astrophys. J.* **559**, 507(October 2001).

42. X. Fan, M. A. Strauss, R. H. Becker, R. L. White, J. E. Gunn, G. R. Knapp, G. T. Richards, D. P. Schneider, J. Brinkmann and M. Fukugita, *Astron. J.* **132**, 117(July 2006).

43. L. Hui and Z. Haiman, *Astrophys. J.* **596**, 9(October 2003).

44. H. Trac, R. Cen and A. Loeb, *Astrophys. J.* **689**, L81(December 2008).

45. S. R. Furlanetto and S. P. Oh, *Astrophys. J.* **701**, 94(August 2009).

46. F. Haardt and P. Madau, *Astrophys. J.* **746**, p. 125(February 2012).

47. J. A. Muñoz, S. P. Oh, F. B. Davies and S. R. Furlanetto, *ArXiv e-prints* (October 2014).

48. S. R. Furlanetto and A. Mesinger, *Mon. Not. R. Astron. Soc.* **394**, 1667(April 2009).

49. A. Mesinger and S. Furlanetto, *Mon. Not. R. Astron. Soc.* **400**, 1461(December 2009).

50. G. D. Becker, J. S. Bolton, P. Madau, M. Pettini, E. V. Ryan-Weber and B. P. Venemans, *ArXiv e-prints* (July 2014).

51. G. D. Becker and J. S. Bolton, *Mon. Not. R. Astron. Soc.* **436**, 1023(December 2013).

52. G. Hinshaw, D. Larson, E. Komatsu, D. N. Spergel, C. L. Bennett, J. Dunkley, M. R. Nolta, M. Halpern, R. S. Hill, N. Odegard, L. Page, K. M. Smith, J. L. Weiland, B. Gold, N. Jarosik, A. Kogut, M. Limon, S. S. Meyer, G. S. Tucker, E. Wollack and E. L. Wright, *Astrophys. J. Supp.* **208**, p. 19(October 2013).

53. J. Dunkley, R. Hlozek, J. Sievers, V. Acquaviva, P. A. R. Ade, P. Aguirre, M. Amiri, J. W. Appel, L. F. Barrientos, E. S. Battistelli, J. R. Bond, B. Brown, B. Burger, J. Chervenak, S. Das, M. J. Devlin, S. R. Dicker, W. Bertrand Doriese, R. Dünner, T. Essinger-Hileman, R. P. Fisher, J. W. Fowler, A. Hajian, M. Halpern, M. Hasselfield, C. Hernández-Monteagudo, G. C. Hilton, M. Hilton, A. D. Hincks, K. M. Huffenberger, D. H. Hughes, J. P. Hughes, L. Infante, K. D. Irwin, J. B. Juin, M. Kaul, J. Klein, A. Kosowsky, J. M. Lau, M. Limon, Y.-T. Lin, R. H. Lupton, T. A. Marriage, D. Marsden, P. Mauskopf, F. Menanteau, K. Moodley, H. Moseley, C. B. Netterfield, M. D. Niemack, M. R. Nolta, L. A. Page, L. Parker, B. Partridge, B. Reid, N. Sehgal, B. Sherwin, D. N. Spergel, S. T. Staggs, D. S. Swetz, E. R. Switzer, R. Thornton, H. Trac, C. Tucker, R. Warne, E. Wollack and Y. Zhao, *Astrophys. J.* **739**, p. 52(September 2011).

54. O. e. a. Zahn, *Astrophys. J.* **756**, p. 65(September 2012).

55. M. McQuinn, S. R. Furlanetto, L. Hernquist, O. Zahn and M. Zaldarriaga, *Astrophys. J.* **630**, 643(September 2005).

56. A. Mesinger, M. McQuinn and D. N. Spergel, *Mon. Not. R. Astron. Soc.* **422**, 1403(May 2012).

57. C. L. Carilli, R. Wang, X. Fan, F. Walter, J. Kurk, D. Riechers, J. Wagg, J. Hennawi, L. Jiang, K. M. Menten, F. Bertoldi, M. A. Strauss and P. Cox, *Astrophys. J.* **714**, 834(May 2010).

58. D. J. Mortlock, S. J. Warren, B. P. Venemans, M. Patel, P. C. Hewett, R. G. McMahon, C. Simpson, T. Theuns, E. A. Gonzáles-Solares, A. Adamson, S. Dye, N. C. Hambly, P. Hirst, M. J. Irwin, E. Kuiper, A. Lawrence and H. J. A. Röttgering, *Nature* **474**, 616(June 2011).

59. J. S. Bolton, M. G. Haehnelt, S. J. Warren, P. C. Hewett, D. J. Mortlock, B. P. Venemans, R. G. McMahon and C. Simpson, *Mon. Not. R. Astron. Soc.* **416**, L70(September 2011).

60. S. Cantalupo, C. Porciani and S. J. Lilly, *Astrophys. J.* **672**, 48(January 2008).

61. F. B. Davies, S. R. Furlanetto and M. McQuinn, *ArXiv e-prints* (September 2014).

62. P. Madau, A. Meiksin and M. J. Rees, *Astrophys. J.* **475**, p. 429(February 1997).

63. M. Valdés, C. Evoli, A. Mesinger, A. Ferrara and N. Yoshida, *Mon. Not. R. Astron. Soc.* **429**, 1705(February 2013).

64. S. A. Wouthuysen, *Astron. J.* **57**, p. 31 (1952).

65. G. B. Field, *Proceedings of the Institute of Radio Engineers* **46**, p. 240 (1958).

66. G. B. Field, *Astrophys. J.* **129**, p. 536(May 1959).

67. C. M. Hirata, *Mon. Not. R. Astron. Soc.* **367**, 259(March 2006).

68. G. B. Field, *Astrophys. J.* **129**, p. 551(May 1959).

69. K. Ahn, P. R. Shapiro, I. T. Iliev, G. Mellema and U.-L. Pen, *Astrophys. J.* **695**, 1430(April 2009).

70. L. N. Holzbauer and S. R. Furlanetto, *Mon. Not. R. Astron. Soc.* **419**, 718(January 2012).

71. D. Tseliakhovich and C. Hirata, *Phys. Rev. D* **82**, p. 083520(October 2010).

72. N. Dalal, U.-L. Pen and U. Seljak, *J. Cos. Astroparticle Phys.* **11**, p. 7(November 2010).

73. E. Visbal, R. Barkana, A. Fialkov, D. Tseliakhovich and C. M. Hirata, *Nature* **487**, 70(July 2012).

74. S. P. Oh, *Astrophys. J.* **553**, 499(June 2001).

75. S. R. Furlanetto, *Mon. Not. R. Astron. Soc.* **371**, 867(September 2006).

76. I. F. Mirabel, M. Dijkstra, P. Laurent, A. Loeb and J. R. Pritchard, *Astron. Astrophys.* **528**, p. A149(April 2011).

77. T. Fragos, B. D. Lehmer, S. Naoz, A. Zezas and A. Basu-Zych, *Astrophys. J.* **776**, p. L31(October 2013).

78. J. Mirocha, *Mon. Not. R. Astron. Soc.* **443**, 1211(September 2014).

79. A. Fialkov, R. Barkana and E. Visbal, *Nature* **506**, 197(February 2014).

80. F. Pacucci, A. Mesinger, S. Mineo and A. Ferrara, *Mon. Not. R. Astron. Soc.* **443**, 678(September 2014).

81. J. M. Shull and M. E. van Steenberg, *Astrophys. J.* **298**, 268(November 1985).

82. S. R. Furlanetto and S. Johnson Stoever, *Mon. Not. R. Astron. Soc.* **404**, 1869(June 2010).

83. A. Mesinger, A. Ewall-Wice and J. Hewitt, *Mon. Not. R. Astron. Soc.* **439**, 3262(April 2014).

84. A. R. Parsons, A. Liu, J. E. Aguirre, Z. S. Ali, R. F. Bradley, C. L. Carilli, D. R. DeBoer, M. R. Dexter, N. E. Gugliucci, D. C. Jacobs, P. Klima, D. H. E. MacMahon, J. R. Manley, D. F. Moore, J. C. Pober, I. I. Stefan and W. P. Walbrugh, *Astrophys. J.* **788**, p. 106(June 2014).

85. J. R. Pritchard and A. Loeb, *Phys. Rev. D* **82**, p. 023006(July 2010).

86. J. D. Bowman and A. E. E. Rogers, *Nature* **468**, 796(December 2010).

87. T. C. Voytek, A. Natarajan, J. M. Jáuregui García, J. B. Peterson and O. López-Cruz, *Astrophys. J.* **782**, p. L9(February 2014).

88. J. O. Burns, J. Lazio, S. Bale, J. Bowman, R. Bradley, C. Carilli, S. Furlanetto, G. Harker, A. Loeb and J. Pritchard, *Advances in Space Research* **49**, 433(February 2012).

89. P. A. Shaver, R. A. Windhorst, P. Madau and A. G. de Bruyn, *Astron. Astrophys.* **345**, 380(May 1999).

90. A. Datta, R. Bradley, J. O. Burns, G. Harker, A. Komjathy and T. J. W. Lazio, *ArXiv e-prints* (September 2014).

91. G. Paciga, T.-C. Chang, Y. Gupta, R. Nityanada, J. Odegova, U.-L. Pen, J. B.

Peterson, J. Roy and K. Sigurdson, *Mon. Not. R. Astron. Soc.*   **413**, 1174(May 2011).

92. G. Paciga, J. G. Albert, K. Bandura, T.-C. Chang, Y. Gupta, C. Hirata, J. Odegova, U.-L. Pen, J. B. Peterson, J. Roy, J. R. Shaw, K. Sigurdson and T. Voytek, *Mon. Not. R. Astron. Soc.*   **433**, 639(July 2013).

93. M. Zaldarriaga, S. R. Furlanetto and L. Hernquist, *Astrophys. J.*   **608**, 622(June 2004).

94. M. F. Morales, J. D. Bowman and J. N. Hewitt, *Astrophys. J.*   **648**, 767(September 2006).

95. M. McQuinn, O. Zahn, M. Zaldarriaga, L. Hernquist and S. R. Furlanetto, *Astrophys. J.*   **653**, 815(December 2006).

96. A. Liu and M. Tegmark, *Phys. Rev. D*   **83**, p. 103006(May 2011).

97. H. Vedantham, N. Udaya Shankar and R. Subrahmanyan, *Astrophys. J.*   **745**, p. 176(February 2012).

98. A. Datta, J. D. Bowman and C. L. Carilli, *Astrophys. J.*   **724**, 526(November 2010).

99. M. F. Morales, B. Hazelton, I. Sullivan and A. Beardsley, *Astrophys. J.*   **752**, p. 137(June 2012).

100. A. R. Parsons, J. C. Pober, J. E. Aguirre, C. L. Carilli, D. C. Jacobs and D. F. Moore, *Astrophys. J.*   **756**, p. 165(September 2012).

101. A. Liu, A. R. Parsons and C. M. Trott, *Phys. Rev. D*   **90**, p. 023018(July 2014).

102. A. Liu, A. R. Parsons and C. M. Trott, *Phys. Rev. D*   **90**, p. 023019(July 2014).

103. B. J. Hazelton, M. F. Morales and I. S. Sullivan, *Astrophys. J.*   **770**, p. 156(June 2013).

Discussion

J. Carlstrom Is there another integral constraint from the kinematical SZ effect due to patchy reionization where there are already strong observational limits?

S. Furlanetto I think it is potentially a very interesting probe. I think the big question now is about how much contamination there is from the cross-correlation between dust and thermal SZ at lower redshift. If that could be sorted out, I think we would end up with a very stringent constraint, and the way this works is basically that the velocity of the bubbles that last during ionization imprints this secondary fluctuation on the CMB. The simplest thing you could think at, out of this constraint, is the limit on the duration of reionization. If this limit is as stringent as it might be, then that would turn out to be a big problem. In light of the fact that the model Richard showed from the galaxy that we know, it is relatively hard to reionize the universe and that would take a long time and, given the high optical depth that we have from WMAP, you really need to stretch out reinoziation to make everything consistent and so, if you get some upper limit on duration of reionization, that is potentially very useful and I am very much looking forward to seeing whether ACT and SPT are going with that. I only neglect to talk about it for reason of time.

R. Wijers As a non expert I was puzzled by the relationship between your first two conclusions. It appears to me you are saying galaxy evolution is poorly understood but reionization is better understood. So I kind of miss what is the saving grace that makes you understand the reionization better when you don't understand the sources of your reionization?

S. Furlanetto Given a model for galaxies, you can predict how reionization should have unfolded relatively easily, at least until the very end stages of the process, and that is useful because then you can use the pattern of ionized bubble as a probe of the galaxy evolution process. We cannot predict from a theoretical standpoint when reionization would have occurred or what sources were responsible for but we can use reionization as a tool to understand the sources.

R. Blandford Do you think from the physics perspective, that clumping is relatively well understood, given all the uncertainties in the galaxy history?

S. Furlanetto I would agree that this is much less of a concern than for instance the escape fraction and solar populations. During the bulk of reionization these dense clumps that are hard to understand are not so significant, so the recombination is not very far above what you would get just with uniform medium. It is only at the very end stage when the transition starts to the normal Lyman-α forest that we think that at lower redshift becomes the hard problem and you have to worry about all these details I mentioned. But at the same time that is a very interesting problem because this is how you go from this exotic early phase to the Universe that we do understand. That transition is worth studying even if it is a small part of reionization.

Prepared comment

T. Abel: The First Stars are Massive

Over the past two decades firm predictions about the nature of primordial star formation in the pristine early Universe have been made. Numerical algorithm development and particularly the advent of adaptive mesh refinement for the use in cosmology have enabled studies of unprecedented dynamic range. We now have a solid appreciation of the interplay of chemistry, cooling, hydrodynamics, and gravity in forming the objects that host the very first stars. Figure 1 gives representative results and highlight the complex interplay of the relevant physical processes. The inefficient H_2 cooling which keeps the high redshift analogues of molecular clouds at about 200 Kelvin rather than 10 Kelvin in nearby star forming regions fundamentally alters how star formation proceeds and in particular leads to very high accretion rates onto the first protostars. The other key aspect of H_2 cooling is that it leads to a density independent cooling time at densities above 10^4cm^{-3} which leads to a slower collapse as compared to present day star formation. Very massive stars with masses between 30 and 300 solar masses may be formed[a] where the range is to be understood as a theoretical uncertainty encapsulating our inability to calculate much further after the first photo-star forms as the timescales of the system prohibits one to calculate a significant fraction of the time over which the protostars grow. In simulations that capture a few years of the tens of thousand years of the expected accretion time[b] some very lumpy accretion is found. At the same time central objects grow at just the rates seen in the earlier AMR simulations. The excitement now is about the question of whether the solar mass lumps formed in the massive accretion disks around the very massive first stars could be flung out and whether some lower mass stars can form and survive. Observational searches have not found any pristine metal free stars in our Galaxy consistent with the picture that all primordial stars are massive. However, whether a small mass fraction of material participating in primordial star formation could leave a few low enough mass stars to survive and not be enriched by accretion and nearby supernovae certainly should motivate further searches of the Milky Way's stellar populations. The fossil record may well hold important clues about structure formation in the very early Universe.

Discussion

S. White This is a very nice coherent story which is now being developed over more than 10 years, with many ratifications from many beautiful simulations.

---

[a]Tom Abel, Greg L. Bryan, Michael L. Norman 2002, The Formation of the First Star in the Universe, Science, 295, 93A.

[b]Thomas Greif et al. 2012, Formation and Evolution of Primordial Protostellar Systems, MNRAS, 424, 399.

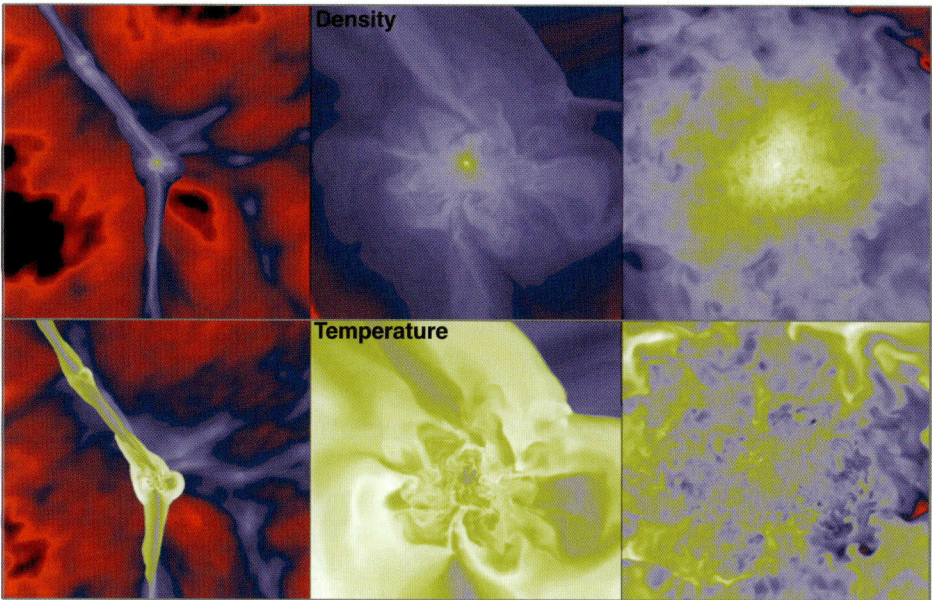

Fig. 1.   Cosmological adaptive mesh refinement simulations of the first objects in the Universe. The left column shows both density (top) and temperature (bottom) for a thin slice through a hundred thousand solar mass object at kilo parsec scales. The middle column zooms in to about 200 parsec on a side and the right column focuses in further by another factor of hundred smaller in scale. The mini-halos are bounded by viral shocks with approximately a thousand degree Kelvin post-shock temperature while in the very center a cooling flow develops with gas cooling to approximately 200 degree Kelvin by rotational and vibrational excitation cooling of molecular hydrogen.

That is all very convincing. One of the reasons it is so convincing is that there is no observational evidence to support it or confuse it. We hope eventually we will be able to see this but there is at least one area which is relevant where there has been a great deal of observational advances, that is in studying the metal poor stars and the hope was looking at the metal poor stars in our galaxy we would see the abundance patterns produced by population III stars. We now find stars down with less than a billionth of the heavy elements of the Sun at least for the elements like iron and they have very specific abundances. My impression is there is still no real link between the particular abundances that are found and the prediction of the first stars and perhaps to some extent this is because the predictions from the first stars are not yet very precise for its properties but I think this is an interesting avenue. I would be interested to hear what are the prospects for proceeding in this way because seeing the objects directly is not so easy.

R. Ellis I think this is a very important point. People have looked at the abundance variation in nearby dwarf galaxies, very old objects where there were no star formation for a very long time. And you are right that the patterns of enrichment from say pair-instability supernovae have been predicted and compared with those abundances patterns and they do not agree. That does not necessarily means a discrepancy between early enrichment if the bulk of the enrichment was done in an enriched environment itself. This comes back down to the point of how quickly does an object self-enriched and what is the abundance and mass of material enrichment that comes from normal SN.

S. White I agree with that but I think it is clear that the abundance found in this extremely metal poor stars are not those from normal SN, so it is from something strange and I think we have not sort out yet what these objects were. These objects are extremely old, so they could plausibly have been enriched in the first phase but we cannot be that precise.

T. Abel I think it is interesting that these extremely metal poor stars were in fact all carbon rich and oxygen rich and it is only just now that we are finding stars where the total metal content is really small, 10^{-4} or so. It is particularly fascinating to think where the 10^{-4} total metallicity stars formed because from the calculation, even with non pair-instability SN, with just regular SN only giving one solar mass of mass yield, these small first objects are enriched to a higher level, $10^{-2} - 10^{-3}$ right away. So it becomes an even stronger puzzle actually how 10^{-4} ones could be made. This very simple naive version of a closed box chemical evolution, where the lowest metallicity stars are the next stars that form after metal free ones, it can't possibly hold.

M. Rees I just would like to mention the effect pointed out by Hirata et al which is relevant to some of this. They pointed out that at recombination the baryons would be left from the streaming motion relative to the DM of a few kilometers per second and that velocity is enough to affect the efficiency of the infall of baryons into the first mini-haloes. That therefore has the interesting effect that the formation of the 1st generation of population III stars in the mini-haloes would be modulated on a scale which is actually very large, the scale of baryonic acoustic oscillations. This is slightly good news for high-z 21 cm, because if there is any fluctuation in the heating rate at high redshifts, that would be modulated on this very large scale even though the overall $\delta\rho/\rho$ on that very large scale is very small at high redshifts, so that's slightly good news for 21cm.

M. Zaldarriaga Indeed, there has been a lot of efforts trying to make predictions including these kind of effects for 21cm and also trying to see if there is any remnant of this in the clustering of later galaxies.

R. Wijers One interesting point, coming back to the GRB and metallicity discussion, even though the selection of sight-lines by GRB is of course very unbiased, one thing that is striking in this metallicity question is that we have been finding that, in the nearby to medium distant Universe, they prefer low metallicity regions but then if you go to higher redshift, the metallicity refuses to go further down. It seems to stick at about 10^{-2} solar. So even though these are presumably among the most massive objects, therefore the first ones to explode, it seems that they come from such intense star forming regions that there has always been already some enrichment. So once again you may see a trend where it is very hard to see very metal poor things.

G. Efstathiou I just would like to make the remark about the constraints on the optical depth from CMB. These are very difficult measurements and I did not really appreciate just how difficult they were until we tried to do that with Planck. A reasonable thing to do is to take WMAP as an upper limit. And it is probably wise to do the same for Planck when Planck comes out. If you do that then I think that you cannot exclude a model where reionization comes late and very quickly and then a lot of these tensions just disappear.

M. Zaldarriaga Indeed, this is a very interesting topic. Probably David Spergel has something to say about how to interpret WMAP measurement. I think it would be unfortunate if all what we get from this kind of measurements is an upper limit.

Prepared comment

G. de Bruyn The LOFAR EoR Project

The LOFAR EoR project is one of several projects that aim to make a statistical detection of the redshifted 21cm line signals from the young Universe, and in particular the evolution of its power spectrum with redshift. Observations with LOFAR, a phased array consisting of 62 stations with each 384 dipoles, centred in the Northern part of the Netherlands began in earnest in December 2012. Thusfar we have acquired about 600 hours of integration on each of two windows. The amount of data is staggering (2PB) and we have a large team of people working on various tasks, from calibration and imaging to simulations of signals and instrumental errors. We are partly guided here by theory, efforts that are coordinated by my colleague Saleem Zaroubi, also present at this meeting. Our initial efforts have concentrated on a field centred on the North Celestial Pole. It is a lovely field that, as seen from the 53° geographic latitude of LOFAR, can be observed, in principle, at good sensitivity every night of the year ! The feeble signals, however, are swamped by those from the foregrounds which, even on the scales that we can probe well, from 3–30 arcmin, are at least 1000 × brighter. It is only thanks to their assumed spectral smoothness that we have a chance to detect signal from the EoR.

On the one slide that I am allowed to show there are various images and tables. They summarise the issues that we are up against and I will guide you through some of it. The starting point for this brief tour is a low resolution image (see Figure 1). It is confusion limited in the inner 10°, where the stations have good sensitivity. Many thousands of the brightest discrete sources, radio galaxies and

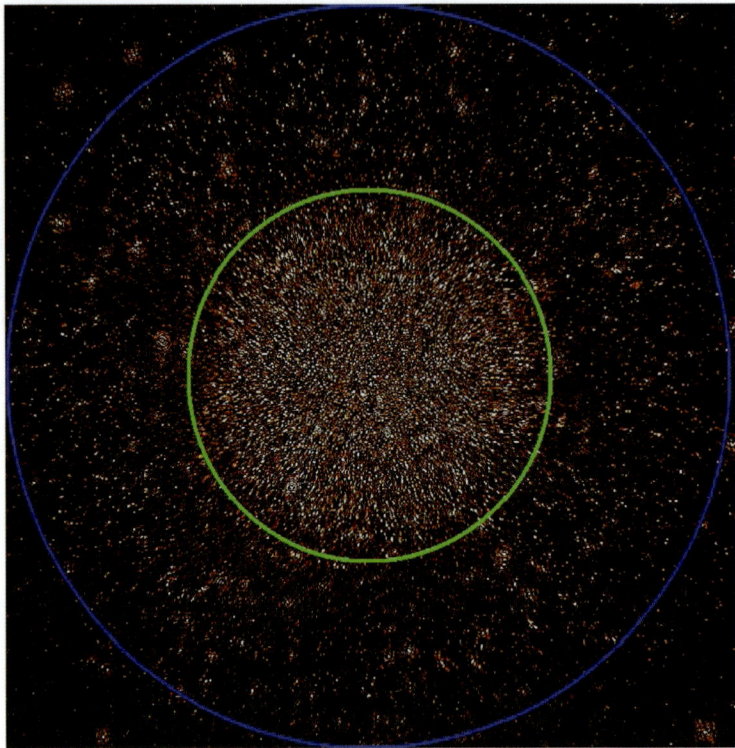

Fig. 1.   A low-resolution wide-field LOFAR image of the 115–175 MHz emission from the North Celestial Pole, after removing thousands of bright sources. The green and blue circles have diameters of 10° and 20°, respectively. Many of the brighter sources at the edges of the field are distorted and require direction-dependent calibration to be properly removed.

quasars, have already been subtracted from the data. The point I want to make however, in showing this image, is that there are also thousands of sources well beyond the first null of the station beam. In fact sources can be detected from all over the sky and their complicated frequency-dependent sidelobes are well above the signals we are interested in. Our approach is to remove these sources, and their sidelobes, as best as we can. This obviously requires higher resolution data which we also have available.

Although the redshifted HI signals are detectable only on the shorter baselines we use the long baselines to model and remove the multitude of bright discrete sources. In addition we use the long baselines to calibrate the stations within the LOFAR core. That is, we use 'additional' information to remove calibration errors without corrupting the desired signals that are buried in the noise. We also detect diffuse Galactic emission. In total intensity this emission appears to be weaker than that of residual discrete sources. Besides foregrounds, and their sidelobes, there are at least two other complications that effect the final data quality. Long baselines are effected by rapid ionospheric phase variations which make the apparent sky change within as well as between nights. A potentially more complicated effect is due to the diffuse polarisation from the Galaxy. Its strength is higher than expected (several K in brightness temperature) and if one cannot properly correct for instrumental polarisation (leakage) the polarized signals could contaminate the total intensity emission where the HI signals reside. Thusfar we have not been limited by these effects, or so it seems. The end result, after calibration and source removal, will be a 3-D cube of images as a function of frequency (typically 300 frames × 0.2 MHz). The remaining frequency-smooth foregrounds are then filtered making use of the fact that HI signals have a correlation length in frequency which is less than a few MHz. Once this filter has been applied we can form power spectra, either from the images themselves or from the sky-subtracted visibility data.

The total frequency range covered by LOFAR, from 115–185 MHz, will allow us to set limits on the signal between redshifts of 6.5 to 11.5. Preliminary results based on 170 hours of data have been obtained for redshifts ranging from 7.8 (0.5) 10.3 and on angular scales corresponding to scales of 0.03 –0.3 Mpc^{-1} suggest that we are still a factor of 5-10 above the expected signals. Part of this large factor is due to systematic errors which we are slowly beginning to understand. With new data to come in the seasons ahead, improved calibration algorithms as well as new processing resources (factor 5–10 faster) we expect to get at least a factor 2–3× deeper in 2015 which will bring us close to a detection.

Discussion

M. Zaldarriaga Perhaps we can also hear from Ue-Li Pen who has been involved on some of the current constraints from the 21cm and he could tell us a bit about the other efforts that are going on.

U. Pen I think you did a very nice job summarizing the ongoing experiments including GMRT, PAPER and others. I would like to take a slightly broader view on the experiments that are there and upcoming, in particular Steven already alluded to the basic picture that is emerging theoretically and we just heard from Gerr about the very nice progress on the redshift ∼ 10 observations of the 21cm structure. The challenge, as we are all appreciating now, is the very bright foregrounds, factors of order ten thousands times bigger new signals, anticipated to be spectrally smooth, so in principle feasible, but of course it is a very challenging process. This ratio of signal to foreground, actually have three sweet spots, so the one we have been hearing about is a redshift 10 where of course the Universe has this big patchy contrast where we hope to see these bubbles. We also saw on Steven's plot that at redshift 20, when we saw the evolution in this movie, that at redshift 20 there were actually the biggest signal in that simulation because in that simulation the Universe was in absorption and the 21cm line in absorption can have much more optical depth and much more contrast in temperature at redshift 20, if there was such a phase of Lyman-α pumped absorption before the X-ray heat brought lots of uncertainties. But potentially it's sweet spot and as Martin reminded us there is this Tseliakhovich/Hirata effect which again may, especially at this high redshift like 20, it's not clear that there is really a smoking big signal because of that and there are other propagation effects that the mean free path of photons may or may not preserve the intrinsically large modulation of mini-haloes, but there is a theoretical possibility that the signal would be much larger because the regions could be large, so there is a potential and if that it is the case there are other telescopes that exist or nearing completion like the FAST telescope in China that in the 70 MHz might probe into this higher redshift 21cm window. One could think about maybe even Arecibo, very challenging in that TV band, so we should think very broadly here that there are theoretical opportunities to probe structures in different sweet spots. One more sweet spot that is not cosmic dawn is redshift 1 intensity mapping where we already have the same foreground challenge and have made already substantial progress, making detection and building telescopes.

S. Kulkarni A technical question. It is not obvious to me why using a single dish telescope like FAST is advantageous in this game?

U. Pen For the challenge of removing foregrounds against 21cm, the very best case to date of a real detection is the Greenbank telescope at redshift 1, where we did achieve this 10000 to 1 dynamic range. I agree with you that in searching what happened we also thought of the GMRT (Giant Meterwave Radio Telescope) that did not achieved that. Interferometers have obviously the advantage that they are easier to calibrate but, at the end the fact is that sideline structure is much cleaner in single dish. You do not have there this oscillating structure that mixes spatial and frequency modes and that is what makes this analysis on a single dish much easier. If you had the choice to have a 1 km size dish that would be probably much easier to deal than with an interferometer.

G. de Bruyn I think I would always go for an interferometer array, of course with large filling factor a single dish of a kilometer would be great, but I would like to point out that that concept is almost like a filled array, but interferometer have so much advantages over single dishes even though you talk about 10000 to 1 spectral dynamic range. Doing that at low frequencies, where your sidelobe structures are much higher there, they easily reach 30 db rather than 40 or 50 dB, I don't think it would be easy, the kind of thing that you do at redshift 1, to do it at redshift 10.

S. Mukhanov Could you tell about future perspective? What do you expect to get from 21cm for cosmology because there were talks that you can make better primordial non-gaussianity than with CMB from 21cm, so until which range of z can we go and how accurately would you do it along the most optimistic expectations?

M. Zaldarriaga I am sure there would be a lot to say about that, but the chance that what we would say would be true is pretty small, so let us shift the discussion to some of the issues related to the CMB. What a year it has been for the CMB! We went from a 7 sigma clear detection of B-mode polarization to an upper limit on tau in a couple of months. It would be nice to get some perspective on how much we might expect τ from WMAP to be shifted around and also on the other source of information about CMB from secondaries.

D. Spergel I mostly know what I just heard from George but I will also add some comments from Planck papers. One of the many things that Planck has taught us is typically that 353 GHz maps give us a much better picture of what dust polarization is doing. With WMAP 9-year result our best value was about 0.088 $\pm$ 0.013 and the statement in the Planck paper is a shift down by about a sigma when corrected by 353 GHz and to me that seems very plausible and it actually goes in the direction of making a lot of pieces fit together better. If you look at the amplitude of fluctuations measured from clusters or from large scale structures you won't tend to get values of σ_8 around .79-.80. With the standard WMAP optical depth from the Planck results you get a value of σ_8 of about 0.82. So the optical depth coming down by 1-1.5 sigma brings those into a limit. Also looking at Richard's plot you got to look at the lower end of the WMAP optical depth, if you pull down by a sigma or so the pieces fit together better. So we will know more I hope in December when we see the Planck results but I would not be surprised by an outcome where the optical depth shifts down to say 0.07. That said I think this shows the importance of trying to make precision measurements of optical depth and I think one thing I am excited about for the future is the possibility that NASA in the next explorer round will select a mission like PIXIE. To remind you what PIXIE would do is make a low l polarization measurement and going after GW at the low multipoles, with an experiment with effectively 400 frequency channels to give us very good control on foregrounds, so to be able to measure the μ and y distortions that Richard mentioned briefly. That would tell us also a great deal about the early Universe back to redshift about a million with the μ distortion, look for signals of GW and give us really a precision measurement of optical depth through an experiment designed to go after low l polarization, and to be able to get better constraints on the reionization history that way. And I think one of the things we have seen on the progress side in the last year is the BICEP measurement has shown that ground based experiments can make precision measurement of the sky at fairly low multipoles, at $l \sim 40$ there is a very nice Bicep measurement and regardless of the interpretation those are really precise measurements, we must congratulate them. At the same time POLARBEAR, SPTpol, are all making intermediate to high polarization measurements from the ground, so we can perhaps delegate to space the responsibility to make these very low l measurements and think about ways with relatively small mission we could go after those low l polarization features in combination with ground based measurement and have the opportunity for the reionization question of having really precision constraints on optical depth and better constraints on reionization history and also the ability, if r is greater than 10^{-3}, to get constraints on how to measure the B-modes at those scales.

M. Kamionkowski So if one experiment does a measurement and another experiment does another one and they differ by 1 sigma it is not a big deal but Planck and WMAP are looking at the same sky and a lot of the information on tau comes from low l and I am wondering if we know whether this one sigma shift is in the noise or if something comes from high l but Planck gets something WMAP does not get, or whether the measurement at low l is different.

D. Spergel For polarization, unlike temperature, WMAP is completely noise limited, it is not cosmic variance (except maybe for a few of the m's), and I think we will see what the Planck results are but some of this is also what the dust polarization correction is at 60 GHz. We had thought that the crossover between dust and synchrotron is closer to 90, maybe a little bit higher, but the dust polarization is higher than that, so the crossover looks like to be at about 70 GHz, that makes a bigger contribution than we had estimated.

G. Efstathiou I wanted to pick up your introductory comment Matias. The BICEP polarization measurement are targeting recombination bumps in the BB spectrum at multipoles between 50 and 200 or so. The tau measurements come from a very restricted set of multipoles, 2 to 7. By multipole 8 there is nothing there. And so the issue I think is not foreground or noise but has to do with instrumental systematics. It is not related to BICEP.

J. Dunkley To come back to Mark's question. We reported in the Planck paper of last year that if you clean up the WMAP map with the Planck dust map, then it can shift the τ value by about 1 sigma.

M. Zaldarriaga An interesting question is if we could measure τ in a completely different way, by direct tracing of reionization?

E. Komatsu I would like to comment on the issue of systematics. For WMAP it is a relatively clean measurement and I think it is doing a better job than Planck. Noise is much lower for Planck, but for systematics WMAP is cleaner. As David said, for our τ value, I think the main source of uncertainty is indeed dust. The best approach to me is to use Planck's dust map, to clean WMAP. That would be the most robust measurement of τ.

J.-L. Puget I do not agree with that. To explain a bit better what George alluded to, we are going to get, on the very low l and for the τ measurement, certainly a measurement that will have an uncertainty that will be at least 2 or 3 times better than WMAP. Regarding systematics vs foreground, when we do the cleaning on the dust, on the low l, when we look at sky cuts, we see residuals when we go to 70 or more %. Below that it is very stable. So from that point of view, we know already that we can clean the dust and synchrotron in the Planck data to a very good level. We are limited by systematics but, it is not a great secret, we will release something in November, thus on a very short time scale. We might start with an upper limit on tau, but eventually we will have a much better measurement than WMAP.

E. Komatsu I am confused. I think you do agree with me that at the moment the limiting factor for Planck is systematics and the WMAP limiting factor is dust polarization. I think that is a true statement.

R. Bond I was going to point to Jean-Loup Puget to give the official word on the state of Planck's Compton depth τ, and he and George Efstathiou have now done this; the τ issue is all very exciting obviously. I would like to make another two points concerning τ. You may know the saga of the SPIDER balloon experiment. We are going to fly this year and our target is the tensor-to-scalar ratio r. We were going to have a test flight a number of years ago which never materialized, on which we wanted to go for τ on a balloon and try to do as well as Planck. So there is a possibility of having an intermediate set of experiments, long before a new satellite experiment like PIXIE, if τ is not fully resolved with Planck. My second point is that, having personally been thinking for over 3 decades reionizing the universe theoretically, I would like to express a little bit of the frustration we have had because, worse even than dealing with the CIB, there are so many dials that we have with the theoretical modeling. At some level, it was expressed today that there was some sort of an attractor, with the theoretical community was in very good agreement about the outcome of the simulations, but really that's largely because it is doing the same thing, in my view. That is to say, the attractor is human driven and not driven by the science. As we confront the emerging period of great data, it behooves us to have a good theoretical prior to begin with, namely that we have a good assessment of the wide range of possibilities. This implies we need a wide spread of mock simulations. What we have already seen is good, but I think the practitioners would admit we are only scratching the surface and ultimately the only way we will be able to go forward is through a very intense crosstalk between the emerging data and the simulations because the astrophysical and radiation environment of the early universe in the first reionization epoch is just mind bogglingly complicated. Issues include whether there is a first reionization, then partial or complete recombination, then another burst of ionization, etc. These are just some of the possibilities among many scenarios that have been on the table for a very long time and they remain possibilities today. So what is the way forward? For one thing, we are tuning or calibrating semi-analytical calculations, sometimes using N-body calculations and sometimes using approximate methods, with 3D radiative transport simulations, which is somewhat problematic since what we are dealing with is extra-galactic HII regions in a very complex environment about which we know little and are non-trivial to treat numerically. The calculational advances have been tremendous, but we have been doing 3D radiative transfer for a fairly short time so the cross talk between the calibration of N-body/semi-analytic with something without a totally solid foundation frustrates our ability to deal well with the theoretical uncertainty. The corollary is one of great long-term promises for our subject, with a lot of work has to be done, especially by the new young people entering theory helping the cause. That is, we have a big future in reionization work, not just experimentally, but definitely also theoretically.

Prepared comment

P. Madau: Intergalactic Absorption as a Probe of Cosmic Dawn

We have heard some beautiful talks about cosmic microwave background radiation data, which yield a detailed portrait of the Universe when it was less than 400,000 years old. Following recombination, the ever-fading blackbody radiation cooled below 3000 K and shifted first into the infrared and then into the radio, and the smooth baryonic plasma that filled the Universe became neutral. The Universe then entered a "dark age" that persisted until the first cosmic structures collapsed into gravitationally-bound systems, and evolved into stars, galaxies, and black holes that lit up the Universe again. The history of the Universe during and soon after these crucial formative stages is recorded in the all-pervading intergalactic medium (IGM), which is the dominant reservoir of the baryonic material left over from the Big Bang. Throughout the epoch of structure formation, the IGM becomes clumpy and acquires peculiar motions under the influence of gravity, and acts as a source for the gas that gets accreted, cools, and forms stars within galaxies, and as a sink for the metal enriched material, energy, and radiation they eject.

Observations of absorption lines in quasar spectra at redshifts up to 6.5 have provided invaluable insight into the chemical composition and thermodynamic state of intergalactic gas and the primordial density fluctuation spectrum of some of the earliest formed cosmological structures. They have also challenged our understanding of the cosmic reionization process:

- A residual neutral fraction is detected in quasar spectra at both low and high redshifts, revealing a highly fluctuating medium with temperatures characteristic of photo-ionized gas. The statistics of the fluctuations are well-reproduced by numerical hydrodynamics simulations within the context of standard ΛCDM cosmological scenarios. The absorption lines predicted by these simulations appear to be substantially narrower than measured, however.
- The detection of metal systems within the diffuse IGM shows that it was enriched by the products of stellar nucleosynthesis early in its history, demonstrating an intimate connection between galaxy formation and intergalactic gas. The details of this enrichment process are still not firmly established.
- Transmitted flux is observed in the Lyman-alpha forest up to redshift 5 or so, suggesting that re-ionization had largely ended by that point. The data, particularly near $z \simeq 5.6 - 5.8$, require fluctuations in the volume-

weighted hydrogen neutral fraction that are a factor of 3 or more beyond those expected from density variations alone. These fluctuations are most likely driven by large-scale variations in the mean free path to ionizing radiation, consistent with expectations for the final stages of inhomogeneous hydrogen reionization.

– The inferred nearly constant value of the hydrogen photoionization rate over $2 < z < 5$ appears to be in stark contrast with the rapidly evolving quasar and star-forming galaxy ionizing emissivities over the same redshift interval.

– The thermal history of the IGM, reflecting the timing and duration of reionization as well as the nature of ionizing sources, remains poorly understood. Measurements of the IGM temperature evolution from redshift 4.5 to 3 are not consistent with the monotonic decrease with redshift expected after the completion of hydrogen reionization. The observed increase in temperature over this interval requires a substantial injection of additional energy, perhaps from the photoionization of singly ionized helium.

Despite a lot of effort, a complete, satisfactory theory of hydrogen and helium reionization in the Universe is still lacking. Hopefully, before the next Solvay Astrophysics and Cosmology meeting, new advances in observations and theory will produce a complete description of the evolution of the IGM over cosmic history.

Discussion

G. Efstathiou There has been some controversy in the literature about the equation of state of the galactic medium and whether there is a temperature jump at redshift 3. What is your view on that now?

P. Madau Some people have inferred an increase of temperature of the diffuse IGM at $z = 3$, which is usually explained as extra heating from radiative transfer effects during helium reionization. That's ok and I think you can reproduce that. Now the only problem is that the same simulation that produces that increase in temperature also produces an optical depth to Lyman-α scattering of HeII which is increasing rapidly as you go back in redshift, say 2.5 and 3 and recent data, from HST show that optical depth is in fact not changing as fast as simulations would predict. So there seems to be some tension between the increase in temperature that you infer from the hydrogen forest and the relatively flat evolution of the HeII optical depth that is seen in HST data.

M. Kamionkowski I remember a few years ago there was some prospect to use neutral oxygen on absorption to measure the ionization fraction. Is there any progress on that?

P. Madau I am not aware of that.

S. Furlanatto The basic idea is that OI and HI are in charge exchange equilibrium, so one is neutral, so is the other one. There have been low ionization metals discovered at redshifts up to 6 but they don't appear to be a good probe of reionization, they seem to be a good probe of small galaxies. They are very interesting. There have huge fluctuations in their abundance along line of sight and it is not clear how the CII and CIV are related but so far the picture is more consistent with a continuation of the metals that are in the Lyman-α forest rather than something that is related to reionization.

E. van den Heuvel It is just because of my ignorance but I was wondering about the source of the ionizing radiation, which are used in these models. Are those single star populations, because recently it has become clear that basically all massive stars are binaries or triples so they are multiple, and that changes the evolution of stellar populations very dramatically. You produce for example lots of Wolf-Rayet stars, helium stars, which have much more UV radiation, you produce lots of X-ray binaries, so I am wondering what sources of radiation are used for calculating this reionization?

R. Ellis So this is a very important point. Our understanding of the evolution of the top of the main sequence is, as you say, is evolving, and we are learning a lot more about binary stars and systems that would stay on the main sequence longer. So the answer to the question is that the calculations that we do to estimate the contribution of ionizing photons do not take into account these effects and so in some sense young stellar populations, like ones that we are inferring exist in redshifts 7 and 8 galaxies, are likely to underestimate by some factor the ability to generate ionizing photons. So the code that we need to implement should take this into account.

C. Kouveliotou What is then the important mass range for you in your assumption of the IMF (initial mass function)? Is it the very massive sources or is it more the intermediate mass range, where stars are probably much more numerous?

R. Ellis Yet again, that is another assumption that we have to make about these early populations, is that the initial mass function is not different from the one we observe in star forming galaxies today, but you are right, there is an uncertainty. If the IMF was skewed to high mass stars in very low metallicity systems then again, this would be helpful, it would generate more ionizing photons than we actually calculate in our models.

P. Madau But there is no evidence from the colors of those high-redshift galaxies that they require any fancy popIII or top-heavy IMF, isn't?

R. Ellis That is right. And in fact, in one of my slides I showed as an illustration how the colors of these galaxies are consistent with having relatively normal IMF but of course there are so many variables here, there is dust extinction, etc, so there is still a lot of ambiguity in the interpretation of the colors.

M. Rees It is widely said that the early population of pure hydrogen-helium stars and then the IMF may change when heavy elements cooling comes in. I wonder if people have thought enough about how the first heavy elements diffuse, because it could be that they do not diffuse very efficiently and therefore there could be rather more pristine high mass stars forming than you would imagine. It seems to be assumed that the heavy elements will diffuse within the mini-halos and that doesn't seem to me to be a very reliable assumption. I wonder if you could comment on that?

T. Abel Mixing generically in these calculations is something that we just start to appreciate as a community. Take some of the old calculations that I showed today. Ten years later we can afford a 100 times more resolution and we start noticing some fluid instability directly in the calculations. My feeling right now is that we are learning that we generate magnetic fields very rapidly on small scales and they will have a play to say of how mixing will work. We are certainly going in this direction as our numerics will finally have Reynolds number sufficient to start capture the turbulence and fluid instabilities we anticipate so that we can describe mixing. Every indication currently is that mixing is indeed quite fast.

M. Zaldarriaga We can now switch to discussing some of the effects of reionization at low redshifts.

Prepared comment

C. Frenk: Theoretical Insights into Reionization from Gasdynamic Simulations

We know that Hydrogen must have been reionized some time between the epoch of (re)combination and the present. This process is one of the remaining major unsolved problems in studies of galaxy formation: when exactly, how and by whom was the Universe reionized? These are difficult questions to answer because we have as yet no observations of photons produced when the Hydrogen was still neutral. The best constraints on the epoch of reionization come from polarization measurements of the cosmic microwave background (CMB) radiation and indicate that Hydrogen was reionized at around $z_{re} \sim 11$.[a]

Observers are naturally wont to extrapolate from the objects they think they can see at redshifts 6-10 and speculate whether they might have identified the population responsible for the production of the bulk of the UV photons that reionized the Universe. There is no convincing answer as yet. While we look forward to deeper observations (possibly with the next generation of telescopes), simulations can serve as a useful guide of what we might expect to find. Such simulations need to follow the evolution of the gaseous component of the universe under the gravitational action of the dark matter and must include the panoply of processes known to play a role in galaxy formation: radiative cooling of gas, star formation and evolution, the formation of supermassive black holes, feedback effects on the cooling gas produced by energy returned during stel-

---

[a]The Planck Collaboration (2014, A&A 571, A16) quote $z_{re} = 11.1 \pm 1.1$.

lar evolution (such as supernovae explosions) and AGN phenomena, radiative transfer, magnetic fields, etc.

Hydrodynamic simulations of galaxy formation in a cold dark matter (CDM) universe have a long history but the latest generation of simulations have achieved an impressive level of realism. An example is the "Evolution and assembly of galaxies and their environment" (Eagle) project carried out by the Virgo Consortium.[b] These simulations follow galaxy formation in volumes of up to $(100\text{Mpc})^3$ and include all the processes mentioned above, except radiative transfer and magnetic fields. The parameters of the simulations were fixed by requiring as good a match as possible to the local galaxy stellar mass function and to their distribution of sizes. Without any further adjustments, the simulations reproduce well the observed evolution of the galaxy stellar mass function out to redshift $z = 7$ and, somewhat less accurately, the star formation history out to that redshift.[c]

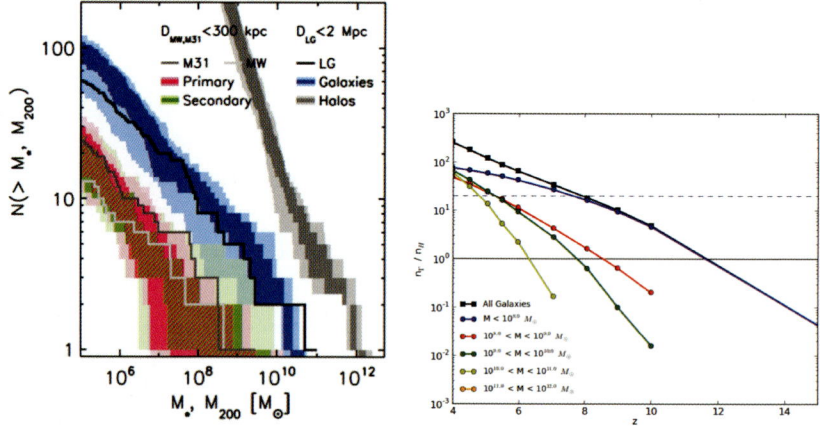

Fig. 1.   Left *(from Sawala et al. 2014, arXiv:1412.2748):* stellar mass functions of satellites within 300 kpc of each of the two main Local Group galaxies (red and green) and of galaxies within 2 Mpc of the Local Group barycentre (blue) averaged over twelve simulations of the Local Group. The dark colour-shaded areas bound the 16th and 84th percentiles; light shaded areas indicate the full range. The grey area corresponds to the mass function of all dark matter halos within 2 Mpc. Lines show the *measured* stellar mass function of the satellites of the Milky Way (light grey) and M31 (dark grey), and of every known galaxy within 2 Mpc from the Local Group barycentre (black). Right *(courtesy of M. Furlong):* number of ionizing photons per H atom produced by stars as a function of redshift in the Eagle simulation. Different colour lines correspond to different galaxy mass ranges.

In CDM simulations there is an interesting interplay between the physics of the small galaxies whose radiation reionize the Universe and the physics of present

[b]Schaye, J. et al. 2015, MNRAS, 446, 521.
[c]Furlong, M. et al. 2014, arXiv:1410.3485.

day dwarf galaxies such as those that have become satellites of the Milky Way. There is no direct evolutionary connection between these two populations (the descendants of the high redshift galaxies responsible for reionization are preferentially found today in clusters of galaxies) but rather an indirect connection through the process of supernovae feedback. Supernovae feedback is inimical to galaxy formation: by blowing out gas, it inhibits the formation of galaxies in small mass halos (or, more precisely, in halos with shallow potentials).

Efficient supernovae feedback is essential to ensure that only the largest of the huge number of subhalos present today in the halo of galaxies like the Milky Way succeed in making a galaxy since only a few tens of satellites are observed. But by suppresing galaxy formation at high redshift, efficient supernova feedback can kill off the sources of UV photons needed to reionize the Universe! With the high redshift of reionization inferred from the Planck CMB data, this becomes an acute problem.

As shown on the left panel of the figure, the Eagle simulations reproduce the luminosity function of satellites measured in the Milky Way and Andromeda, as well as the luminosity function of dwarfs in the Local Group neighbourhood. This is a remarkable success of these simulations. But the flipside is that rather extreme conditions are then required for enough UV photons to be produced to reionize the Universe by $z \simeq 11$. This is illustrated on the right panel of the figure which gives the number of ionizing photons per Hydrogen atom produced by stars forming in galaxies of different mass at different redshift. How many of these photons manage to escape the galaxy in which they are produced so they can be deployed to reionize the Universe (the "escape fraction") is not known. And, as a rule of thumb, typically one photon is wasted due to Hydrogen recombinations in the host galaxy.

If the escape fraction has a value around 0.1, as is often assumed, then, according to the figure, the Universe would only be reionized by $z \simeq 8$ (after taking into account the factor of two due to recombinations). A much larger escape fraction, above 0.5, is required to reionize the Universe by $z \simeq 11$ (with most of the photons coming from galaxies with $M_* < 10^8 M_\odot$). Whether such large values are attainable in practice is unclear, but the simulations show that there is little wiggle room if both reionization and the abundance of Milky Way satellites are to be explained in a CDM universe.

Discussion

M. Zaldarriaga Of course for a long time this question about the number of faint galaxies around us has been a big source of willingness to modify the ΛCDM and add all kinds of bells and whistles to the dark matter sector.

C. Frenk I think that those are very simplistic views that ignore all the complexities of galaxy formation. Actually they are not that complex, it is just reionization that suppresses the efficiency of galaxy formation in these small halos, together with supernovae feedback.

R. Wijers Again I am not an expert in galaxy formation, but I know something about stars. I was interested to hear this comment that apparently we now no longer believe that the high redshift IMF needs to be top-heavy. Could somebody maybe precise a little bit why that has changed? What has changed in the data? And perhaps there is still a large margin of uncertainty here?

C. Frenk I think that for reionization the largest uncertainty is the escape fraction for the ionizing radiation and so you can change the IMF and tune it to have more UV radiation produced but you still don't know what is the fraction that is going to escape. So I think that this and the fact that there is no evidence in the core of the high redshift galaxies for different IMF has moved the community away from trying to explain reionization with strange population III or other things. But it does not mean that it is not happening.

J. Ostriker Just a comment. As practitioners of this game of making galaxies we just say how difficult it is to do, especially at the low mass end. At the high mass end, you do not seem to need many extra wheels and knobs to make things that look more or less like galaxies. But at the low mass end, where the feedback from star formation is extreme in driving the matter out of halos. What you calculate is highly uncertain. Of course if you normalize it to the observations, then you get out what you put in but I would say from looking at things that we do not know very well how to deal with the low mass end right now.

C. Kouveliotou I just would like to go back to a question that might be relevant here, which is the question of star formation rate as estimated from gamma ray bursts. I wanted to make qualification. I do agree with Tsvi that possibly we shouldn't be using GRB to measure the average star formation rate in galaxies in high redshifts, however there is a qualification here. You do use GRB to measure star formation rate but this probes particular properties of the regions of star formation where GRB reside, namely areas with clusters of very high mass. So when you go to the high redshifts, where you see this discrepancy, probably what one can state is that GRB probe regions of star formation rate of very massive stars in clusters. Whether that is useful in order to use in your simulations or not remains to be seen but I would be surprised if I saw that they agree with the average star formation rate of the galaxies, especially when in these redshifts you cannot actually distinguish between the region where the GRB reside and the entire galaxy when you make a measurement.

P. Madau I had a question for Richard. Rather than looking directly for dwarfs at redshift 7 or 8 which is very challenging, there seems to be a lot of information now coming out from resolved star populations. You can do color-magnitude diagrams of dwarfs in the local group, with relatively high precision, and those type of observations tell us exactly what is the fraction of stars that those dwarfs where forming, which is typically 20 or 30% at high redshifts, so I guess my question is if that kind of information has been folded in these reionization calculations?

R. Ellis I think that is what Carlos was heading at. The answer is yes. There is a lot of information in the stellar population and chemical composition that can be useful and indeed there are conferences where the two communities can pat nose. I think that for quantitative work, the difficulty is that only a subset of these early objects make it to the present day. So one can imagine many biases in analyzing a local residual population and inferring absolute properties of the history of the whole object by the one that happened to have survived. So I think the answer is, yes, it is useful, but I do not think that the connection is mature enough yet but it is a tool that people are using.

S. White Just a quick comment. As Carlos has referred to, the most extreme of the small objects, the so-called ultra-faint galaxies, which are the dwarfs that have been found recently, and just in the last year, the color-magnitude diagrams for five or six of these faint dwarfs have been published. It is striking that all of them are consistent with having all their stars forming during the reionization period.

Session 4

Dark Matter

Chair: *Simon White*, Max-Planck-Institut fuer Astrophysik, Garching, Germany
Rapporteurs: *Laura Baudis*, University of Zürich, Switzerland and *Neil Weiner*, New York University, USA
Scientific secretaries: *Glenn Barnich*, Université Libre de Bruxelles, Belgium and *Gianfranco Gentile*, Universiteit Gent and Vrije Universiteit Brussel

S. White Good morning everybody, this morning's session is on dark matter. Dark matter has been with us for about 80 years or so now. It took the first 40 years for the astrophysicists to get interested in it. Most looked on it as a curiosity. I think the idea that dark matter was a particle came to us in the 70s. That got the particle physicists interested in it. The program this morning is to start with the particle physics aspects, and I have two rapporteurs who are going to tell us what the experimental status for looking for dark matter is and current ideas about what it might be. We will have the discussions before coffee based on their talks about what dark matter might be, how we might find out what it is and whether we could see something perhaps from accelerators, and then after coffee the discussion will shift to the more astrophysical aspects of what dark matter could be, how we could find out more about it than what we know. So, that is the plan. I would like to start with the first contribution, our first rapporteur is Laura Baudis from Zurich who is going to tell us about the more experimental aspects of trying to look for dark matter other than by its gravitation.

Rapporteur Talk by L. Baudis: Dark Matter Detection: Experimental Overview

Abstract

One major challenge of astrophysics and particle physics is to decipher the nature of dark matter. I discuss experimental searches for two well-motivated classes of candidates: axions and weakly interacting massive particles (WIMPs). Dark matter axions can be detected by exploiting their predicted coupling to two photons, where the highest sensitivity is reached by experiments using a microwave cavity permeated by a strong magnetic field. WIMPs could be observed via direct and indirect detection experiments, and through production at accelerators. I will show the reach and limitations of direct searches in the laboratory, and compare with indirect searches and with recent input from the LHC. I will discuss selective technologies for the future, addressing their complementarity as well as their main sources of backgrounds. I will end with a set of questions that are to stimulate the ensuing discussion.

Keywords: Dark matter, axions, WIMPs, direct detection, indirect detection, collider searches.

1. Introduction

After decades of increasingly precise astronomical observations, we have unequivocal evidence that the majority of the material that forms galaxies, clusters of galaxies and the largest observed structures in the universe is non-luminous, or dark. This conclusion rests upon accurate measurements of galactic rotation curves, measurements of orbital velocities of individual galaxies in clusters, cluster mass determinations via gravitational lensing, precise measurements of the cosmic microwave background acoustic fluctuations and of the abundance of light elements, and upon the mapping of large scale structures. In addition, cosmological simulations based on the ΛCDM model successfully predict the observed large-scale structures in the universe. In this model, which so far provides the only paradigm that can explain all observations, our universe is spatially flat and composed of $\sim$5% atoms, $\sim$27% dark matter and $\sim$68% dark energy.[1,2]

The first quantitative case for a dark matter dominance of the Coma galaxy cluster was made as early as 1933 by the Swiss astronomer Fritz Zwicky.[3] He found that the velocity dispersion of individual galaxies in this galaxy rich cluster was far too large to be supported by visible matter alone. In the 1970's, Vera Rubin and collaborators, and Albert Bosma measured the rotation curves of spiral galaxies and also found evidence for a "missing mass" component. In the early 80's, evidence for this non-luminous matter was firmly established on galactic scales, once rotation curves of galaxies well beyond their optical radii were measured in radio emission of neutral hydrogen.[4,5]

Since then, our understanding of the total amount of dark matter and its overall distribution deepened, but we still lack the answer to the most basic question: what is the dark matter made of? One intriguing answer is that it is made of a new particle, yet to be discovered. Instantly, more questions arise: what are the properties of the particle, such as its mass, interaction cross section, spin and other quantum numbers? Is it one particle species, or many? Is it absolutely stable, or very long-lived? Here I will discuss searches for two particular classes of dark matter particle candidates: QCD axions with masses in the range 1 μeV - 3 meV and weakly interacting massive particles (WIMPs) with masses in the $\sim$0.3 GeV-100 TeV range.

2. Axions

The axion is one of the most promising hypothetical particles proposed to solve the dark matter puzzle. Originally, axions were introduced by Peccei and Quinn (PQ) as a solution to the strong-CP problem in QCD. They postulated a global U(1) symmetry that is spontaneously broken below an energy scale f_a. While the original PQ axion with f_a around the weak scale was soon excluded, so-called invisible axion models are still viable. The axion mass is given in terms of f_a as:

$$m_a \simeq 6 \cdot 10^{-6} \text{eV} \, \frac{10^{12} \text{ GeV}}{f_a}. \tag{1}$$

Cold axions are produced during the QCD phase transition, but the exact mass of the axion is not known. Constraints from laboratory searches and from astrophysical observations restrict the mass range to $\sim$1 μeV - 3 meV. All axion couplings are inversely proportional to f_a, which is also called the axion decay constant. For dark matter searches, the axion coupling to two photons, dictated by the following Lagrangian, is of particular interest:

$$\mathcal{L}_{a\gamma\gamma} = -g_\gamma \frac{\alpha}{\pi f_a} a(x) \vec{E}(x) \cdot \vec{B}(x) = -g_{a\gamma\gamma} a(x) \vec{E}(x) \cdot \vec{B}(x), \tag{2}$$

where $a(x)$ is the scalar axion field, $\vec{E}(x)$ and $\vec{B}(x)$ are the electric and magnetic fields of the two propagating photons, α is the fine structure constant and g_γ is a model-dependent coefficient of order one ($g_\gamma = 0.36$ and $g_\gamma = -0.97$ in the DFSZ and KSVZ models, respectively).

Although the coupling of the axion to electromagnetism is inferred to be extremely weak, and the spontaneous decay life-time of an axion to two real photons is vastly greater than the age of our universe, axions can be detected via the inverse Primakoff effect, where the axion decay is accelerated through a static, external magnetic field. In this field, one photon is replaced by a virtual photon, and the other maintains the energy of the axion, namely its rest mass plus its kinetic energy:

$$E \simeq m_a c^2 + \frac{1}{2} m_a c^2 \beta^2. \tag{3}$$

The ADMX experiment[6] exploits the axion detection scheme proposed by Sikivie, based on the Primakoff effect: in a microwave cavity permeated by a strong magnetic field, the axion-photon conversion is enhanced when the resonant frequency f of the cavity equals the axion rest mass:

$$f \simeq \frac{m_a c^2}{h} \tag{4}$$

where h is Planck's constant. Assuming an axion mass of $5\,\mu\mathrm{eV}$ and an expected velocity dispersion of $\Delta\beta \sim 10^{-3}$ in our galaxy, the spread in the axion energy is expected to be $\sim 10^{-6}$ or $\sim 1.2\,\mathrm{kHz}$. The resonant cavity is tunable, and the axion decay signal is to be detected by observing the proper modes at a given frequency. The expected axion-to-photon conversion power is:

$$P = g_{a\gamma\gamma}^2 \frac{\rho_a}{m_a} B_0^2 V C \min(Q_L, Q_a)$$

$$= 4 \times 10^{-26}\mathrm{W} \left(\frac{g_\gamma}{0.97}\right)^2 \frac{\rho_a}{0.5 \times 10^{-24}\mathrm{g\,cm}^{-3}} \frac{m_a}{2\pi(\mathrm{GHz})} \tag{5}$$

$$\times \left(\frac{B_0}{8.5\,\mathrm{T}}\right)^2 \frac{V}{0.22\,\mathrm{m}^3} C \min(Q_L, Q_a),$$

where ρ_a is the local axion density, B_0 is the strength of the static magnetic field, V is the cavity volume, C is a mode-dependent cavity form factor, and $\min(Q_L, Q_a)$ is the smaller of either the cavity or axion quality factors. The axion signal quality factor is $Q_a = 10^6$, the ratio of their energy to the energy spread. The signal power is thus expected to be exceedingly weak, and the cavity and amplifiers are cooled to very low temperatures to minimize thermal noise.

The new ADMX experiment is currently being assembled at the University of Washington. It will use a tuneable cavity with a higher quality factor, and a lower intrinsic noise, due to a dilution refrigerator and quantum-limited SQUID amplifiers. Staring in 2015, ADMS will test a large fraction of the predicted parameter space for the QCD axion as a dark matter candidate, as shown in Figure 1. In addition, an R&D program for a next-generation experiment (ADMX-HF), that can probe the theoretically allowed higher mass region (10-100 GHz) is ongoing.

3. WIMPs: Overview

Weakly interactive massive particles (WIMPs), which would have been in thermal equilibrium with quarks and leptons in the hot early universe, and decoupled when they were non-relativistic, represent a generic class of dark matter candidates.[7] Their relic density can account for the dark matter density if the annihilation cross section σ_{ann} is around the weak scale:

$$\Omega h^2 \simeq 3 \times 10^{-27}\mathrm{cm}^3\mathrm{s}^{-1}\frac{1}{\langle\sigma_{\mathrm{ann}}\mathrm{v}\rangle} \tag{6}$$

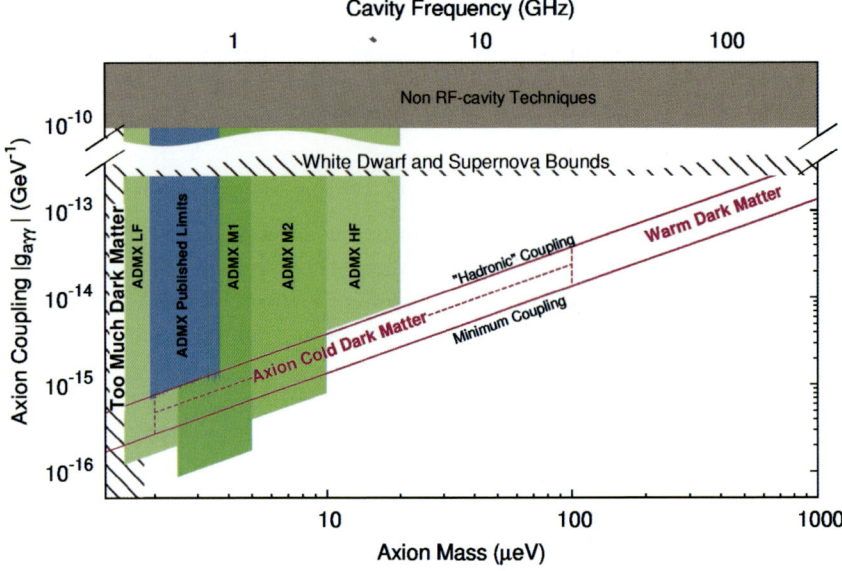

Fig. 1.   Current limits and projected reach for the various stages of the ADMX experiment.

where v is the relative velocity of the annihilating WIMPs, and the angle brackets denote an average over the WIMP thermal distribution. $\langle \sigma_{ann} v \rangle = 3 \times 10^{-26} \mathrm{cm}^2 \mathrm{s}^{-1}$ gives the correct relic density, and is often considered as the benchmark annihilation cross section value. Concrete examples for WIMPs are the lightest superpartner in supersymmetry with R-parity conservation,[8] and the lightest Kaluza-Klein particle, for instance the first excitation of the hypercharge gauge boson, in theories with universal extra dimensions.[9]

Perhaps the most intriguing aspect of the WIMP hypothesis is the fact that it is testable by experiment. WIMPs with masses around the electroweak scale are within reach of high-energy colliders and of direct and indirect dark matter searches.[10] Direct detection experiments search for a scattering signal between the WIMP and an atomic nucleus, while indirect detection experiments aim to detect annihilation products of dark matter particles, such as neutrinos, gamma rays, antiprotons and positrons above the astrophysical background. Colliders look for dark matter production in high energy particle collisions. At the LHC, the signature is missing energy, as the WIMP leaves the detectors unobserved, accompanied by a jet, a photon, a Z etc, that are required for tagging an event. These approaches are complementary to one another, and to astrophysical probes that can test non-gravitational interactions of dark matter.[11] Although no conclusive evidence for WIMPs in any of these detection channels exists, it is likely that we will be faced with detections in multiple approaches within this decade.

4. Direct WIMP Detection

The insight that WIMPs can be detected by elastic scattering off nuclei in an earth-bound detector goes back to Goodman and Witten,[12] following the suggestion of Drukier and Stodolsky[13] to detect solar and reactor neutrinos by exploiting their elastic neutral-current scattering of nuclei in a detector made of superconducting grains embedded in a non-superconducting material. The study was extended by Drukier, Freese and Spergel[14] to include a variety of cold dark matter candidates, as well as details of the detector and the galactic halo model. They also showed that the Earth's motion around the Sun produces an annual modulation in the expected signal. On the theoretical side, there was much progress in refining all aspects entering the prediction of scattering event rates: from detailed cross section calculations in specific particle and nuclear physics models, to refined dark matter halo models that take into account uncertainties in the local WIMP density, in their mean velocity and velocity distribution, as well as in the galactic escape velocity. Progress has been tremendous on the experimental side: in developing new technologies that yield an increasing amount of information about every single particle interaction, in applying these technologies to detectors with masses soon to reach the ton-scale, and in fighting the background noise such that levels below 1 event per kg and year have now been reached.

The differential rate for elastic scattering can be expressed as:[15]

$$\frac{dR}{dE_R} = N_T \frac{\rho_{dm}}{m_W} \int_{v_{\min}}^{v_{\max}} d\vec{v}\, f(\vec{v})\, v\, \frac{d\sigma}{dE_R} \tag{7}$$

where N_T is the number of target nuclei, ρ_{dm} is the local dark matter density in the galactic halo, m_W is the WIMP mass, $\vec{v}$ and $f(\vec{v})$ are the WIMP velocity and velocity distribution function in the Earth frame and $d\sigma/dE_R$ is the WIMP-nucleus differential cross section. The nuclear recoil energy is $E_R = m_r^2 v^2 (1 - \cos\theta)/m_N$, where θ is the scattering angle in the center-of-mass frame, m_N is the nuclear mass and m_r is the reduced mass. The minimum velocity is defined as $v_{\min} = (m_N E_{th}/2m_r^2)^{\frac{1}{2}}$, where E_{th} is the energy threshold of the detector, and $v_{\max}$ is the escape velocity in the Earth frame.

The simplest galactic model assumes a Maxwell-Boltzmann distribution for the WIMP velocity in the galactic rest frame, with a velocity dispersion of $\sigma \approx 270\,\mathrm{km\,s^{-1}}$ and an escape velocity of $v_{esc} \approx 544\,\mathrm{km\,s^{-1}}$.[16,17] For direct detection experiments, the mass density and velocity distribution at a radius around 8 kpc are most relevant. State of the art, dark-matter-only cosmological simulations of Milky Way-like halos find that the dark matter mass distribution at the solar position is smooth, with substructures being far away from the Sun. The local velocity distribution of dark matter particles is likewise found to be smooth, and close to Maxwellian.[18] The efforts to measure the mean density of dark matter near our Sun are extensively reviewed in:[19] the Milky Way seems to be consistent with featuring a spherical dark matter halo with little flattening in the disc plane, and $\rho_{dm}=0.2-0.56\,\mathrm{GeV\,cm^{-3}}$.[19] The largest source of uncertainty in ρ_{dm} comes from the baryonic contribution to the local dynamical mass.

General treatments of the WIMP-nucleon cross section in non-relativistic, effective field theory approaches,[20–23] identify a full set of operators that describe the potential interactions of dark matter particles, and relate these to nuclear response functions. These are of great interest because, in principle, more information can be extracted from direct detection experiments in case of a signal detection, assuming a certain variety of targets is employed in the search. In general, the coherent, spin-independent part dominates the WIMP-nucleus interaction (depending however on the characteristics and composition of the dark matter particle) for target masses with $A{\geq}30$,[8] and is used as a benchmark to compare results from various experiments. Nuclear form factors, expressed as a function of the recoil energy, become significant at large WIMP and nucleus masses, leading to a suppression of the differential scattering rate at higher recoil energies.

As an example, for a WIMP and target nucleus mass of 100 GeV, and an elastic scattering cross section from the nucleus of $\sigma_{WN} \sim 10^{-38}\mathrm{cm^2}$, the rate for elastic scattering can be expressed as:

$$R \sim 0.13 \ \frac{\text{events}}{\text{kg year}} \left[\frac{A}{100} \times \frac{\sigma_{WN}}{10^{-38}\,\mathrm{cm^2}} \times \frac{\langle v \rangle}{220\,\mathrm{km\,s^{-1}}} \times \frac{\rho_0}{0.3\,\mathrm{GeVcm^{-3}}} \right]. \quad (8)$$

Figure 2 shows the expected differential nuclear recoil spectrum for various WIMP target materials, a WIMP mass of 100 GeV and an assumed WIMP-nucleon scattering cross section of $1 \times 10^{-47}\mathrm{cm^2}$.

5. Signatures and Backgrounds

A dark matter particle with a mass in the GeV−TeV range has a mean momentum of a few tens of MeV and an energy below 50 keV is transferred to a nucleus in a terrestrial detector. Expected event rates range from one event to less than 10^{-3} events per kg detector material and year. To observe a WIMP-induced spectrum, a low energy threshold, an ultra-low background noise and a large target mass are essential. Specific experimental signatures from a particle populating our galactic halo come from the Earth's motion through the galaxy. It induces both a seasonal variation of the total event rate[14,24] and a forward-backward asymmetry in a directional signal.[25,26] The annual modulation of the WIMP signal arises because of the Earth's motion in the galactic rest frame, which is a superposition of the Earth's rotation around the Sun and the Sun's rotation around the galactic center. Since the Earth's orbital speed v_{orb} is much smaller than the Sun's circular speed, the amplitude of the modulation is small, of the order of $v_{\mathrm{orb}}/v_c \simeq 0.07$. A stronger signature would be given by the ability to detect the axis and direction of the recoil nucleus. Since the WIMP flux in the lab frame is peaked in the direction of motion

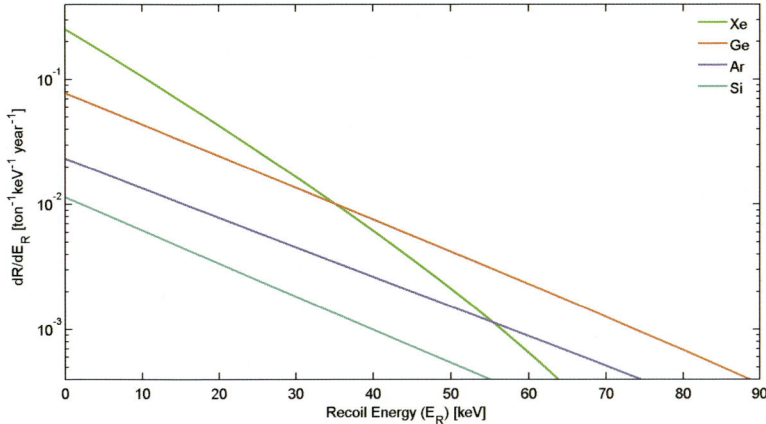

Fig. 2. Expected differential nuclear recoil spectra for Xe, Ge, Ar and Si as WIMP target materials. A WIMP mass of 100 GeV and a spin-independent WIMP-nucleon cross section of 1×10^{-47} cm^2 are assumed.

of the Sun, namely towards the constellation Cygnus, the recoil spectrum is peaked in the opposite direction. The forward-backward asymmetry yields a large effect of the order of $\mathcal{O}(v^E_{orb}/v_c) \approx 1$. Thus fewer events, namely a few tens to a few hundred, depending on the halo model, are needed to discover a WIMP signal compared to the case where the seasonal modulation effect is exploited.[26,27] The experimental challenge is to build massive detectors capable of detecting the direction of the incoming WIMP at low energy deposits.

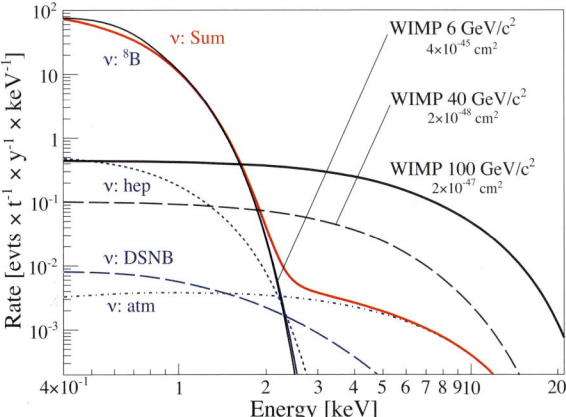

Fig. 3. The differential nuclear recoil spectrum from coherent scattering of neutrinos (red) from the Sun, the diffuse supernova background (DSNB), and the atmosphere (atm), compared to the one from WIMPs for various masses and cross sections (black). For both plots, the nuclear recoil signals are converted to an electronic recoil scale (see[28]) and a nuclear recoil acceptance of 50% is assumed. Figure from.[28]

The main background sources are the environmental radioactivity, radio-impurities in the detector construction and shield material, neutrons from (α, n) and fission reactions, cosmic rays and their secondaries, activation of detector materials during exposure at the Earth's surface, as well as sources intrinsic to the target materials. The ultimate backgrounds might come from neutrino-induced nuclear recoils from coherent neutrino-nucleus scatters. The coherent scattering rate from ^{8}B solar neutrinos will provide an irreducible background for low-mass WIMPs, limiting the cross section sensitivity to $\sim 4 \times 10^{-45} \, \mathrm{cm}^2$ for WIMPs of $6 \, \mathrm{GeV}/c^2$ mass. Atmospheric and diffuse supernova neutrinos will dominate the rate for WIMP masses above $\sim 10 \, \mathrm{GeV}/c^2$ and cross section below $10^{-49} \mathrm{cm}^2$.[28–32] Figure 5 shows the expected background spectrum from coherent scattering of solar, atmospheric and diffuse supernova background neutrinos in a detector using xenon as target material. It also displays the expected differential recoil spectra for WIMPs with three different masses and cross sections.

6. Direct Detection Experiments

In an earthbound detector, the kinetic energy carried away by the scattered atomic nucleus is transformed into an ionization signal, in scintillation light or lattice vibration quanta (phonons). The simultaneous detection of two observables yields a powerful discrimination against background events, which are mostly interactions with electrons, as opposed to WIMPs and neutrons, which scatter off nuclei. Good timing and position resolution will distinguish localized energy depositions from multiple scatters (highly improbable from WIMPs) within the active detector volume and in addition allow to *in situ* measure the neutron background.

Existing upper limits and projected sensitivities for the spin-independent WIMP-nucleon interactions as a function of the WIMP mass are summarized in Figure 6. In spite of observed anomalies in a handful of experiments, that were interpreted as due to WIMPs (albeit not consistently), we have no convincing evidence of a direct detection signal induced by galactic dark matter. The low mass region is accessible by experiments with very low energy threshold and/or lighter target nuclei, while the higher mass region is probed by experiments with very low backgrounds and high target masses. The most constraining upper limits for WIMP masses above $6 \, \mathrm{GeV}$ come from the LUX experiment.[36] The reached 90% upper C.L. cross section for spin-independent WIMP-nucleon couplings has a minimum of $7.6 \times 10^{-46} \, \mathrm{cm}^2$ at a WIMP mass of $33 \, \mathrm{GeV}/c^2$, thus confirming and improving upon the earlier XENON100 results.[35]

Cryogenic experiments operated at sub-Kelvin temperatures feature the highest sensitivities at WIMP masses in the range 0.3-$10 \, \mathrm{GeV}/c^2$, considering their low energy threshold ($<10 \, \mathrm{keV}$), excellent energy resolution ($<1\%$ at $10 \, \mathrm{keV}$) and the ability to highly differentiate nuclear from electron recoils on an event-by-event basis. Their development had been driven by the exciting possibility of performing a calorimetric energy measurement down to very low energies with unsurpassed en-

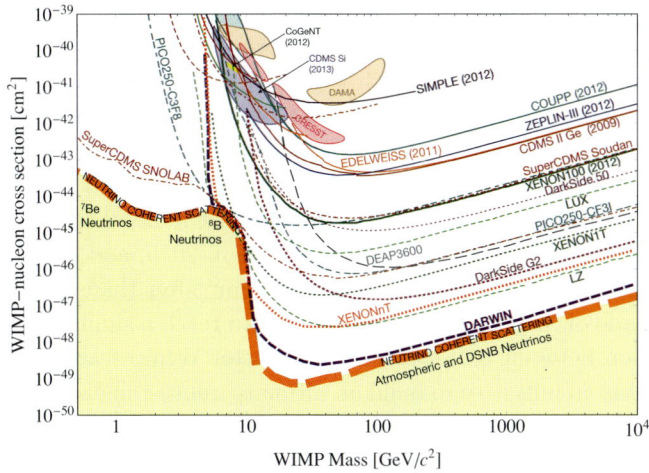

Fig. 4. Spin-independent WIMP-nucleon scattering results: Existing limits from the ZEPLIN-III,[33] XENON10,[34] XENON100,[35] and LUX[36] experiments, along with projections for DarkSide-50,[37] LUX,[36] DEAP3600,[38] XENON1T, DarkSide G2, XENONnT, LZ,[39] and DARWIN[40] are shown. DARWIN is designed to probe the entire parameter region for WIMP masses above ~6 GeV/c^2, until the neutrino background (yellow region) will start to dominate the recoil spectrum. Experiments based on the mK cryogenic technique such as SuperCDMS[41] and EURECA[42] have access to lower WIMP masses. Figure adapted from.[43]

ergy resolution. Because of the T^3-dependence of the heat capacity of a dielectric crystal, at low temperatures a small energy deposition can significantly change the temperature of the absorber. The change in temperature is measured either after the phonons reach equilibrium, or thermalize, or when they are still out of equilibrium, or athermal, the latter providing additional information about the location of an interaction in the crystal.[44] The CDMS,[45] CRESST[46] and EDELWEISS[47] experiments are successful implementations of these techniques. The SuperCDMS experiment, using 50 kg of Ge and Si detectors, has been approved for construction at SNOLAB. The new cryostat will be able to house up to 400 kg of detectors, and discussions are ongoing with EURECA (CRESST+EDELWEISS) towards a common project at this deep site.

Liquid noble elements such as argon and xenon offer excellent media for building non-segmented, homogeneous, compact and self-shielding dark matter detectors. Liquid xenon (LXe) and liquid argon (LAr) are good scintillators and ionizers in response to the passage of radiation, and the simultaneous detection of ionization and scintillation signals allows to identify the primary particle interacting in the liquid. In addition, the 3D position of an interaction can be determined with sub-mm (in the z-coordinate) to mm (in the x-y-coordinate) precisions in a time projection chamber (TPC). These features, together with the relative ease of scale-up to large masses, have contributed to make LXe and LAr powerful targets for WIMP searches.[48] The dual-phase LXe TPCs ZEPLIN-III,[33,49] XENON10,[34,50]

XENON100[35,51,52] and LUX[36] are successful implementations of the technique (also the LAr projects!). The most stringent limits for spin-independent WIMP-nucleon couplings at WIMP masses above 6 GeV come from XENON100 and LUX, currently running experiments. XENON1T, the next step of the XENON program, is a dual-phase xenon TPC with total (active) mass of 3.3 t ($\sim$2 t), in construction at LNGS. Completion and commissioning are expected for mid and late 2015, respectively. The next phase in the LUX program, LUX-ZEPLIN (LZ), foresees a 7 t LXe detector; construction is expected to start in 2015, and operation in 2016, with the goal of reaching a sensitivity of 2×10^{-48}cm^2 after three years of data taking.[39] The upgrade of XENON1T, XENONnT, is to reach a similar sensitivity, with planned operation between 2018-2021. DARk matter WImp search with Noble liquids (DARWIN) is an initiative to build an ultimate, multi-ton dark matter detector at LNGS,[40,53] to probe the spin-independent WIMP-nucleon cross section down to the 10^{-49} cm^2 region. It would thus explore the experimentally accessible parameter space, which will eventually be limited by irreducible neutrino backgrounds.

Various other technologies are employed to search for WIMPs - scintillating crystals (DAMA/LIBRA, KIMS, ANAIS and DM-Ice), superheated liquid detectors (PICASSO, COUPP, SIMPLE and lately PICO), Ge-ionization detectors (CoGeNT, CDEX, etc). In addition, a strong R&D program for detectors capable of measuring the direction of the recoiling nucleus is in progress (DRIFT, DMTPC, MIMAC, NEWAGE). In the near future, a common approach for building a large directional detector will be proposed.

As soon as direct evidence for a dark matter signal has been firmly established, the efforts will shift towards inferring the mass, cross section and possibly the spin of the dark matter particle. To reconstruct these properties with a certain accuracy, based on statistical considerations alone, exposures of several ton-years and multiple targets are required even for a cross section as large as 10^{-45}cm^2, increasing to several tens of ton-years for a cross section of 10^{-46}cm^2.[54] If we consider a 50 GeV WIMP as an example, its mass can be reconstructed with a 1-σ accuracy of about 5% when using a combination of data from xenon, germanium and argon experiments and assuming fixed astrophysical parameters. Allowing for 1-σ uncertainties in the local density, circular and escape velocity of 0.1 GeV cm^{-3}, 30 km s^{-1} and 33 km s^{-1}, respectively, increases the 1-σ accuracy of the mass determination to about 10%.[55,56] If the astrophysical parameters are left to vary in a broad range, data from dark matter experiments alone, when using multiple targets, can constrain the local circular velocity at least as accurately as it is currently measured,[55] see also Figure 5.

6.1. *Direct Detection: Future*

An overview of the evolution of upper limits on the spin-independent cross section as a function of time, together with projections for the future, is shown in Figure 6. The rate of progress was slower during the first decade shown here, and mostly

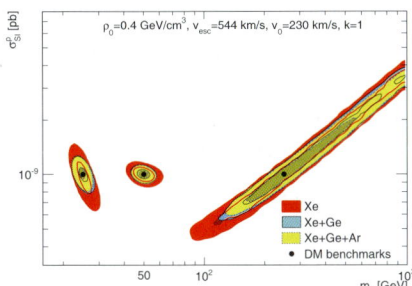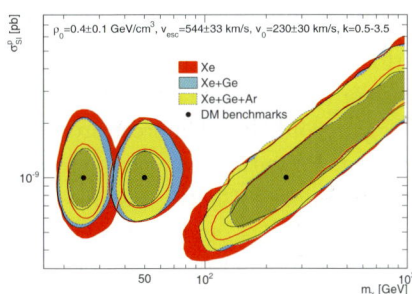

Fig. 5. Reconstruction capabilities of combined Xe+Ge and Xe+Ge+Ar data sets as joint 68% and 95% posterior probability contours in the $m_\chi - \sigma_{SI}^p$ plane for the three WIMP masses ($m_\chi = 25, 50, 250\,\text{GeV/c}^2$) with fixed astrophysical parameters (left) and considering astrophysical uncertainties (right). Figure from.[55]

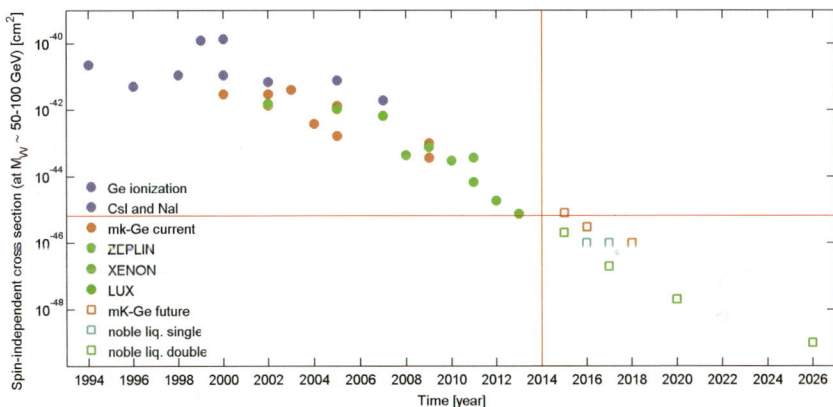

Fig. 6. Existing upper limits on spin-independent WIMP-nucleon cross sections (for WIMP masses around 50-100 GeV) from various direct detection techniques (filled circles), along with projections for the future (open squares), as a function time. Part of the data was retrieved from dmtools.[57]

dominated by low-background Ge experiments searching for neutrinoless double beta decay. The rate increased with the advent of cryogenic mK-detectors and of homogeneous noble-liquid detectors. These techniques are capable of fiducialization and can distinguish between electronic and nuclear recoils on an event-by-event basis. Looking at the recent progress, from ~2002-2013, we notice that the sensitivity increased by about one order of magnitude roughly every two years. The predictions of reachable sensitivities until ~2030 seem reasonable if the current trend is extrapolated into the future, assuming not only larger and lower background detectors, but that known backgrounds can be reduced at the same rate as demonstrated during the last few years.

7. Comparison with Indirect Detection

WIMPs could also be detected by observing the radiation produced when they annihilate or decay.[58] The flux of annihilation products is proportional to $(\rho_W/m_W)^2$,
thus regions of highest interest are those expected to have an enhanced WIMP concentration. Potential signatures are high energy neutrinos from the Sun's core and
from the Galactic Centre, gamma-rays from the Galactic Centre and from dwarf
spheroidal galaxies, and positrons, antiprotons and antideuterons from the galactic halo. The predicted fluxes depend on the particle physics model delivering the
WIMP candidate and on astrophysical input such as the dark matter halo profile,
the presence of sub-structure and the galactic cosmic ray diffusion model, the latter
being particularly relevant for the propagation of charged particles.

Several existing observational anomalies have been interpreted as signatures for
dark matter annihilation in our galaxy, or in extragalactic dark matter halos,[59,60] to
be discussed by Neal Weiner in this volume. Among these are a line feature around
130 GeV in the Fermi-LAT data, a rising positron fraction in the PAMELA and
lately AMS data, an excess of $\sim$1-3 GeV gamma rays from the region surrounding
the Galactic Centre. There is clearly no single WIMP capable of explaining all the
data: WIMP masses from $\sim$30 GeV to several TeV are required, and there is a
strong demand for more data. It will come from the continued operation of Fermi
and AMS, by the current generation of atmospheric Cherenkov detectors and the
future CTA, as well as by existing and future neutrino experiments. In addition,
WMAP and Planck data provide direct constraints on dark matter annihilation
that takes place during the era of recombination, for WIMP annihilation could give
rise to a sufficient number of energetic particles to impact the observed anisotropies
in the cosmic microwave background.[61]

It is most straightforward to compare direct detection results with those from
searches for high-energy neutrinos from the Sun, as the WIMP-proton cross section plays a crucial role, initiating the capture process. WIMPs with orbits passing
through the Sun can scatter from nuclei and lose kinetic energy. If their final velocity is smaller than the escape velocity, they will be gravitationally trapped and will
settle to the Sun's core. Over the age of the solar system, a sufficiently large number
of particles can accumulate and efficiently annihilate, whereby only neutrinos are
able to escape and be observed in terrestrial detectors. The primary annihilation
spectrum is model dependent, the range of possible models is usually considered by
assuming 100% branching into channels with different characteristics: the so-called
hard channel, where the WIMPs annihilate into W^+W^- and the so-called soft channel, with annihilation into $b\bar{b}$. Typical neutrino energies are $1/3$–$1/2$ of the WIMP
mass, thus well above the solar neutrino background.

Observation of high energy neutrinos from the direction of the Sun would thus
provide a clear signature for dark matter in the halo. The annihilation rate is set
by the capture rate, which scales with $(m_W)^{-1}$ for a given halo density. Thus,

annihilation and direct detection rates have the same scaling with the WIMP mass. However, the probability of detecting a neutrino by searching for muons produced in charge-current interactions scales with E_ν^2, making these searches more sensitive at high WIMP masses when compared to direct detection experiments. The best technique to detect high-energy neutrinos is to observe the upward-going muons produced in charged-current interactions below the detector. To distinguish neutrinos coming from the Sun's core from backgrounds induced by atmospheric neutrinos, directional information is needed. The direction of the upward-going muon and the primary neutrino direction are correlated, the rms angle scaling roughly with $\sim 20(E_\nu/10\,\mathrm{GeV})^{-1/2}$.

Two types of detectors are used to search for high-energy neutrino signals, with no excess above the atmospheric neutrino background reported so far. In the first category are large underground detectors, such as SuperKamiokande, while the second type are dedicated to neutrino telescopes, employing large arrays of PMTs deep in glacier ice or in water, such as IceCube, ANTARES and the future KM3NET. These experiments detect the Cerenkov light emitted when muons move with speeds larger than the velocity of light in water/ice, with ~ 1 ns timing resolution. The PMT hit pattern and relative arrival times of the photons are used to reconstruct the direction of the incoming particle, which is correlated with the direction of the neutrino. Neutrino telescopes have higher energy thresholds (in the range 50-100 GeV), but their effective area is much larger, thus compensating for the lower fluxes predicted for heavy WIMPs.

The strongest limits on high-energy neutrinos coming from the Sun are placed by IceCube[62] and Super-Kamiokande. Figure 7 shows the upper limits on spin-dependent WIMP-proton cross section as a function of the WIMP mass (in the hard and soft annihilation channels) from IceCube, in comparison with results from direct detection experiments.

Although no conclusive evidence for a dark matter particle from indirect detection experiments exists, these searches now start to probe annihilation cross sections around the value for a simple thermal relic, $\langle \sigma v \rangle \approx 3 \times 10^{-26} \mathrm{cm}^3 \mathrm{s}^{-1}$, albeit annihilation cross sections below this value are certainly possible. As experimental sensitivities are poised to drastically improve during this decade, we urgently need a deeper understanding of the astrophysical backgrounds and of the dark matter distribution on galactic scales.

8. Input from the LHC

Another avenue to search for WIMPs (χ) is to look for their production at the LHC: $pp \to \chi\bar{\chi}$. While the presence of dark matter particles is not directly observable in a detector at a collider, it can be inferred from their recoil on standard model particles. So far, there is no evidence for dark matter from LHC searches. In particular, there is no evidence for supersymmetry, or for any other new theoretical model motivated by solving the naturalness problem.[64] In more model-independent dark

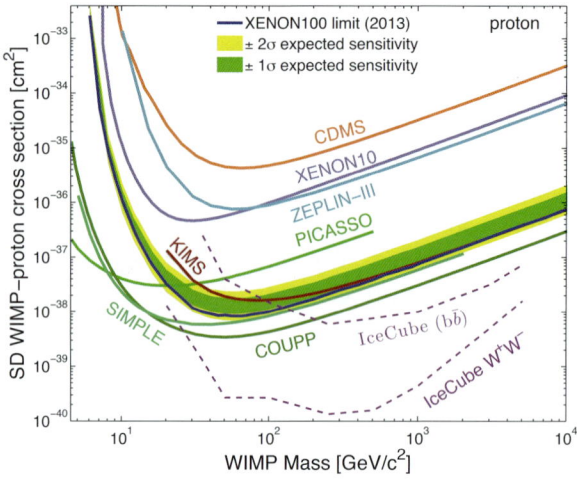

Fig. 7.   IceCube [62] upper limits on the spin-dependent, WIMP-proton cross section, in the hard $(W^+W^-, \tau^+\tau^-$ for WIMP masses $<80.4\,\mathrm{GeV}/c^2)$ and soft $(b\bar{b})$ annihilation channels. Also shown are results from direct detection experiments. Figure from.[63]

matter searches, the unknown interactions between dark matter and standard model particles are usually described by a set of effective operators. The expected signature is then missing transverse energy accompanied by a so-called mono-object (denoted by X: a photon, a single jet, a Z, etc) required to tag the event: $pp \to \chi\bar{\chi} + X$. Presently, the collider searches using mono-jets or mono-photons accompanied by missing transverse energy remained fruitless.[65–67] In addition, unlike the case for direct detection, the effective field theory (EFT) approach is not always valid, for instance when the energy scale probed by the effective operators is smaller than the energy of the partons taking part in the collision.[68,69]

To allow a meaningful comparison with direct detection experiments, benchmark scenarios in simplified models were proposed (see e.g.[68–70] and references therein). One approach is to classify possible mediators of the interaction between the dark matter and the standard model particles. For particles exchanged in the s-channel, the mediators must be electrically neutral, with spin 1 or 0, while for particles exchanged in the t-channel, the mediator can be a colour triplet. An example for an s-channel mediator is a new massive spin-one vector boson, Z', from a broken U(1)' gauge symmetry. However, the mediator of interactions between the WIMP and quarks could also be the Z boson or the Higgs boson.

As an example which nicely illustrates the complementarity between collider and direct searches, Figure 8 (from[69]), shows the vector (g_V^{DM}) and axial-vector (g_A^{DM}) coupling versus mass regions for fermionic dark matter that couples to the Z. Shown are constraints from the LHC, from direct detection (exemplified by the LUX 2013 results) and from the Z-invisible decay width. Also shown is the predicted thermal relic abundance via Z-coupling annihilation, and the predictions for LHC14.

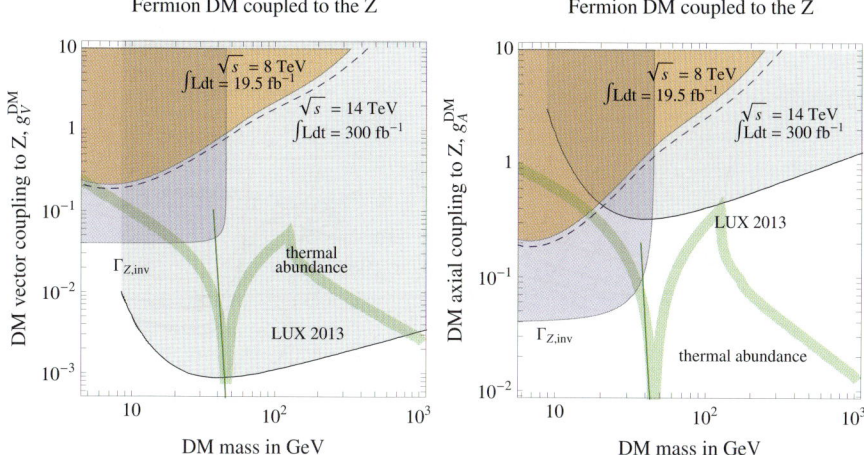

Fig. 8. Regions of the dark matter mass and vector (g_V^{DM}) and axial-vector (g_A^{DM}) couplings to the Z, along with constraints from the LHC, from direct detection (the LUX 2013 results) and from the Z-invisible width constraint. Also shown is the thermal relic abundance via Z-coupling annihilation. Figure from.[69]

Another example for an alternative to the EFT interpretations is to characterise searches for dark matter via simplified models, that are constructed using four parameters only: the mass of the WIMP, the mass of the mediator and the couplings of the mediator to the WIMP, as well as to quarks.[68,70] Assuming that the WIMP is a Dirac fermion and that the mediator couples to all quarks with the same strength, it is straightforward to compare collider and direct detection searches. For the exchange of a vector mediator, the LHC mono-jet searches have better sensitivity at WIMP masses below ∼5 GeV. For axial-vector mediators, the LHC has greater sensitivity than direct searches at WIMP masses below ∼200 GeV, as shown in Figure 9 (from[70]). The figure shows projected spin-dependent and spin-independent sensitivities in the WIMP-nucleon cross section versus WIMP mass for LHC14 (and various couplings) and a large liquid xenon detector (7 t LXe, exemplified by LZ), as well as a Si+Ge detector such as SuperCDMS. The region below which coherent neutrino-nucleus scattering will start to dominate the event rates in direct detection experiments is also shown.

In summary, the current generation of collider searches can probe very low WIMP masses, and up to masses of ∼1 TeV, while direct detection experiments can probe mass regions up to 100 TeV and even above. In terms of the WIMP couplings to standard model particles, colliders are superior for axial-vector couplings, while direct searches show much larger sensitivities than colliders for spin-independent scattering.

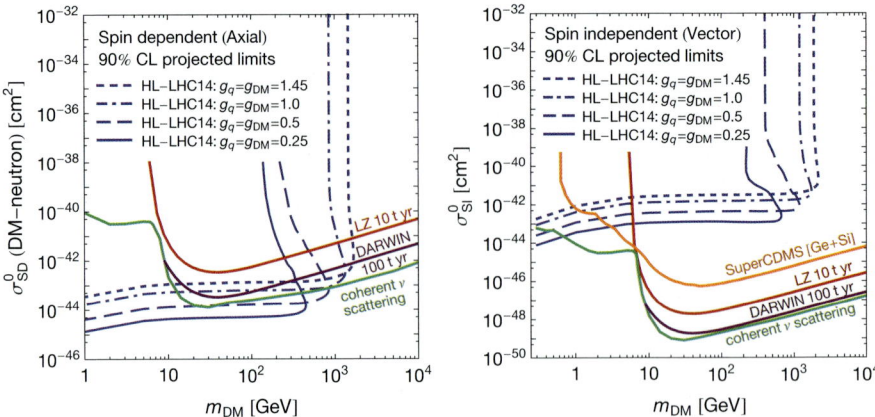

Fig. 9.   Projected spin-dependent (left) and spin-independent (right) 90% CL limits for LHC14 in various coupling scenarios (blue lines) and for direct detection experiments (exemplified by large liquid xenon detectors, LZ (red line) and DARWIN (magenta line), and by a Si+Ge detector, SuperCDMS (orange line). The region below which coherent neutrino-nucleus scattering will start do dominate the event rates in direct detection experiments is also shown (green curve). Figure from.[70]

9. Conclusions

Although invisible, dark matter is five times as abundant as normal matter in our Universe, its identity, more than eighty years after its postulation by Fritz Zwicky in its modern form, remains a secret. Uncovering the nature of this dominant form of matter is thus one of the grand challenges of modern physics. Considering the immense progress in a large range of experiments operated at the Earth's surface, in space or deep underground, it seems likely that we will be faced with multiple and hopefully consistent discoveries within the next couple of years. The immediate next goal will be to reveal the detailed properties of dark matter particles, such as their mass, spin and couplings to ordinary matter, and to shed light on their phase-space distribution in the Milky Way's halo. This could herald the start of a new field, dark matter astronomy.

10. Questions

I append a set of questions that are to stimulate the discussion. Perhaps some of these will have been answered by the next Solvay meeting in astrophysics.

1. Can we convincingly discover dark matter by direct and indirect searches given the uncertainties on their phase-space distribution in the Milky Way, and the astrophysical background?

2. To which degree will we be able to constrain the astrophysical observables (local dark matter density, spacial distribution, velocity distribution) relevant for a detection in the near future?

3. Can we control and reduce the background levels in next-generation laboratory experiments by another 2-3 orders of magnitude?

4. Can we achieve much lower energy thresholds and higher target masses, also for directional detectors?

5. How will we accurately calibrate larger and larger detectors (with self-shielding), can we use neutrinos?

6. What is required to overcome the limitations caused by the 'irreducible' backgrounds from atmospheric neutrinos?

7. What do we require as a definitive proof that we observed galactic particle dark matter, and not a new source of backgrounds?

8. How many particle species make up the dark matter? What are their masses, spins and couplings to Standard Model particles?

9. Are our search strategies adequate to answer above questions, and the set of experiments sufficiently diverse?

10. What are the limitations of the ΛCDM model, and how to solve the problems at small scales?

11. What can we learn about warm and cold dark matter from astrophysical probes, and how well will these constrain the self-interaction strength in the future?

References

1. P. Ade *et al.*, *Astron.Astrophys.* (2014).
2. P. Ade *et al.*, *Astron.Astrophys.* (2015).
3. F. Zwicky, *Helv.Phys.Acta* **6**, 110 (1933).
4. A. Bosma, *Astron.J.* **86**, p. 1825 (1981).
5. V. Rubin, D. Burstein, J. Ford, W.K. and N. Thonnard, *Astrophys.J.* **289**, p. 81 (1985).
6. G. Rybka, *Phys.Dark Univ.* **4**, 14 (2014).
7. B. W. Lee and S. Weinberg, *Phys.Rev.Lett.* **39**, 165 (1977).
8. G. Jungman, M. Kamionkowski and K. Griest, *Phys.Rept.* **267**, 195 (1996).
9. D. Hooper and S. Profumo, *Phys.Rept.* **453**, 29 (2007).
10. G. Bertone, D. Hooper and J. Silk, *Phys.Rept.* **405**, 279 (2005).
11. S. Arrenberg, H. Baer, V. Barger, L. Baudis, D. Bauer *et al.* (2013).
12. M. W. Goodman and E. Witten, *Phys.Rev.* **D31**, p. 3059 (1985).
13. A. Drukier and L. Stodolsky, *Phys.Rev.* **D30**, p. 2295 (1984).
14. A. Drukier, K. Freese and D. Spergel, *Phys.Rev.* **D33**, 3495 (1986).
15. J. Lewin and P. Smith, *Astropart.Phys.* **6**, 87 (1996).
16. A. M. Green, *Mod.Phys.Lett.* **A27**, p. 1230004 (2012).
17. M. C. Smith *et al.*, *Mon. Not. Roy. Astron. Soc.* **379**, 755 (2007).
18. M. Vogelsberger, A. Helmi, V. Springel, S. D. White, J. Wang *et al.*, *Mon.Not.Roy.Astron.Soc.* **395**, 797 (2009).
19. J. Read (2014).
20. J. Fan, M. Reece and L.-T. Wang, *JCAP* **1011**, p. 042 (2010).
21. A. L. Fitzpatrick, W. Haxton, E. Katz, N. Lubbers and Y. Xu, *JCAP* **1302**, p. 004 (2013).

22. R. J. Hill and M. P. Solon, *Phys.Rev.Lett.* **112**, p. 211602 (2014).
23. N. Anand, A. L. Fitzpatrick and W. Haxton, *Phys.Rev.* **C89**, p. 065501 (2014).
24. K. Freese, J. A. Frieman and A. Gould, *Phys.Rev.* **D37**, p. 3388 (1988).
25. D. N. Spergel, *Phys.Rev.* **D37**, p. 1353 (1988).
26. C. J. Copi, J. Heo and L. M. Krauss, *Phys.Lett.* **B461**, 43 (1999).
27. C. J. Copi and L. M. Krauss, *Phys.Rev.* **D63**, p. 043507 (2001).
28. L. Baudis, A. Ferella, A. Kish, A. Manalaysay, T. M. Undagoitia *et al.*, *JCAP* **01**, p. 044 (2014).
29. L. E. Strigari, *New J.Phys.* **11**, p. 105011 (2009).
30. A. Gutlein, C. Ciemniak, F. von Feilitzsch, N. Haag, M. Hofmann *et al.*, *Astropart.Phys.* **34**, 90 (2010).
31. J. Billard, L. Strigari and E. Figueroa-Feliciano, *Phys.Rev.* **D89**, p. 023524 (2014).
32. A. Anderson, J. Conrad, E. Figueroa-Feliciano, K. Scholberg and J. Spitz, *Phys.Rev.* **D84**, p. 013008 (2011).
33. D. Y. Akimov, H. Araujo, E. Barnes, V. Belov, A. Bewick *et al.*, *Phys.Lett.* **B709**, 14 (2012).
34. J. Angle *et al.*, *Phys.Rev.Lett.* **100**, p. 021303 (2008).
35. E. Aprile *et al.* (2012).
36. D. Akerib *et al.* (2013).
37. M. Bossa, *JINST* **9**, p. C01034 (2014).
38. M. Boulay, *J.Phys.Conf.Ser.* **375**, p. 012027 (2012).
39. D. Malling, D. Akerib, H. Araujo, X. Bai, S. Bedikian *et al.* (2011).
40. L. Baudis, *J.Phys.Conf.Ser.* **375**, p. 012028 (2012).
41. R. Agnese *et al.* (2014).
42. H. Kraus, E. Armengaud, C. Augier, M. Bauer, N. Bechtold *et al.*, *PoS* **IDM2010**, p. 109 (2011).
43. P. Cushman, C. Galbiati, D. McKinsey, H. Robertson, T. Tait *et al.* (2013).
44. N. Booth, B. Cabrera and E. Fiorini, *Ann.Rev.Nucl.Part.Sci.* **46**, 471 (1996).
45. Z. Ahmed *et al.*, *Science* **327**, 1619 (2010).
46. G. Angloher, M. Bauer, I. Bavykina, A. Bento, C. Bucci *et al.*, *Eur.Phys.J.* **C72**, p. 1971 (2012).
47. E. Armengaud *et al.*, *Phys.Lett.* **B702**, 329 (2011).
48. E. Aprile and L. Baudis (2010).
49. D. Y. Akimov, *Nucl.Instrum.Meth.* **A623**, 451 (2010).
50. E. Aprile *et al.*, *Astropart.Phys.* **34**, 679 (2011).
51. E. Aprile *et al.*, *Phys.Rev.Lett.* **107**, p. 131302 (2011).
52. E. Aprile *et al.*, *Astropart.Phys.* **35**, 573 (2012).
53. L. Baudis, *PoS* **IDM2010**, p. 122 (2011).
54. C. Strege, R. Trotta, G. Bertone, A. H. Peter and P. Scott, *Phys.Rev.* **D86**, p. 023507 (2012).
55. M. Pato, L. Baudis, G. Bertone, R. Ruiz de Austri, L. E. Strigari *et al.*, *Phys.Rev.* **D83**, p. 083505 (2011).
56. J. L. Newstead, T. D. Jacques, L. M. Krauss, J. B. Dent and F. Ferrer, *Phys.Rev.* **D88**, p. 076011 (2013).
57. http://dmtools.brown.edu/.
58. J. L. Feng, *Ann.Rev.Astron.Astrophys.* **48**, 495 (2010).
59. J. Buckley, D. Cowen, S. Profumo, A. Archer, M. Cahill-Rowley *et al.* (2013).
60. J. Conrad (2014).
61. D. P. Finkbeiner, S. Galli, T. Lin and T. R. Slatyer, *Phys.Rev.* **D85**, p. 043522 (2012).
62. M. Aartsen *et al.*, *Phys.Rev.Lett.* **110**, p. 131302 (2013).

63. E. Aprile *et al.*, *Phys.Rev.Lett.* **111**, p. 021301 (2013).
64. K. Olive *et al.*, *Chin.Phys.* **C38**, p. 090001 (2014).
65. G. Aad *et al.*, *Phys.Rev.* **D90**, p. 012004 (2014).
66. G. Aad *et al.*, *JHEP* **1304**, p. 075 (2013).
67. S. Chatrchyan *et al.*, *JHEP* **1209**, p. 094 (2012).
68. O. Buchmueller, M. J. Dolan and C. McCabe, *JHEP* **1401**, p. 025 (2014).
69. A. De Simone, G. F. Giudice and A. Strumia, *JHEP* **1406**, p. 081 (2014).
70. S. Malik, C. McCabe, H. Araujo, A. Belyaev, C. Boehm *et al.* (2014).

Discussion

S. White Thank you Laura. Neal is going to tell us next about theory, but I thought that maybe someone wants to have some discussion of the experimental aspects of dark matter searches first, maybe some questions come up. Scott, I can see you want to say something here.

S. Tremaine According to your projections of a factor of 10 in increase in sensitivity every 2 years, you are about 4 orders of magnitude away from the neutrino background, so in 8 years you will be at the neutrino background. At that point should we give up the search for WIMPs or what changes in strategy and direction do you think that the community should consider at that point?

L. Baudis This is a very good question. Of course, we are trying to think about it. It might not be 8 years, it might be 10 years, but still it will come up at some point. One of the strategies would be, of course, to observe the direction of the incoming WIMP, because you could then at least determine whether it comes from the Sun or not. You would then even have some handles on atmospheric neutrinos, but this is very very challenging because if you have not seen anything until then, building a detector, so there is quite some R&D on directional detectors, but to measure the direction of the incoming WIMP is connected to the recoil direction so you have to be able to measure it and also the sense, this is what is called "head versus tail", so you cannot use the type of experiments that I have shown so far. You really need a low-pressure gas that would mean a very little target mass. So building a 20-ton scale-detector that is in the form of a low-pressure gas will be quite challenging. And I am also not sure if it would even get the funding if we have not seen any dark matter. So maybe the answer is: if we have seen some sign for dark matter until then, and if the rate is high enough, then we can think of building directional detectors and we can even think of background subtraction and so on. I think we will be clever enough to figure out how to deal with this background. The question is, of course, what to do if you have not seen anything down to that point.

T. Piran I just want to clarify the situation about the DAMA results. Do I understand correctly from your talk that the result is established, they see the oscillations...

L. Baudis It is established that they see a variation in the event rate observed in the detector. But where the variation comes from, what the source is, is not clear.

T. Piran But the region that would explain this with dark matter is ruled out by another experiment.

L. Baudis By many other experiments, yes.

T. Piran And at the same time there is no theoretical explanation, which is accepted, which explains that they see.

L. Baudis Absolutely, yes. People have tried, yes, you are totally right.

T. Piran I just wanted to see that I understand the paradox.

L. Baudis Absolutely. Still, there are attempts to measure or disprove or prove the same signal with the same type of detectors, namely sodium and iodide. But you are perfectly right.

S. White A question from Mitch before we move on.

M. Begelman In the speculative approaches being encouraged here by those of us who are ignorant in the field, could someone comment about recent speculations in the literature that quantum interference effects could be important in the CDM structure on small scales and could this have any effect on detectability and strategies?

L. Baudis I am not sure I understood the question.

N. Weiner I do not think it would be very relevant for what she is talking about, because their fairly classical description of the particle dark matter at these densities, for WIMPs, is appropriate.

S. White There is the question of whether the small-scale classical structure of the distribution could be interesting, for example structure in the velocity distribution, or other kinds of structure. We will come back to that in the astrophysics session.

S. White Now Neal Weiner is going to give us a survey of theoretical possibilities from the particle physics point of view. As I think you will see, there are plenty of candidates.

Rapporteur Talk by N. Weiner: Current Directions in Dark Matter Theory

Abstract

While dark matter has been convincingly shown to be present from its gravitational influence, so far its particle physics nature has proved elusive. I will explore various motivations for dark matter, and the models that have arisen. I will review some recent claims of signals consistent with dark matter and consider their theoretical implications.

1. Introduction: Discovering the Nature of Dark Matter

The 20th century saw tremendous success in the progress made by physics. From the properties of the atom and nucleus to the earliest moments of the universe to its long-term fate, a huge range of questions were answered.

At the same time, new questions were opened. One of the most central of those is the nature of dark matter. Dark matter is ubiquitous and the dominant form of matter in the universe. Evidence for its existence comes from a wide range of sources, (see for instance, a discussion by Murayama[1]), and should be considered at this point, the conservative explanation for the universe we observe, rather than an exotic one.

At the same time, we have no conclusive evidence as of yet as to its nature. In this endeavor, we find ourselves in a peculiar situation. We are searching for something, which we - by definition - are not certain of how we should be looking for it. Without a theoretical model, we cannot say what would be a sign for dark matter. Signals such as anomalous nuclear recoils, excesses in cosmic rays, or tiny signals from microwave cavities, even if you just lucked upon them would not seem as evidence of dark matter without some idea as to what dark matter is. Thus, theory input is essential just to begin to look for dark matter.

Unfortunately or fortunately, theory guidance is broad. For instance, in principle, the mass scale of dark matter could be anything. The natural particle physics scales one might consider would range from 10^{-33} eV (the mass of a particle with Compton wavelength the current horizon size) to 10^{19} GeV (the scale at which gravity becomes strong). Within that, there are only a few additional points of note. A boson should be more massive than roughly 10^{-22} eV, so that its Compton wavelength can fit into known dwarf galaxies and galaxy cores,[2] while a Fermion should be heavier than roughly 100 eV (the Tremaine-Gunn bound[3]), so that the fermionic phase space can contain enough dark matter inside of galaxies.

Beyond that, we have little guidance. Indeed, dark matter could even be much heavier, if it is a composite state (such as a nucleus or a planet). We have theoretical motivations that single out the weak scale for thermal relics, and the proton mass

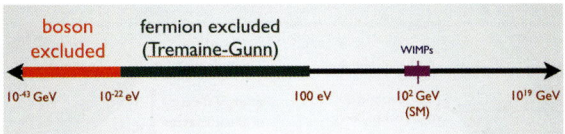

Fig. 1.   The mass range for dark matter particles. Courtesy S. Rajendran.

scale if the similarities between dark and baryonic matter are not an accident. Beyond those tantalizing suggestions, the field is wide open.

The question one encounters is then: how should one approach developing dark matter theories?

I should say that what follows is my own, personal impression of things. A different person would give you different, perhaps very different, perspective on how to approach dark matter theory. But I am here, and I can only comment on my opinion. With that caveat, I will proceed.

Roughly speaking, most models of dark matter arise from one of three approaches.

- **"Top down"** - In this approach, one begins with a theory motivation, for instance the hierarchy problem, the strong CP problem, and so on. One then develops a model (e.g. the MSSM or an axion). Within the model, one looks for a stable, neutral particle that might serve as dark matter (i.e. the LSP or axion)
- **"Bottom up"** - Motivated often by specific experimental anomalies, one constructs a model to explain it. Such a model might be embedded within a larger framework (such as supersymmetry or extra dimensions), modifying the implications of the usual signals of those theories (such as the stable LKP of extra dimensional models). From there, one makes predictions for other experiments.
- **"Phenomenological"** - Without a specific signal, one attempts to develop a model with a given property, such as a mass range, a direct detection signal (such as electron recoils), or cosmological sign (such as changes to large scale structure). One then can further consider the prospect of testing such a scenario in existing, future, or hypothetical experiments.

The feature that almost all these models have in common is that they give *some* hope of experimental implication. This is almost by design - a dark matter model that cannot be tested does not offer much hope in the way of actually learning what dark matter is. This does not guarantee, of course, that dark matter *is* actually discoverable (or at least, not by us, not right now), but our only hope of doing so is to consider as many testable possibilities as we can.

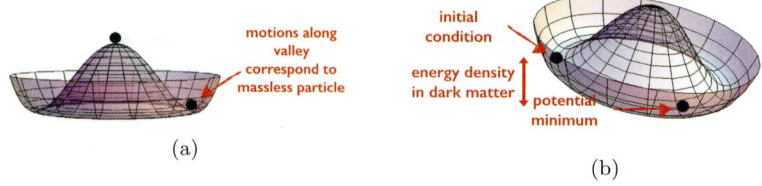

(a)

(b)

Fig. 2.   A qualitative sketch of the dynamics of an axion field. (a) After spontaneous symmetry breaking, a flat (massless) direction persists which corresponds to the axion field. (b) At low temperatures, QCD effects correct this and generate a potential. The initial location of the axion field then corresponds to some initial energy.

2. The Axion

One of the most popular models around today is the axion. The axion is probably not as popular today as the WIMP is, but it's important not to confuse popularity with a determination of which idea is better. Popularity has more to do with what one can do with a model, and WIMPs have a tremendous number of directions to go, with broad direct and indirect detection programs, as well as the role within collider physics and the connection to many models of weak scale physics. While axions do have interesting implications, their experimental signals are more limited, with the only established technique being the Sikivie cavity currently employed by the ADMX experiment.[4] Notably, there has been a resurgence of interest here, with novel new proposals.[5–7]

Nonetheless, my discussion here will parallel that of the field, with more emphasis placed on weak-scale models than axions. I would encourage you not to confuse that emphasis with endorsement, any more than I would the number of publications in the field.

The axion is a pseudo-goldstone boson arising from a global symmetry breaking. The assumption is that the symmetry breaking occurs at some scale f_a, after which there is some flat direction in the theory. If that flat direction is identified with the QCD θ parameter, then the QCD θ parameter will become dynamical.

This is important, because the QCD θ (really $\bar{\theta}$) contributes to CP violating observables, namely the neutron EDM. Constraints from this require $\bar{\theta} \lesssim 10^{-10}$.[8] In making the $\bar{\theta}$ parameter dynamical, one can imagine physics that would relax this to its minimum.

And, indeed, this is precisely what occurs. The $U(1)_a$ symmetry is anomalous, and QCD effects generate a potential that drives the axion to cancel the $\bar{\theta}$ parameter. The resulting axion mass is

$$m_a \sim \frac{m_\pi f_\pi}{f_a} \sim 0.6 \, \text{meV} \times \left(\frac{10^{10} \, \text{GeV}}{f_a} \right). \tag{1}$$

Because of the initial value of the axion, before it drifts down to its minimum, there is some energy density in the axion field. It is this energy we identify with the dark matter energy density of the universe. If the axion starts far from its minimum, there is a great deal of energy in dark matter. If it starts near its minimum, there is very little.

And so, in thinking about this, we are immediately led to two, separate thermal histories for the axion. In the first scenario, the Peccei-Quinn (PQ) symmetry breaks after inflation. In this case, parts of the universe has large initial values. In other parts of the universe it is small. On average, however, it is uniformly distributed around the initial values and the abundance is totally calculable, and the dark matter abundance is known.[a] On the other hand, if the PQ symmetry breaks *before* inflation, then there is a single value that covers the entire universe.

In this case, it is not clear how much dark matter there is. For small axion masses and large initial axion values, there could be far more dark matter than we observe in the universe. But, if we just happened to live somewhere with a tiny initial value, it would be consistent with the data. If one argues, such as argued by Tegmark et al[10] that too much dark matter is anthropically selected against, then perhaps the small axion value is just a selection effect. This "anthropic axion" has become an important idea in thinking about dark matter.

And for this reason, there was added importance to the question of primordial B-modes. While B-modes give us the prospect of knowing about the early universe, they also would tell us that the initial energy of the universe was quite high. Because fields fluctuate by amounts $\delta\phi \sim H$ in the early universe, if the inflation scale is high, the fields fluctuate more. Those fluctuations can imprint themselves on the universe as isocurvature pertubations which are strongly constrained. Indeed, it has been argued that if primordial B-modes are observed, the anthropic axion would be excluded.[11] In this way, modern cosmology can teach us about the existence or absence of very light fields. For now, however, we must wait for such a detection.

3. A Thermal Relic

A very simple alternative hypothesis for dark matter is to take a cue from the thermal nature of the early universe. It is quite reasonable to imagine that the dark matter was in thermal equilibrium with us at some early time.

Thus, we begin with the assumption that dark matter is in thermal contact with the Standard Model bath, and then at some temperature $T < M_{\mathrm{DM}}$, it decouples. (Chemical and kinetic decoupling need not happen simultaneously.) The picture is then quite simple, once $T < M_{\mathrm{DM}}$, the number density of dark matter begins to drop precipitously, due to the $exp(-M_{\mathrm{DM}}/T)$ Boltzmann suppression. When decoupling occurs, that is the amount of dark matter we have left. If the annihilation cross section is large, the dark matter will stay in equilibrium until late times, and it will be underabundant. If the cross section is small, it will depart equilibrium early and it will be overabundant. In between, one will find the correct abundance, and, simultaneously, the correct cross section.

---

[a]This is actually overly simplified, as it turns out that strings from the phase transition can be the dominant source of axions,[9] but misalignment still gives approximately the right scales.

As such, in studying thermal dark matter, we learn precisely one number, namely the annihilation cross section of dark matter in the early universe. Specifically,

$$\frac{\Omega_{DM}}{0.1} \simeq \frac{3 \times 10^{-26} \text{cm}^3 \text{sec}^{-1}}{\langle \sigma v \rangle} \approx \frac{\alpha_{EM}^2}{(200 \text{GeV})^2} \frac{1}{\langle \sigma v \rangle}. \tag{2}$$

The fact that the scale of the thermal WIMP cross section corresponds to approximately the weak scale is referred to as the "WIMP miracle." Namely, that the weak scale, in the form of the hierarchy problem, suggests new particles in that regime. Simultaneously, thermal freezeout suggests particles at the same scale. This coincidence bolsters each motivation separately. Note that we needn't actually assume a solution to the hierarchy problem - we can simply say that the weak scale is an interesting scale in a numerological sense, and the WIMP miracle would still be intriguing.

Of course, we must proceed with some humility here. We have probably several orders of magnitude to play in the cross section (200 GeV → 1TeV, for instance, a few 4π's out front, replacing $\alpha_{EM} \to \alpha_s$), and we would still find the overall coincidence compelling. Thus, this coincidence is a possible indication, but we should be careful in its application.

Let us proceed under the assumption that dark matter is a thermal relic. The next question is: how does it annihilate into the standard model? It can annihilate directly into any standard model particles, including quarks, leptons, gauge bosons or the Higgs boson (if kinematically possible).

But there is another important alternative that has gained a great deal of interest of late, which goes by a variety of monikers, including "hidden sector models," "dark sector models," "exciting dark matter," "secluded dark matter," and others. For simplicity, I will refer to these scenarios as "hidden sector models." The idea is quite simple: one can assume that dark matter has some interaction of its own. This idea is not radical: in the standard model, left handed fields have interactions right handed fields do not, quarks have interactions that leptons do not, so it it quite reasonable to imagine that dark matter has interactions that standard model fermions do not.

Under that assumption, it is possible for dark matter to annihilate into its new force carrier, ϕ. (The force carrier also carries many names, including A', ϕ, U, X and others. I will use A' when referring to a vector but ϕ more generally.) This force carrier generally picks up very weak couplings to standard model fields through a mixing, either with the photon or the Higgs boson. Under the annihilation $\chi\chi \leftrightarrow \phi\phi$, χ is in equilibrium with ϕ, while ϕ maintains equilibrium separately, such as via $\phi e \leftrightarrow \gamma g$. In this scenario, freezeout proceeds normally in its numerical details, but the implications for our detection are quite different, as we shall see.

3.1. *Some Comments on SUSY*

One cannot discuss the thermal WIMP without discussing the most common WIMP, namely, the Lightest Super Partner (LSP) in Supersymmetry. The LSP is often

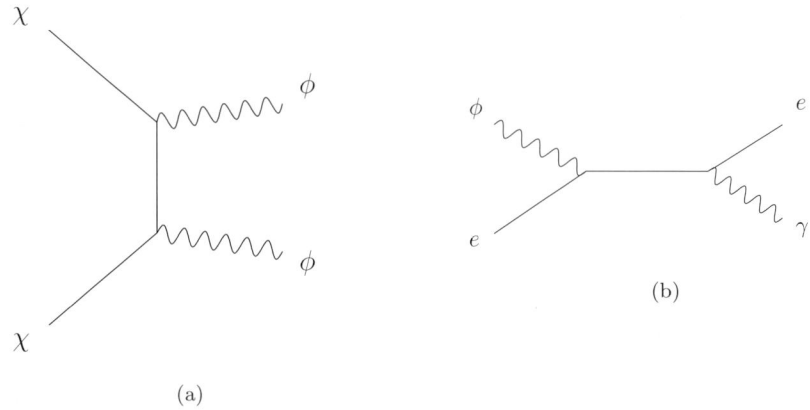

Fig. 3.   Dark matter freezeout and equilibrium via annihilation into a hidden sector particle.

stable and neutral, and as such gives SUSY a natural dark matter candidate.

The phenomenology of the LSP is quite broad, there are a number of different possibilities for its thermal history depending on the rest of the SUSY spectrum. Its collider, cosmic ray and collider signals can vary depending on what its makeup is. It is an important candidate that serves in many ways as a benchmark WIMP model.

At the same time, we must be careful not to generalize too much from it, as the neutralino has important features that may not be general. It is Majorana (conventionally), and does not have any even approximate symmetries associated with it aside from R-parity. It couples to the SM only through the fields present (so gauge bosons, Higgs bosons and off-shell squarks and sleptons). It does not have cosmologically significant self interactions.

Moreover, the "WIMP miracle" was supposed to have been a two-way street. The anticipation was that the search for dark matter (in the form of direct and indirect) would show signs of the weak scale at the same time colliders would find evidence of new particles at the similar scale. Together, they would inform us of the more complete theory of the weak scale. So far, this has not been the case.

Through the 8 TeV run of the LHC there has been no clear sign of new physics. Generic SUSY searches have pushed the mass for colored partners into the several hundred GeV range, or above a TeV for gluinos. While the possibility of SUSY to be discovered in the next run of the LHC, it is fair to say that many of the expected scenarios have been excluded. At a minimum, I would say, we should be careful about our theoretical prejudice. After all, if our ideas of the nature of SUSY were wrong, should we be confident in our ideas of what dark matter will look like? I include within this possibility the idea that SUSY might still be found, but simply be a different looking character than what we had anticipated twenty years ago.

As a consequence, it is important to keep an open mind about dark matter models, and not to hew too closely to the neutralino in our ideas of how to look for it.

Adding to this is the fact that almost any model can be embedded into Supersymmetry. Supersymmetry, we must remember, is not, itself, a model, but a property of theories. Essentially all of the new models discussed in the past several years are easily combined with SUSY, and, in many cases, benefit from it. Thus, for my remaining time, we can safely separate these questions.

3.2. *Some Comments on Direct Detection*

The "direct" search for WIMPs has been underway for nearly three decades, motivated by the seminal work of Goodman and Witten[12] who suggested that nuclear recoil searches could be capable of discovering relic WIMPs. Since then, there has been a steady progression of the sensitivity of these experiments. Recently, rather than slowing, the pace of the progress has accelerated, largely due to the success of liquid noble detectors, such as XENON and LUX. The present reach of LUX, currently the most sensitive experiment, is to constrain a maximum cross section of $7.6 \times 10^{-46} \text{cm}^2$ for particles with mass of 33 GeV (the mass to which it has the strongest limits).[13]

A common refrain from people is this: after all this time, we still have not seen a WIMP. How worried should we be? Is the WIMP hypothesis already squeezed.

In my opinion, the answer to this is no, and we should not worry (yet). To understand this, we can consider three basic models: a Dirac SU(2) doublet, also known as a heavy Dirac neutrino or pure Higgsino, a mixed SU(2)-doublet/singlet Majorana state, also known as a neutralino, and a Majorana SU(2) triplet, also known as a pure Wino.

If we calculate the cross section for the first case, we see that its interaction with ordinary matter is mediated by the Z-boson, and find the WIMP-neutron cross section to be $\sim 10^{-39} \text{cm}^2$. Such a cross section was ruled out long ago, by experiments such as the Heidelberg-Moscow experiment,[14] and is far, far above the current limits.

In contrast, if we consider the second model, because the particle is Majorana, there are no spin-independent interactions mediated by the Z-boson. For this model, the dominant cross section is via Higgs exchange. The Higgs, however, has a much smaller coupling to neutrons than Z-bosons, nearly 10^{-3} smaller, meaning that we expect cross sections at the scale of 10^{-44}cm^2 and smaller. Thus, at the moment, the experiments are probing the "Higgs-mediated" scattering of dark matter.

If we consider the third model, there are no tree level Z or Higgs couplings, and all interactions are mediated by loops. There, we expect scattering cross sections more at the 10^{-47}cm^2 level.[15,16] There is nothing about this third model that I would not consider a "WIMP," and yet it's quite a challenge to discover via direct detection. This particular model, especially if a thermal relic, is likely excluded from cosmic ray searches,[17,18] but still drives home that we cannot rely on direct detection, alone, any more than we can rely on indirect detection or collider physics alone to answer the question about dark matter.

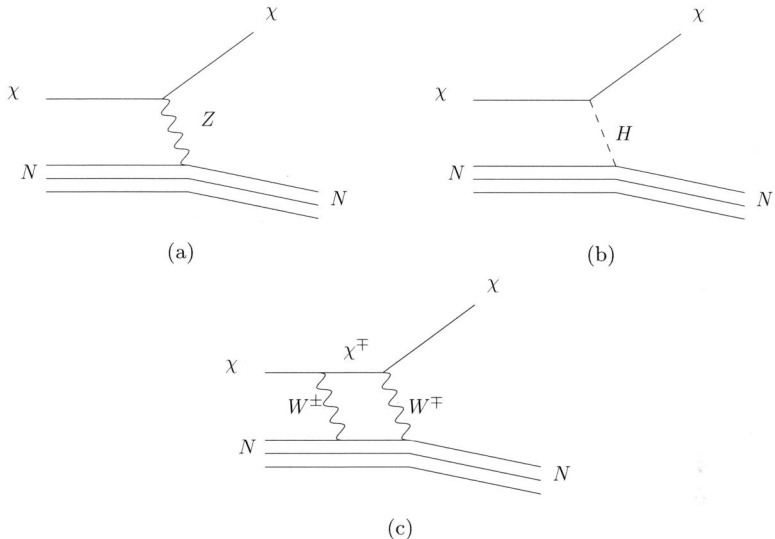

Fig. 4. Scattering of the WIMP via (a) Z exchange, (b) Higgs exchange or (c) loop exchange. In general loop exchange also involves loops generating Higgs exchange, with this just one example of a loop process.

The critical point is this: we are currently in the era where we can answer the question: does the dark matter talk at a significant level to the Higgs boson? If yes, we will see it, if not, there are still other compelling scenarios to be probed.

3.3. *Some Comments on Indirect Detection*

Direct detection has as well made tremendous leaps in recent years, especially with the tremendous progress made by the Fermi Gamma-Ray Space Telescope. While we shall discuss the claims of excesses attributed to dark matter shortly, it is worth considering the results first as the tremendous constraints that they are on what dark matter might be.

I will only discuss these constraints in a limited capacity, but two limits of note, which are quite complementary are the Fermi limits from dwarf galaxies, and the HESS search for gamma ray lines. (HESS is a large, ground based, atmospheric Cerenkov telescope that looks for the showers produced from cosmic rays impacting the upper atmosphere in the TeV energy range.)

Dwarfs are important targets to look for dark matter because they are not expected to be significant sources of gamma rays on their own. Thus, if we were to see a signal from some, especially if the spectra were identical, would give strong evidence for dark matter. At the moment, we have only limits.[b] These limits, however, are quite strong.

---

[b]Subsequent to the Solvay conference, claims for an excess in a newly discovered dwarf, Reticulum II have appeared, but this remains an open question.[19,20]

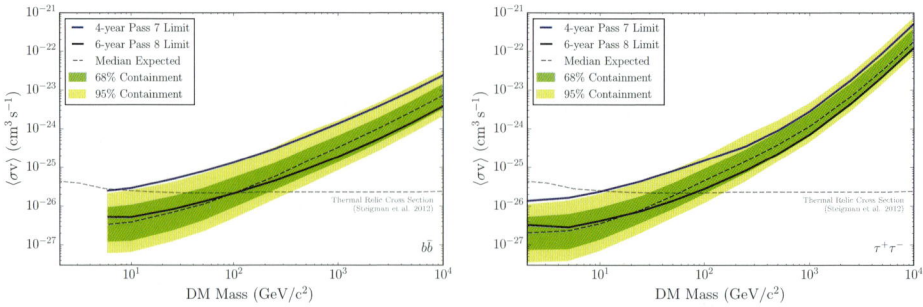

Fig. 5.   Fermi limits on dwarf galaxies. Taken from Ackermann et al.[21]

After much effort within Fermi, the absence of a clear signal has set important limits on dark matter models. In particular, these searches constrain models below ~ 100 GeV which annihilate today with an approximately thermal cross section. (Importantly, if the cross section drops with velocity, these constraints are not significant, with current velocities in the 10^{-3} range. Still, there are many models where the cross section is *not* velocity dependent.)

At a much higher energy we have the HESS experiment looking for lines. If dark matter annihilates directly $\chi\chi \to \gamma\gamma$, one expects a monochromatic line in the gamma ray spectrum. The absence of such lines places constraints on a number of models. In particular, it places important constraints on Wino dark matter,[17,18] which is so hard to see in direct detection. Over much of the mass range, including the 2 TeV needed for a thermal Wino, these limits can exclude the possibility of Wino dark matter. Indeed, if a 2 TeV Wino were the LSP, we would also be in trouble on the LHC front, as it would imply the other super partners were even *heavier*, and possibly out of reach. This drives home the importance of these complementary approaches.

Finally, one can employ the CMB to place constraints on dark matter annihilation. Pointed out originally by Padmanabhan and Finkbeiner,[22] and studied recently by others,[23–26] dark matter injects energy into the CMB during the era of recombination (even if it is out of equilibrium). This perturbs the CMB temperature and polarization spectrum and can be constrained. Recent constraints from Planck[27] are the strongest to date and constrain many scenarios of recent interest, including those that might produce observable positron signals.

4. Anomaly Driven Dark Matter

A major effort in the past $\sim$10 years has been a development of "anomaly driven" dark matter models. These are models that were either created or gained much of their steam from their ability to explain some signal via a dark matter interpreta-

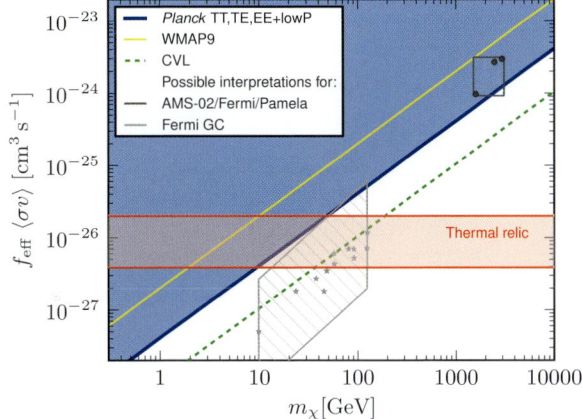

Fig. 6. Limits on dark matter annihilation during the era of recombination from the Planck satellite.[27]

tion. Various anomalies have come and gone but the ones that are still potential candidates (without any assessment of probability) are

– The annual modulation in the DAMA experiment
– The MeV-energy positrons observed (via a 511 keV line) by INTEGRAL
– The GeV+ energy positrons (and electrons) observed by PAMELA/Fermi/AMS
– The GeV-scale gamma ray excess from the galactic center
– The unexplained 3.5 keV x-ray claimed present in clusters, M31 and possibly the Milky Way

I will, in turn, discuss scenarios for these different anomalies.

I should begin with a disclaimer: I believe the anomalies, even ones that are ultimately not explained by dark matter, are often a good thing. The reason for this is simple: before we attempt to explain an anomaly, we often have a number of preconceptions (a.k.a. conventional wisdom) about how dark matter should or should not act. When pressed by an anomaly, we challenge those preconceptions directly, and often find that they are not nearly as general as we expected. This has happened frequently in the past ten years, in the form of the Sommerfeld enhancement (boosts to the dark matter cross section), leptophilic models, light ($\lesssim GeV$) dark matter, and self-interacting dark matter models, just to name a few.

In essence, our searches for dark matter rely on us looking under the lampost. Anomalies push us to expand our lampost, and even if it turns out that a particular anomaly does not arise from dark matter, it may help us realize pieces that go into the dark matter puzzle.

A major consequence of many of these anomalies of recent years is the more broadly considered possibility that the dark sector has a richer physics than we had commonly entertained before. In contrast to the relatively inert neutralino, there

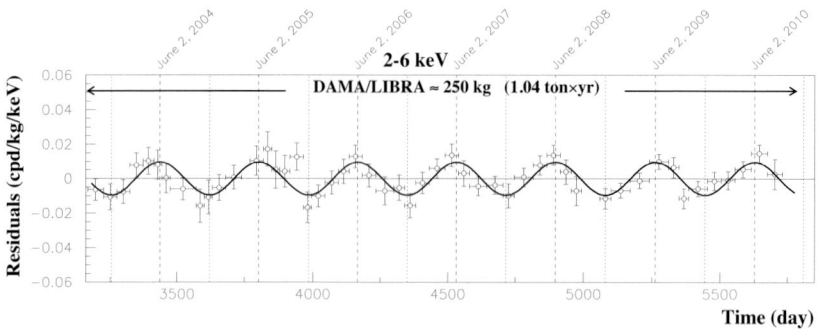

Fig. 7.   Modulation observed in the DAMA experiment, from Bernabei et al.[28]

has been great discussions of dark interactions, leading to signals in cosmic rays, direct detection, collider physics, and cosmological structures. Furthermore, there has been an explosion of interest in "dark forces," in analog to the forces of the standard model, and the means to look for them in controlled experiments.

4.1. *A Brief Comment on DAMA*

The DAMA annual modulation has been one of the most perplexing of all the anomalies. It has persisted for fifteen years, through an upgrade of the DAMA experiment, still without resolution. The basic signal is simple: an observation of a $\sim 1\%$ modulation in the low energy, singlet hit events in a NaI crystal detector. Dark matter is expected to produce such a signal, with the signal rising and falling as the Earth's motion around the sun adds constructively and destructively with the orbital motion of the sun through the milky way.

At this point, there is certainly no obvious consensus on what the background explanation would be. At the same time, no dark matter model in the literature (that I know of) can explain the excess, without appealing to ignorance (e.g. that energy calibrations are such that the LUX experiment has a large rate outside the region they show[29]). As a consequence, the situation is quite confusing.

There is a very basic reason that it is difficult to find explanations of DAMA (exotic or not). While the modulation is $O(1-2\%)$, much of the background is actually understood. Indeed, perhaps $10--20\%$ of the signal could be of some source other than the expected background. As a consequence, this $\sim 1-2\%$ modulation is really more like a $\sim 5-20\%$ modulation on this new component.[30,31] It is quite difficult to get things to modulate so profoundly, although some dark matter models naturally do.[32]

It would be unwise to confuse our inability to find an explanation with the absence of one, however. I consider, for instance, the model "Luminous Dark Matter".[33] In it, dark matter upscatters to an excited state in the Earth, then enters the detector and decays back down, emitting a photon at ~ 3.3 keV. This can lead

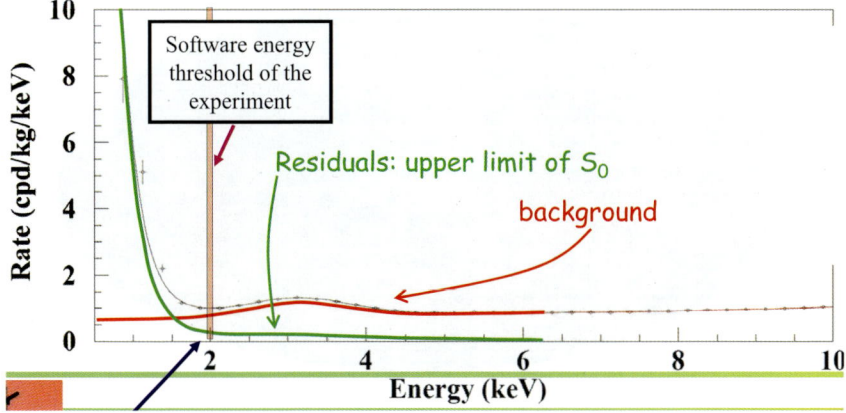

Fig. 8.   Model of background at DAMA, from talk by F. Nozzoli at TAUP 2009.

to a large modulation amplitude, and with a characteristic feature that it is *not* a nuclear recoil, and so would generally be vetoed by many experiments. At this point, it seems (even by inspection) to be inconsistent with what is observed at the XENON and LUX experiments, so this model is probably not precisely the explanation. However, the fact that such a qualitatively different model can arise a decade after the initial DAMA result suggests there may be other new ideas waiting to be had, and raises the question: what else are we missing?

4.2. *A Brief Comment on INTEGRAL*

Of the anomalies that have motivated recent work, a second is the *low energy* positron production seen in the galactic center by INTEGRAL. The observation of a prominent 511 keV line from the galactic center gives evidence of copious positron production. The basic arguments that motivate a dark matter interpretation are (1) that it is higher than is expected from conventional sources (by roughly an order of magnitude), and (2) that the morphology is (largely) spherical, while conventional sources would be expected to be disk-like. While there has been a sizable back and forth about conventional sources, such as LMXBs (see for instance arguments pro[34] and con[35]), in the spirit of considering anomaly mediated dark matter, let me focus on the dark matter interpretations.

The essential issue that a dark matter model must confront is that these positrons are very low energy,[36] produced with energies at most $\sim$ MeV. Thus, the MeV scale must appear somewhere in the problem. The most natural approach is to imagine that this is the scale of dark matter, and this is the approach that was taken by Boehm and Fayet,[37] who constructed models of MeV dark matter, annihilating directly into $e^{+}e^{-}$ via a light gauge boson. Another natural approach is to imagine the dark matter is MeVscale but decays, rather than annihilating, such as in axino dark matter.[38] These are both reasonable models, to be sure.

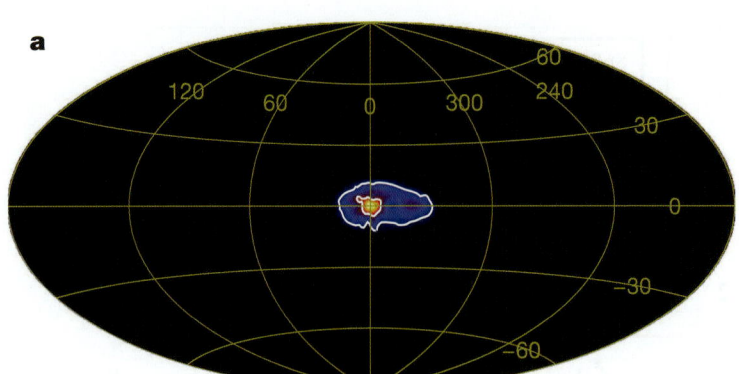

Fig. 9.   Sky distribution of the 511 keV radiation observed by INTEGRAL.[34]

However, there is another possibility: namely that we identify this with the kinetic energy of the WIMP. In a scenario called "exciting dark matter" (XDM),[39] it was proposed that dark matter might interact via a "light" mediator ϕ (i.e. $m_\phi \ll M_{\mathrm{DM}}$) which would allow collisional excitations into an excited state χ^*, which would decay back emitting e^+e^- pairs. (One can consider excitations into a charged state χ^+, but without the long range interaction it is a challenge to explain the excess.[40]) Such a possibility is interesting because it retains the possibility of WIMP type dark matter producing keV $-$ MeV scale signatures.

A simple model of this type involves dark matter χ and a light mediator ϕ. The mediator ϕ can interact with the standard model via a small mixing, for instance via $\phi h h^\dagger$ if ϕ is a scalar, or via a field strength mixing $\phi_{\mu\nu} F_Y^{\mu\nu}$ with hypercharge if a vector. The natural annihilation of this model is of the dark force type already described, where $\chi\chi \to \phi\phi$. When ϕ decays, it produces boosted standard model particles, and thus, in spite of having no *direct* interaction with the standard model, dark matter is capable of producing high energy cosmic ray signals.

4.3. A Slightly Longer Comment on Positrons

While the low energy positrons seen by INTEGRAL yielded some interest, it paled in comparison to the excitement of the field following the announcement by the PAMELA experiment of a sizable excess in positrons at 10+ GeV energies.[41] This was confirmed subsequently by Fermi[42] and AMS.[43]

It is intriguing to look at the situation prior to the announcement by PAMELA. There had been a previous claim by the balloon-borne HEAT experiment of an excess in positrons at slightly lower energies.[44] Anti-matter is a natural signal of dark matter annihilation (as is matter, but with less anti-matter in the universe, it provides generally a higher signal/background). But the expected signals were

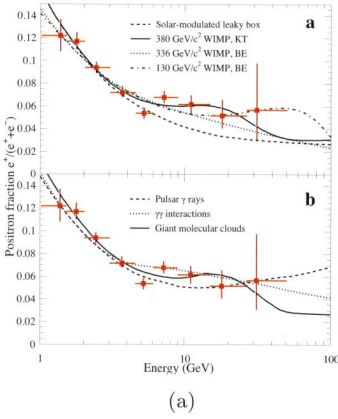

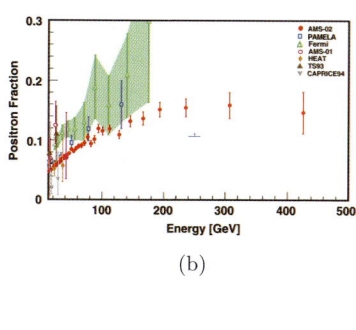

Fig. 10. Searches for cosmic ray positrons from HEAT[44] (a) and PAMELA/AMS[43] (b).

generally quite mild, as evidenced by the candidate theory curves that HEAT compared to in 1999. Thus, when the PAMELA results arrived, it was shocking, because while there was a clear rise in the positron spectrum, it was in striking contrast to expectations. It was, in some sense, too good to be dark matter.

Now, prior to PAMELA and more so subsequently, there have been challenges to how well we can interpret positron signals as evidence for physics beyond the standard model. For instance, both pulsars[45] as well as changes to propagation[46,47] have been claimed as possible explanations for the excess, among other astrophysical proposals. At the same time, while the dark matter explanation has not been definitively excluded, it has had many opportunities to show us that it is the correct explanation (via gamma ray signals, effects on the CMB, or in direct detection), and so far, it has not availed itself of any of those opportunities.

That said, we do not want to discount this possibility simply because we have plausible alternatives - we would like to really exclude it. Let us then attempt to understand the sorts of models that would explain it. The astrophysical proposals are certainly viable, but are not my focus here.

If dark matter, the signal should:

– Produce a rate of positrons above that is expected for a standard WIMP
– Produce a harder spectrum of positrons than expected for a standard WIMP
– Simultaneously produce no or few anti-protons, for which the standard models fit reasonably well

One interesting possibility lies in the dark force models already considered.[48,49] If the mediator ϕ is light $m_\phi \lesssim$ GeV, then an annihilation $\chi\chi \to \phi\phi$ produces a boosted ϕ. When it decays, the resulting particles will yield a hard spectrum of cosmic rays, in agreement with the observed spectrum. If ϕ is light compared to the dark matter, and produces a long range force (which, in this context means

~ fm range), the particles will have an enhanced annihilation cross section. This effect, known as the Sommerfeld enhancement, arises due to the attractive force that becomes more important for low velocity particles. As a consequence, the particles today annihilate with a rate higher than that of the early universe.

To achieve this, one needs a particle with mass $m_\phi \lesssim$ GeV, and, consequently, it is kinematically impossible for ϕ to decay to anti-protons. Thus, the simple assumption of a GeV-scale dark force immediately addresses these three points.

These models do lead to other predictions, however: one would expect gamma rays from the galactic center produced radiatively,[50] or a signal in the CMB as described above, which eliminates many if not most scenarios to explain this with annihilating dark matter. Such signals depend on things like the profile of the halo in the inner galaxy, or if there is even a factor of two boost from local substructure, the constraints from the CMB become easily evadable. Nonetheless, there have been opportunities for signals from these models and, as of yet, none of them have shown a positive sign.

Perhaps a more intriguing, and perhaps more important question is: why have we not see the dark force? This particle is light $m_\phi \lesssim$ GeV, and couples to ordinary matter through mixing. Shouldn't we have seen such a particle already?

Intriguingly, the answer is no. Basic standard model backgrounds are adequately high that makes usual collider searches less effective in looking for these states. There are possible search strategies, using high luminosity, low energy experiments.[51] Motivated in part by these models to explain PAMELA, searches have increased in number and reach, dramatically improving the constraints on these particles even in just a few years. In some ways, the search for dark photons as they are called parallels the search for WIMPs, but unlike that search, this is just in its infancy. And while the positron anomaly served as a jumping off point for these searches, as the field has broadly appreciated the role these dark forces can play, and how easily they can appear, these experiments are motivated by far more generally than any individual anomaly.

4.4. *A Signal in the Galactic Center*

One of the most intriguing results in the past few years has been the claim, originally by Goodenough and Hooper,[52] and later many others[53–57] that there is an excess of gamma rays in the inner galaxy at roughly the GeV energy scale. At this point, the signal has been well confirmed by a number of groups, and it is (I believe) safe to say that the model of diffuse gamma rays from the inner galaxy does not explain the full signal we see. Moreover, the residual appears to be spherically symmetric source, falling off roughly as $\sim (1/r^{1.2})^2$.

It is not obvious what the source of this would be, if not dark matter. A common candidate is a new population of millisecond pulsars (MSPs). As pointed out by Hooper et al[55] there are a number of issues with this:

– MSPs are generally thought to be related to LMXBs. In the inner galaxy, IN-

TEGRAL observed 6, while a standard ratio would have suggested more like 100 to explain the signal.

- The LMXBs trace the stellar population, and are not DM-like in their spatial morphology.
- The luminosity function of known MSPs applied to a population with this overall luminosity would suggest that a number of resolvable point sources should have appeared in this region and have not.
- The low energy spectrum rises while for MSPs it is generally flatter.

While these objections are not conclusive - after all, one might object, perhaps it is just a very distinct population of pulsars - it at least gives one enough basis, in my mind, to consider the dark matter interpretation.

Thus, we confront the question: what kind of dark matter would this be? A ~ 35 GeV particle annihilating dominantly into $b\bar{b}$ can fit the data well. This has led to a fairly common refrain of "Dark matter annihilating into $b\bar{b}$ - what could be simpler?"

This, I believe, sweeps a number of issues under the rug. Perhaps the simplest effective theory to achieve this was presented by Ipek, McKeen and Nelson.[58] There, the challenges of such a model are apparent: first, the annihilation must be via a pseudoscalar coupling, but there are no pseudoscalars in the SM that can do this, thus one must be invoked. To allow that pseudoscalar to couple to quarks, we must introduce one that couples to quarks, which involves an additional Higgs doublet. This then allows the annihilation, but with a number of important constraints, both from direct searches as well as flavor changing processes, on these models. In addition, there is now a new hierarchy problem (because of the added light Higgs multiplet), and there cannot be a low energy scalar field, lest the direct detection signals be too large.

I do not mean to imply that this cannot explain the signal - it can. It is just that a significant number of fields and broad constraints on the parameter space get swept away in these simple statements. For me, at least, this motivates consideration of alternative models.

One such alternative is to look to dark force models again.[59–63] The appeal of such models is that by annihilating into a dark force which then decays to the SM, it removes the tensions of having something with significant coupling to the standard model, while still yielding a significant signal. These models, in annihilating to leptons, will also have significant tension with the CMB and likely either heavy dark photons or some annihilation into scalars will be needed to evade those limits.

At the moment, the best prospect to test these models come from dwarf galaxies. Unfortunately, the improvement from statistics alone will not be enough to settle this question. Significant improvements will only come with more dwarfs. The recent limits (and claims of excesses) from recent DES candidate dwarfs drive this point home.[64] Absent that, this may be an anomaly that lingers for some time without definitive conclusion.

4.5. *A Line at 3.55 keV*

A second recent claim that has intrigued many people is that of an unexplained line at 3.55 keV. The initial claims came from two groups.[65,66] Bulbul et al[67] found evidence for the line in a stack of 73 clusters using XMM-Newton, as well as in a stack of Coma, Ophiucus, and Centaurus, and finally detailed XMM and Chandra observations of the Perseus cluster. Bulbul et al[65] also looked at Virgo with Chandra data and found no evidence for the line. Boyarsky et al[66] found evidence for the line in M31 and Perseus with an orthogonal set of XMM data.

These two papers resulted in a flurry of activity, both theoretical and observational.

This particular energy range is of interest because it is the natural scale for sterile neutrino dark matter, which would decay $\nu_s \to \nu\gamma$. While perhaps not precisely in the parameter space that was predicted, the signal as claimed by these papers is quite close to what was expected for sterile neutrino dark matter, which makes these results particularly interesting.

So far, there have been (including the initial papers) a number of claims to see evidence for the new line,[65,66,68] searches in various systems that *don't* see it,[69–73] and some that see an excess but believe the morphology is not correct to be described by dark matter.[74,75] In addition, there has been a debate about whether the atomic line background has been fully included,[67,68,76,76] which I won't go into here. To some extent, this is because the issues should be settled by Astro-H, and if this line is of exotic origin, we shall know soon enough, and we are left to understand (theoretically) what it might be.

Perhaps most challenging for the dark matter interpretation are the null results from dwarf galaxies[72] and stacked galaxies.[73] These are low background environments, in particular, formally exclude this possibility of decaying dark matter by 12σ.[73]

There are alternative dark matter possibilities, however. In particular, we can return to the proposal of XDM,[39] where dark matter collisionally excites to an excited state χ^*, then decays back to the ground state χ but now by emitting a photon, rather than e^+e^-.[77] Such a model offers the possibility to give better agreement to the data by having the signal more concentrated (so not appearing in outer parts of galaxies[73]) and requiring higher velocities to excite (so absent in dwarfs).

The importance of considering such a model is clear - while we can look to Astro-H to settle this, it will still depend on looking at the correct systems in the correct places. We can easily fool ourselves into believing things are excluded if we are grounded in a single model.

4.6. *A Very Brief Comment on the Properties of CDM*

Much discussion has been had as well on whether the properties of observed galaxies are those we expect from CDM. These deviations include cores in dwarf galaxies,[78] the "too big to fail problem",[79] as well as a question as to whether we see enough Milky Way satellite galaxies. One possibility to explain this is the idea of SIDM or "self interacting dark matter".[80]

This discussion would really warrant its own talk, and involves important questions about what physics is needed in simulations before we can make comparisons to data. I would only offer the minor point that the cross sections needed to yield deviations, with $\sigma/M_{\mathrm{DM}} \sim 1\mathrm{cm}^2/\mathrm{g} \sim 1 \mathrm{~GeV}^{-3}$ are quite easy to achieve in particle physics models.

There is the unfortunate collision of acronyms for SIDM (self interacting dark matter) and SIDM (strongly interacting dark matter). Also, the needed cross section is typically $O(\mathrm{barn})$ for weak scale dark matter to affect galaxy properties. Thus, SIDM (in the one sense) is often SIDM (in the other sense). But it is important to remember that one can have very large cross sections without "strong" dynamics if the mediator of the force is light. For instance, the Thomson cross section $\sim 0.7 \times 10^{-24}\mathrm{cm}^2$ is large because the photon is massless and the electron light, not because electromagnetism is strongly interacting.

Similarly, dark matter models with dark forces can naturally be large. Moreover, these forces are naturally velocity dependent,[81–85] which can alleviate tensions between null results[82] which are often in higher velocity environments. At a minimum, in the current context, cosmologically relevant dark matter interactions are easy to come by, and theoretically well motivated.

5. Conclusions

The search for dark matter is one of the most important questions in particle physics, astrophysics and cosmology. Simultaneously, it is one of the most frustrating, because we simply do not know for sure how we should be looking for it. In a nutshell, the search for particle dark matter is, has been and likely will continue to be, messy.

That said, we have learned a tremendous amount about dark matter in recent years, mostly in the form of what dark matter is not. We have great prospects to learn more about dark matter in coming years, for instance: does it interact with the Higgs? Could it be an axion?

But there is a huge variety of dark matter models out there, with thermal WIMPs and axions just being two. Of late, we have begun to consider more actively dark matter models with significant dark interactions. Some might object to this by appealing to Occam's razor, but the response might be this: that the entire universe we see is very complicated - is it really the simplest possibility to assume that the dark sector is just a single particle?

As we search for dark matter, it is quite clear that there will be many red herrings along the way. Indeed, there have already been many, and it is a personal choice about whether this is a type of science you are willing to engage in. For me, personally, I do not view these events as distractions. Often, these events challenge us to broaden our minds.

Anyone in this field has their own path that led them to consider these new ideas. For me, it started with INTEGRAL. This positron excess pushed me to think about dark force models, which, in turn, helped motivate my thinking about PAMELA. The models for PAMELA may not explain positron excesses, but maybe they are relevant for the galactic center. Or perhaps dark matter excitations is not relevant for INTEGRAL, but developing ideas along the way makes it easier for us to think about the X-ray signal from clusters. Or perhaps none of this is dark matter, but the models we have been developing, as a community, will turn out to allow us to understand a different signal, or make new experiments that help us find dark matter. When you have a shortage of lampposts, investing in new lampposts is as important as looking under the ones you have.

So I cannot tell you whether this is a style of physics that will work for you. It is, for better or worse, an important part of thinking about dark matter. The anomalies along the way - you can ignore them, or accept that a lot of time will be spent developing models that may turn out not to be dark matter. What you do with this, that's up to you. It just depends on what gets you out of bed in the morning.

References

1. H. Murayama (2007).
2. W. Hu, R. Barkana and A. Gruzinov, *Phys.Rev.Lett.* **85**, 1158 (2000).
3. S. Tremaine and J. Gunn, *Phys.Rev.Lett.* **42**, 407 (1979).
4. S. Asztalos *et al.*, *Phys.Rev.Lett.* **104**, p. 041301 (2010).
5. P. W. Graham and S. Rajendran, *Phys.Rev.* **D84**, p. 055013 (2011).
6. P. W. Graham and S. Rajendran, *Phys.Rev.* **D88**, p. 035023 (2013).
7. D. Budker, P. W. Graham, M. Ledbetter, S. Rajendran and A. Sushkov, *Phys.Rev.* **X4**, p. 021030 (2014).
8. K. Olive *et al.*, *Chin.Phys.* **C38**, p. 090001 (2014).
9. O. Wantz and E. Shellard, *Phys.Rev.* **D82**, p. 123508 (2010).
10. M. Tegmark, A. Aguirre, M. Rees and F. Wilczek, *Phys.Rev.* **D73**, p. 023505 (2006).
11. P. Fox, A. Pierce and S. D. Thomas (2004).
12. M. W. Goodman and E. Witten, *Phys.Rev.* **D31**, p. 3059 (1985).
13. D. Akerib *et al.*, *Phys.Rev.Lett.* **112**, p. 091303 (2014).
14. L. Baudis, J. Hellmig, G. Heusser, H. Klapdor-Kleingrothaus, S. Kolb *et al.*, *Phys.Rev.* **D59**, p. 022001 (1999).
15. R. J. Hill and M. P. Solon, *Phys.Rev.* **D91**, p. 043504 (2015).
16. R. J. Hill and M. P. Solon, *Phys.Rev.* **D91**, p. 043505 (2015).
17. T. Cohen, M. Lisanti, A. Pierce and T. R. Slatyer, *JCAP* **1310**, p. 061 (2013).
18. A. Hryczuk, I. Cholis, R. Iengo, M. Tavakoli and P. Ullio, *JCAP* **1407**, p. 031 (2014).
19. A. Geringer-Sameth, M. G. Walker, S. M. Koushiappas, S. E. Koposov, V. Belokurov *et al.* (2015).
20. D. Hooper and T. Linden (2015).
21. M. Ackermann *et al.* (2015).
22. N. Padmanabhan and D. P. Finkbeiner, *Phys.Rev.* **D72**, p. 023508 (2005).
23. T. R. Slatyer, N. Padmanabhan and D. P. Finkbeiner, *Phys.Rev.* **D80**, p. 043526 (2009).
24. S. Galli, F. Iocco, G. Bertone and A. Melchiorri, *Phys.Rev.* **D80**, p. 023505 (2009).

25. S. Galli, F. Iocco, G. Bertone and A. Melchiorri, *Phys.Rev.* **D84**, p. 027302 (2011).
26. M. S. Madhavacheril, N. Sehgal and T. R. Slatyer, *Phys.Rev.* **D89**, p. 103508 (2014).
27. P. Ade *et al.* (2015).
28. R. Bernabei, P. Belli, F. Cappella, V. Caracciolo, S. Castellano *et al.*, *Eur.Phys.J.* **C73**, p. 2648 (2013).
29. G. Barello, S. Chang and C. A. Newby, *Phys.Rev.* **D90**, p. 094027 (2014).
30. J. Pradler, B. Singh and I. Yavin, *Phys.Lett.* **B720**, 399 (2013).
31. J. Pradler and I. Yavin, *Phys.Lett.* **B723**, 168 (2013).
32. D. Tucker-Smith and N. Weiner, *Phys.Rev.* **D64**, p. 043502 (2001).
33. B. Feldstein, P. W. Graham and S. Rajendran, *Phys.Rev.* **D82**, p. 075019 (2010).
34. G. Weidenspointner, G. Skinner, P. Jean, J. Knodlseder, P. von Ballmoos *et al.*, *Nature* **451**, 159 (2008).
35. R. M. Bandyopadhyay, J. Silk, J. E. Taylor and T. J. Maccarone, *Mon.Not.Roy.Astron.Soc.* **392**, p. 1115 (2009).
36. P. Jean, J. Knodlseder, W. Gillard, N. Guessoum, K. Ferriere *et al.*, *Astron.Astrophys.* **445**, 579 (2006).
37. C. Boehm and P. Fayet, *Nucl.Phys.* **B683**, 219 (2004).
38. D. Hooper and L.-T. Wang, *Phys.Rev.* **D70**, p. 063506 (2004).
39. D. P. Finkbeiner and N. Weiner, *Phys.Rev.* **D76**, p. 083519 (2007).
40. M. Pospelov and A. Ritz, *Phys.Lett.* **B651**, 208 (2007).
41. O. Adriani *et al.*, *Nature* **458**, 607 (2009).
42. M. Ackermann *et al.*, *Phys.Rev.Lett.* **108**, p. 011103 (2012).
43. L. Accardo *et al.*, *Phys.Rev.Lett.* **113**, p. 121101 (2014).
44. S. Coutu, S. W. Barwick, J. J. Beatty, A. Bhattacharyya, C. R. Bower *et al.*, *Astropart.Phys.* **11**, 429 (1999).
45. D. Hooper, P. Blasi and P. D. Serpico, *JCAP* **0901**, p. 025 (2009).
46. B. Katz, K. Blum and E. Waxman, *Mon.Not.Roy.Astron.Soc.* **405**, p. 1458 (2010).
47. K. Blum, B. Katz and E. Waxman, *Phys.Rev.Lett.* **111**, p. 211101 (2013).
48. N. Arkani-Hamed, D. P. Finkbeiner, T. R. Slatyer and N. Weiner, *Phys.Rev.* **D79**, p. 015014 (2009).
49. M. Pospelov and A. Ritz, *Phys.Lett.* **B671**, 391 (2009).
50. P. Meade, M. Papucci and T. Volansky, *JHEP* **0912**, p. 052 (2009).
51. R. Essig, J. A. Jaros, W. Wester, P. H. Adrian, S. Andreas *et al.* (2013).
52. D. Hooper and L. Goodenough, *Phys.Lett.* **B697**, 412 (2011).
53. D. Hooper and T. Linden, *Phys.Rev.* **D84**, p. 123005 (2011).
54. K. N. Abazajian and M. Kaplinghat, *Phys.Rev.* **D86**, p. 083511 (2012).
55. D. Hooper, I. Cholis, T. Linden, J. Siegal-Gaskins and T. Slatyer, *Phys.Rev.* **D88**, p. 083009 (2013).
56. D. Hooper and T. R. Slatyer, *Phys.Dark Univ.* **2**, 118 (2013).
57. T. Daylan, D. P. Finkbeiner, D. Hooper, T. Linden, S. K. N. Portillo *et al.* (2014).
58. S. Ipek, D. McKeen and A. E. Nelson, *Phys.Rev.* **D90**, p. 055021 (2014).
59. D. Hooper, N. Weiner and W. Xue, *Phys.Rev.* **D86**, p. 056009 (2012).
60. A. Berlin, P. Gratia, D. Hooper and S. D. McDermott, *Phys.Rev.* **D90**, p. 015032 (2014).
61. M. Abdullah, A. DiFranzo, A. Rajaraman, T. M. Tait, P. Tanedo *et al.*, *Phys.Rev.* **D90**, p. 035004 (2014).
62. A. Martin, J. Shelton and J. Unwin, *Phys.Rev.* **D90**, p. 103513 (2014).
63. J. Liu, N. Weiner and W. Xue (2014).
64. K. Bechtol *et al.* (2015).

65. E. Bulbul, M. Markevitch, A. Foster, R. K. Smith, M. Loewenstein *et al.*, *Astrophys.J.* **789**, p. 13 (2014).
66. A. Boyarsky, O. Ruchayskiy, D. Iakubovskyi and J. Franse, *Phys.Rev.Lett.* **113**, p. 251301 (2014).
67. E. Bulbul, M. Markevitch, A. R. Foster, R. K. Smith, M. Loewenstein *et al.* (2014).
68. A. Boyarsky, J. Franse, D. Iakubovskyi and O. Ruchayskiy (2014).
69. S. Horiuchi, P. J. Humphrey, J. Onorbe, K. N. Abazajian, M. Kaplinghat *et al.*, *Phys.Rev.* **D89**, p. 025017 (2014).
70. T. Tamura, R. Iizuka, Y. Maeda, K. Mitsuda and N. Y. Yamasaki, *Publ.Astron.Soc.Jap.* **67**, p. 23 (2015).
71. S. Riemer-Sorensen (2014).
72. D. Malyshev, A. Neronov and D. Eckert, *Phys.Rev.* **D90**, p. 103506 (2014).
73. M. E. Anderson, E. Churazov and J. N. Bregman (2014).
74. O. Urban, N. Werner, S. Allen, A. Simionescu, J. Kaastra *et al.* (2014).
75. E. Carlson, T. Jeltema and S. Profumo, *JCAP* **1502**, p. 009 (2015).
76. T. Jeltema and S. Profumo (2014).
77. D. P. Finkbeiner and N. Weiner (2014).
78. S.-H. Oh, W. de Blok, E. Brinks, F. Walter and J. Kennicutt, Robert C., *Astron.J.* **141**, p. 193 (2011).
79. M. Boylan-Kolchin, J. S. Bullock and M. Kaplinghat, *Mon.Not.Roy.Astron.Soc.* **415**, p. L40 (2011).
80. D. N. Spergel and P. J. Steinhardt, *Phys.Rev.Lett.* **84**, 3760 (2000).
81. J. L. Feng, M. Kaplinghat and H.-B. Yu, *Phys.Rev.Lett.* **104**, p. 151301 (2010).
82. A. Loeb and N. Weiner, *Phys.Rev.Lett.* **106**, p. 171302 (2011).
83. S. Tulin, H.-B. Yu and K. M. Zurek, *Phys.Rev.Lett.* **110**, p. 111301 (2013).
84. S. Tulin, H.-B. Yu and K. M. Zurek, *Phys.Rev.* **D87**, p. 115007 (2013).
85. K. Schutz and T. R. Slatyer, *JCAP* **1501**, p. 021 (2015).

Discussion

S. White Thank you Neal. Now let us discuss dark matter from the particle physics aspects in general. Let us leave out the accelerators for a bit and talk about the things we have heard in these two talks first. You want to make a comment, Werner?

W. Hofmann Well, it is a comment and a question, mostly on indirect detection. Clearly, indirect detection has right now these interesting signatures, but I agree with you that most of them probably have fairly mundane explanations. It may be worthwhile to point out that somewhat contrary to direct detection, indirect detection is at the limit of the nominal cross section. You pointed out the Fermi data. Hopefully in a decade or so with instruments like CTA, we will cover the range up to TeV energies and really, at the nominal cross section, cover the relevant range. Now what I wanted to ask is: indirect and direct detection are complementary in the sense that they cover somewhat different regions of parameter space, but from a pragmatic point of view would you consider direct detection as convincing or would we need to see indirect detection to make sure that what we see on Earth is actually the dark matter which is out there?

N. Weiner I think she is going to say that direct detection is very convincing.

L. Baudis I think that direct detection would be very convincing, I am not saying it should be the only detection that we have, but in some sense it would be not as messy because we do not have to deal with the astrophysics where you have to integrate over the line of sight. But I should comment also on something that you said: you said that if the dark matter does not interact with quarks or gluons, then we would not see it in direct detection, and I would say that it is not true, because you are actually also looking for a particle that for instance scatters off electrons. So we might not have the same sensitivity, because there the backgrounds are somewhat higher, but we are not ignoring that region. Actually, I did not mention that, but we do set limits on Solar axions, on axion-like particles and so on. And this is also why I mentioned not only the coherent neutrino-nucleus scattering as a background, but actually also the Solar pp neutrino and ^{7}Be-neutrino scattering off electrons because those give us a direct background for any searches for whatever you call it: dark matter scattering electromagnetically.

N. Weiner Let me comment on the question directly. I think that they are very complementary, and there are some situations where indirect detection has been hugely important. Let me give you one example: I showed the example of this SU(2) triplet – the so-called wino. For a number of us this is in some sense a nightmare scenario, because for it to be the dark matter it should have a mass around 2.5 TeV, and if that is the lightest superpartner (that means the bottom of your superspectrum is at 2.5 TeV), and it is an electro-weakly charged particle, which is very hard to produce at the LHC. If your colour particles are then a factor of a few above that, you are essentially out of luck, as far as the LHC goes to produce these things. At least, it is very very hard. The cross-section I showed you for direct detection is very hard to see, it is very low: 10^{-47} and this is a heavy particle. But this is a situation where actually indirect detection has already done a tremendous service, because these particles would have a line-like feature that would show up, and actually searches by H.E.S.S. have already – on this particular model – shown that it is incompatible: this, as a thermal relic, should not be the dark matter. There are some things which are similar to it, that are equally hard to find with direct detection, that are currently not constrained by those things, but that future searches would have the capability of looking for. The cross-talk between these experiments is not a panacea because there are some models that just won't show up anywhere, and there are some models that only show up in one place. But there are some situations where we have the set of experiments we have, we have good theorists working on them, and together they can cover a lot more parameter space than they would have covered separately. As you said, the cross-section that they are studying is interesting.

M. Rees The WIMP miracle argument assumes of course that there are equal numbers of particles and anti-particles. Are there any models which allow an asymmetry as for baryons, in which case you could allow larger cross-sections?

N. Weiner There are, and obviously for the sake of time I did not discuss them, but there is a tremendous amount of interest in this. There has been for some time, motivated originally just by the observation that Ω_{baryons} is relatively close to $\Omega_{\text{darkmatter}}$. People considered originally the possibility that dark matter carries baryon number and that the number density of that original baryon number has somehow separated into a dark sector and the observable baryons. More generally, you can very simply have dark matter which carries some primordial conserved number, and then unfortunately it is very hard to have an indirect detection signal, because then you are left with only dark matter and no anti-dark matter, but very often these models still have interesting collider or direct detection signatures.

R. Blandford I will just make a very short comment about the electron/positron ratio as a possible signature of dark matter. You mentioned correctly that an alternative source might be pulsars. I would characterize this as a sort of sin of emission, but there is also a sin of transmission, which is that, as the particles propagate here from a given source, they are scattered by resonant Alfvén waves, and this is a charge-dependent phenomenon. There are more protons than there are electrons controlling the wave turbulence, and so you could have an energy-dependent variation of propagation which would change the e^+/e^- ratio that you measure.

C. Frenk It is a question for both of you. The 3.5 keV line was very exciting because if it is a sterile neutrino, that seems to be a really good particle to have because it would explain leptogenesis and it gives a mechanism to explain the masses of the light neutrinos. So the question is: if it is not decaying but scattering, do you understand why you do not see it from galaxies. Would that still be compatible with a sterile neutrino and part (B) is: are there any other ways one could try and see whether dark matter is a sterile neutrino, or whether sterile neutrinos exist in the right mass range.

N. Weiner If it is scattering then it won't be a sterile neutrino, it has to be something even more exotic than a sterile neutrino. I do not know if you want to comment on the direct searches for sterile neutrinos.

L. Baudis We would not be able to detect them definitely with the detectors that we have now, not with the type of direct detection experiment because the energy that would be deposited is way below our threshold.

C. Frenk Sure, not with those experiments, but are there any class of experiments that could detect these things?

L. Baudis There are what he mentioned – there are some limits and some other class of experiments that would look to produce this type of particles. Not the sterile neutrinos, but heavier. I guess you are talking about the model by Mikhail Shaposhnikov, so you could be able possibly to produce the heavier GeV or MeV type of particles and see those in an accelerator-type experiment. We would detect the dark matter directly, the lightest one.

S. White Maybe at this point we could have Hitoshi tell us something about the accelerator experiments since this keeps coming up.

Prepared comment

H. Murayama Dark Matter and Accelerator

I was asked to give a remark on the topic of dark matter and accelerators. I interpreted the charge as to discuss the relevance of accelerator-based particle physics experiments on discovering the dark matter in the laboratory. The short answer is *we don't know*, but many of us in the community are excited about this possibility. In the rest of this prepared remark, I'll give a longer answer.

First of all, we know so little about dark matter that its possible mass ranges all the way from 10^{-31} GeV to 10^{50} GeV. The lower limit comes from the requirement that the quantum wave function of dark matter should not be too fluffy, namely that the analogue of the Bohr radius would fit in within the galactic scale. The upper bound comes from the search for MACHOs (Massive Compact Halo Objects) with gravitational microlensing. Therefore, there is no empirical argument for the relevance of accelerators on the dark matter question.

Now come theoretical considerations rather than emprical ones. There is no good idea to create dark matter in the evolution of the Universe from the Planck scale, the heaviest particle one can conceive, to the MACHO limit. It is not to say that it is impossible, but highly inconceivable. A part of the reason is that any elementary particle in this mass range can be viewed as microscopic blackholes, which would evaporate too quickly to compose the dark matter in the current Universe.

The most important reason why people (including myself) are excited about discovering dark matter particle at accelerator experiments is the concept of thermal relics. If the dark matter particles are created in the Big Bang, they are likely to annihilate with each other to reduce their abundance. Assuming that the annihliation process takes place with the strengths similar to the Standard Model gauge interactions, the annihilation cross section should be something like $\sigma_{ann}v \sim \pi\alpha^2/m^2$, where α is the analog of the fine-structure constant in the dark matter annihilation process, and m is the mass of the dark matter. In order to obtain the dark matter abundance required by the cosmological observations (e.g. CMB anisotropy), we need $\sigma_{ann}v \sim 10^{-9}\text{GeV}^{-2}$. This is why TeV-scale mass of dark matter is of intense interest.

The dark matter may not annihilate completely, if there is an asymmetry between dark matter and anti-dark matter, just like in the case of the baryon asymmetry in the Universe $n_B/n_\gamma \sim 10^{-9}$. We know nothing about the asymmetry in dark matter, but if we *assume* the asymmetry is of the same order of magnitude, the preferred mass range would be $m_p\Omega_c/\Omega_b \sim 6$ GeV. This is again within the scope of the accelerator-based experiments. If the mass is this light, the search may be done at high-intensity low-energy accelerators rather than at (relatively speaking) low-intensity high-energy accelerators such as LHC. It is the basis of an argument for the so-called "intensity frontier" experiments with low-energy accelerators.

One way to look for dark matter is the traditional missing energy (or transverse energy in the case of hadron machines) signature. By adding all particles you could observe in a collision event, and if there is an imbalance in energy-momentum conservation, you can conclude that "invisible" particles were created in that event. To use this method to place limits on the dark matter production, we obviously need to make assumptions on its production process, and hence the method is inherently model-dependent. Recently a less model-dependent approach based on the so-called "Effective Theory" has become popular to study signals at LHC, but one cannot be completely model-independent. Nonetheless, the method cuts into the parameter space of specific models such as supersymmmetry in a significant way, raising excitements that we will learn something important in the near future.

Given the discovery of the Higgs boson, another exciting avenue is to look for the decay of the Higgs boson into invisible particles. In the Standard Model, simple dimensional analyses suggest that any interactions with non-standard particles are least suppressed for the Higgs boson. This is the concept called "Higgs portal" to the dark sector. The search in this direction would require an electron-positron collider, such as ILC (International Linear Collider).

In addition to looking for dark matter directly, one may also look for the "mediators," which links the standard model particles with the dark matter. It may be a new force-particle such as Z', or new scalars beyond the Higgs boson. In lack of a definitive theory for mediators, the search for the mediators requires

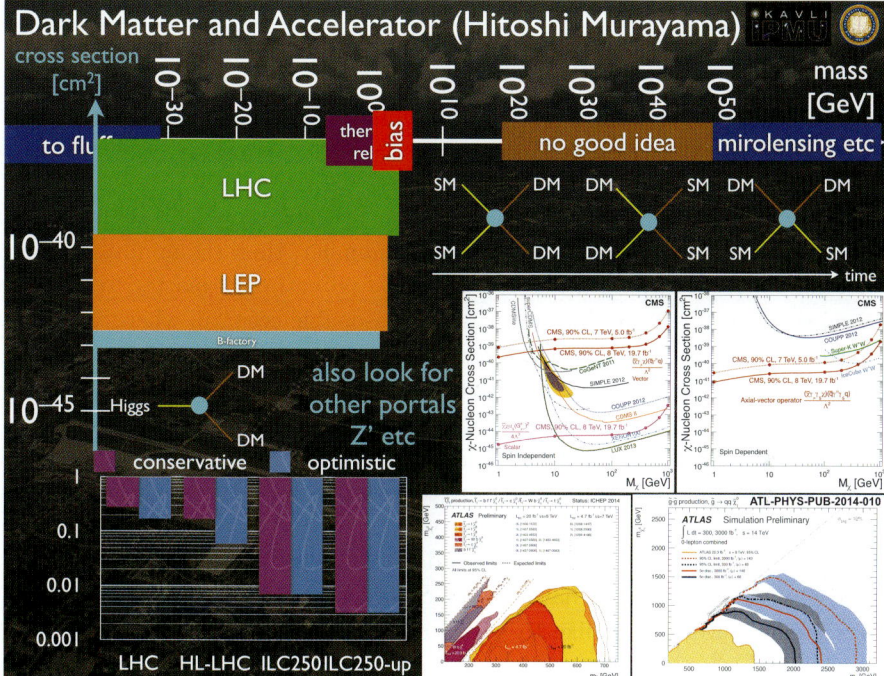

Fig. 1.   The slide presented at the meeting.

casting a wide net.

Laura Baudis in this session discusses the prospect for underground experiments to detect dark matter interactions directly. Most of them are targeted in the same mass range of GeV to TeV as the accelerator-based searches, and their information is complementary.

Since dark matter is mother of all structures in the Universe, from stars to galaxies and clusters of galaxies, we need to understand it. We would like to meet *her* and express our gratitude. We now know for sure that dark matter must exist, yet nobody has met *her*. If the argument based on the thermal relics and/or asymmetry is what Nature decided to use, we are in a good position to hope for either accelerator or underground experiments to see an evidence of new particles relevant to this question.

I hold high hopes that we will have concrete data from these experiments by the next Solvay Conference on this subject.

Discussion

S. White We have more comments about how we can put all these different things together. Ralph has been waiting for a while.

R. Wijers I had a question about the concern that Neal quite rightly raised. Many of the indirect detection experiments for WIMPs assume the nominal standard cross section and you could well be off from that. Are there at least some generic ideas about how far off one could be in the present Universe relative to the early Universe, or is there as much freedom in that correction factor as there is in the types of WIMPs?

N. Weiner There is the example that Martin brought up, which is that, if dark matter actually contains some primordial number, then the annihilation cross section is essentially zero, because you have no anti-dark matter to annihilate against. I am just pointing out a fairly reasonable scenario where you do not have a chance of seeing indirect detection. In the context of models where there is no conserved quantum number, it can be quite small. If the velocity-dependent cross section is an s-wave cross section, the first velocity-dependent cross section is a p-wave that goes like v^2, and so, if there is a reason – and there can be various reasons – why your s-wave cross section is suppressed, then your leading cross section might be p-wave and so you might lose orders of magnitude. And really all you have to lose is a few orders of magnitude and it is very hard to see. But the models that have a large cross section are very reasonable too.

S. Tremaine I would just like to point out that if Neal is correct and the dark sector is as complicated as the baryonic sector, then gravitational wave detectors presumably have at least as good a chance of detecting events from the dark sector as from our sector.

N. Weiner You would have to have some sort of cooling process in the dark sector to get some compact objects. That might be bad, experimentally. But maybe some subcomponent of dark matter could be like that.

M. Kamionkowski I just want to respond to Carlos' question about sterile neutrinos. There is a group from Princeton called PTOLEMY that is starting to think about ways to detect the neutrino background with inverse β-decay detection experiments, and it is reasonable that if those could some day be done, modifications to those might be used to actually look directly for sterile neutrinos. So there is a bigger mass, so the interaction strength is stronger, but that is compensated then by the mixing angle of the sterile neutrino into the neutrinos that you would actually detect. So it is not completely implausible that by the next Solvay conference people will be discussing direct detection of this sterile neutrino dark matter.

S. White Thank you, I can see David Spergel down here.

D. Spergel A question for Neal. If you have this dark photon, and more generically more interactions in the dark sector, is it likely that you will produce another background of particles, basically a dark photon background? We are starting to get interesting constraints on the number of relativistic species from the CMB and that will continue to improve over the next few years. I am wondering whether that will place a useful constraint on things in that sector.

N. Weiner That is actually a very interesting point, because in the simplest models, by design, you make sure that the dark photon is massive enough and decays early enough so you do not have to worry about corrections to the relativistic number of degrees of freedom. But when you branch out and you start thinking, like I said you can make any model with supersymmetry, and when you do that you often have additional particles around that stay around, and those can be in thermal equilibrium, and be extremely light, and be relativistic, so at least in some extensions this would be interesting, because these sectors are in thermal equilibrium until quite late.

S. White We have one hour of discussion now before lunch, which is mostly going to be about astrophysical aspects of dark matter, but I am sure some of the things we have been hearing about the particle aspects will come back again, looked at from different angles. We will start by two short contributions to get things going from Uros Seljak and Carlos Frenk on astrophysical constraints on dark matter from large scale structure and from small scale structure in the properties of galaxies. Uros will discuss the simple linear believable part of the constraint on dark matter from structure. After that Carlos who will talk about all the non-linear complicated parts.

Prepared comments

U. Seljak Dark Matter Constraints from Large Scale Structure of the Universe

I have been asked to summarize what does Large Scale Structure (LSS) of the universe tell us about the dark matter. It can mostly probe the neutrino sector. One example are non-standard neutrinos that could be the dark matter, such as an extra neutrino family. Standard neutrinos would need a mass of around 10eV to match the observed dark matter density, but they would move very fast and suppress small scale structure, and have been ruled out as the dark matter long ago. If you make the neutrinos more massive, and reduce the abundance at the same time, neutrinos become more and more cold and approach cold dark matter in the limit of high mass. In between these two limits there are some unique features and the scenario is called warm dark matter (WDM). The neutrinos can either be fully thermalized or sterile, i.e. weakly coupling through the mixing matrix to the standard neutrinos. Neutrinos free stream and erase their own structures on scales shorter than the free-streaming scale, leading to a suppression of clustering strength on small scales. Ideally this should be searched for on scales which are still linear, so that linear calculations can be used. In the non-linear regime the nonlinear physics and various astrophysical effects make the interpretation more difficult.

At the moment the best observations to search for suppression of clustering on small scales are those of Lyman alpha forest in distant clusters. These observations probe small scales, and because they probe structures up to redshifts of 4, the structures are much more linear than today, making theoretical predictions more reliable. The data show no evidence of WDM, putting strong lower mass limits on these models, and ruling out completely some simple specific scenarios, where absence of X-ray emission in clusters puts an upper limit on the mass. While neutrinos likely cannot be all of the dark matter, they do have mass and are non-relativistic today, so they contribute to the dark matter density. The fraction of neutrino mass density in the dark matter component is expected to be 1% or less. This also leads to a suppression of clustering in LSS data, at a level of about ten times the neutrino fraction, and on scales sufficiently large that it should be observable with the current LSS observations, such as redshift space distortions of galaxies, weak lensing, and cluster abundance. The clustering strength can be compared to that of the cosmic microwave background (CMB) to measure this fraction, and determine the sum of the masses of the three known neutrino families. Of special interest is the so called normal hierarchy scenario, with the minimum sum of the neutrino masses of 0.06eV, versus so called inverted hierarchy, which predicts 0.11eV for the minimum sum of the masses. At the moment combining all the data points to a consistent picture where the sum cannot exceed 0.2eV, with the most likely value somewhere around 0.05-0.10eV. In the near future we will very likely be able to exclude zero mass, but it will take us longer to settle on one of the two scenarios at any relevant significance level. There is currently no evidence that the sum of the masses is above the minimum values above.

A significant uncertainty in this method comes from the optical depth of reionization, which creates a degeneracy with the amplitude of the CMB, and is rapidly becoming the leading source of uncertainty. There is considerable difference in the value of this parameter as reported by WMAP, versus Planck satellite values. The cause of this uncertainty is dust polarization, which has not been measured by the WMAP team. This is the same source of uncertainty which has caused the BICEP2 result on primordial gravity waves to be in doubt. My reanalysis (with Mortonson) of BICEP2, which accounts for dust polarization uncertainty, and includes recent Planck dust polarization constraints on the BICEP2 patch, gives an upper limit $r < 0.09$ at 95% confidence levels. Despite some concerns that have been voiced at this meeting I consider this result to be robust and I expect the joint Planck-BICEP2 analysis, which is underway, to give a similar result (note: subsequent Planck papers in early 2015, which include joint BICEP2-Planck analysis, give the same $r < 0.09$ constraint). An independent measurement of the optical depth would be highly valuable and may come from detailed early dawn 21cm measurements of reionization process at redshifts 6-10.

C. Frenk Astrophysical Clues to the Identity of the Dark Matter

Cold, collisionless particles are the most popular candidates for the dark matter. There are sound reasons for this, both from the points of view of astrophysics and particle physics. The current standard model of cosmology, ΛCDM (where

Λ stands for Einstein's cosmological constant and CDM for cold dark matter), is based on this idea. ΛCDM accounts for an impressive array of data on the structure of the Universe on large-scale scales, from a few gigaparsecs down to a few megaparsecs, where the cosmic microwave background (CMB) radiation and the clustering of galaxies provide clean and well-understood diagnostics. Particle physics provides well-motivated CDM candidates, most famously the lightest supersymmetric particle, which would have the relic abundance required to provide the measured dark matter content of our Universe.

In recent years, an altogether different kind of elementary particle, sterile neutrinos, have emerged also as plausible candidates for the dark matter. These particles are predicted in a simple extension of the Standard Model of particle physics and, if they occur as a triplet, could explain the observed neutrino oscillation rates as well as baryogenesis. In the neutrino Minimal Standard Model one of this triplet would have a mass in the keV range and behave as warm dark matter (WDM). These particles would be relativistic at the time of decoupling and subsequent free streaming would damp primordial density fluctuations below some critical cutoff scale which varies inversely with the particle mass and, for keV particles, corresponds to a dwarf galaxy mass.

CDM and WDM models produce very similar large-scale structure and are thus indistinguishable by standard CMB and galaxy clustering tests. However, as is clearly shown in the figure, they produce completely different structure on scales smaller than dwarf galaxies where the WDM primordial power spectrum is cut off whereas the CDM power spectrum continutes to increase (logarithmically). The search for the identity of the dark matter is currently at a very exciting stage, with tentative, mutually exclusive, claims for detections of CDM[a] and WDM.[b] While eagerly awaiting developments on the dark matter particle detection front, it is incumbent upon astrophysicists to develop tests that could distinguish between the two frontrunner candidates. A glance at the figure suggests that such a test should not be too difficult: the two simulated galactic dark matter halos look completely different. In CDM there is a very large number of small subhalos whereas in WDM only a handful of the most massive ones are present. Surely, it cannot be beyond the ingenuity of observational astronomers to tell us whether the Milky Way resides in a halo like that on the left or in one like that on the right of the figure!

It turns out that distinguishing CDM from WDM merely by counting the number of small satellites orbiting the Milky Way (or similar galaxies) is not possible once galaxy formation is taken into account. This is because there are two physical processes that inhibit the formation of galaxies in small halos: the reionization of Hydrogen in the early universe heats up gas to a temperature higher than the virial temperature of small halos (a few kilometers per second)

---

[a]Hooper, D. & Goodenough, L. 2011, PhL,697,412.
[b]Bulbul, E. et al. 2014, ApJ, 789, 13; Boyarski, A. et al. 2014, PhRvL.113, 1301.

Fig. 1.   N-body simulations of galactic halos in universes dominated by CDM (left) and WDM (right; for a particle mass of 2 keV). Intensity indicates the line-of-sight projected square of the density, and hue the projected density-weighted velocity dispersion. Taken from Lovell et al. (2012).

and, in slightly larger halos where gas can cool, the first supernovae explosions expel the remaining gas. Models of galaxy formation going back over 10 years,[c] and more recent simulations, confirm this conclusion. In fact, if anything, requiring that a galactic halo should contain at least as many subhalos as there are observed satellites in the Milky Way sets a lower limit on the mass of WDM particle candidates.[d] (A similar limit can be placed from the degree of inhomogeneity in the density distribution at early times as measured by the clustering of Lyman-α clouds[e]).

Other tests based on the observed properties of Milky Way satellites also fail to distinguish CDM from WDM. In CDM halos and subhalos have cuspy "NFW" density profiles while in WDM one would expect cores to form in their inner regions. However, detailed calculations show that such cores would be much too small to be astrophysically relevant.[f] Another test is offered by the "too-big-to-fail" argument[g] that CDM N-body simulations predict a population of massive, dense subhalos whose structure is incompatible with the observed kinematics of stars in the brightest Milky Way satellites. WDM models neatly avoid this problem because the inner densities of subhalos are slightly less dense that those of their cold dark matter counterparts (on account of their more recent formation epoch).[h] However, baryon physics produce exactly the same effect in CDM subhalos.[i]

[c]Benson, A. J. et al. 2002, MNRAS, 333, 177.
[d]Kennedy, R. et al. 2014, MNRAS, 442, 2487.
[e]Viel, M. et al. 2013, PhRvD, 88, 3502.
[f]Shao, S. et al. 2013, MNRAS, 430, 2346.
[g]Boylan-Kolchin, M. et al. 2011, MNRAS, 415, L40.
[h]Lovell, M. et al. 2012, MNRAS, 420, 2318.
[i]Sawala, T. et al. 2014, arXiv:1412.2748.

At present, the best strategy for distinguishing CDM from WDM would seem to be the use of gravitational lensing. There are two complementary techniques that could reveal the presence of the myriad small subhalos predicted to exist in the halos of galaxies by CDM simulations: flux ratio anomalies and distortions of giant arcs. Whether these techniques will work in practice is still an open question.

Discussion

S. White Thank you. So let us have comments on astrophysical evidence for the nature of dark matter.

G. Efstathiou I just wanted to stress a point that Uros made. The CMB is sensitive to lensing and, particularly if you have polarization information at high multipoles, the CMB alone is very sensitive to neutrino masses. So you can use the CMB, and you see this in the 2014 Planck papers that you can get very good limits. In the future, with polarization experiments extending to high angular resolutions, it should be possible to detect neutrino masses of .06 eV, just from the CMB.

J. Ostriker A comment on two of the points that Carlos made and their agreement, but strengthening them. One is on the "too big to fail". I do not think there is a problem for an additional reason, and that is satellites spiral in much faster than you would think from dark matter only simulations if you include the baryons, and so their absence is not a problem. The second point is on cores versus cusps. The NFW cusp is very soft, it is $1/r$ in density, so the potential, if $1/r^2$ is only logarithmically divergent, and the potential in the $1/r$ barely increases towards the center, so if you imagine stars orbiting in a $1/r$ potential, they do not go much faster when they come to the center of the "cusp", and so it is just undetectable. So you cannot tell from observing stellar motions in small galaxies whether they have a core or a cusp.

J. Carlstrom You mentioned gravitational lensing for looking at sub-structure. You were talking about looking at galaxy-galaxy lensing, ...

C. Frenk No, I was talking about all the things, like for example the flux anomalies, where you have a multiply imaged quasar that produces ratios of the fluxes of images but not what you expect in a smooth potential. So that is one way, and there are other techniques that can detect in strong lensing, perturbations in the lensing due to the presence of the sub-structure.

J. Carlstrom I just want to comment that we now have these high redshift, very dusty galaxies which are being lensed, and with ALMA it is terrific, you do better than ...

C. Frenk Yes, those are ideal for these tests.

S. White The current situation is that there are a few good cases of multiple imaged distant galaxies. What you need is a distant galaxy with structure, with some extent where you see multiple images, and then what happens is that if there is a small object in the lens, it will affect one of the images and not the other. So you will find that when you try to rectify the images to coincide, there is something which is anomalous. By looking for these anomalies, you can put constraints and find small objects. The current constraints come from a lensing system where the lens itself is at a redshift .6 or .7, I do not remember exactly the number, but it is impressive that the constraint on the mass of the object that is directly detected is similar to the smallest dwarfs in the halo of the Milky Way. By improving this technique, for example by using radio-interferometric data, which can get you to smaller scales, it should be possible to go down at least another order of magnitude. So it should be able to detect the small halos that Carlos claims that may be there, but they would not have any galaxies in them.

N. Weiner I wonder if you can be a bit more quantitative on the limits on warm dark matter down to 2 keV and the prospects for going forward with the theory of these experiments.

U. Seljak If I start first with the Lyman-α forest. Lyman-α forest is limited because it has its own cut-off scale, Jeans scale, and we are reaching that limit. So, anything that has a free streaming scale which is shorter than the Jeans scale, we are not going to be able to probe it. So we are almost at that limit already. So I do not think that from Lyman-α forest we can improve those limits a lot better than what we have right now.

N. Weiner What about from these other techniques?

C. Frenk There comes obviously a point where, if the mass is so big, it is effectively not warm but cold, and you cannot tell the difference from an astrophysics point of view. I think for the masses that are currently allowed from the Lyman-α forest, there is still a lot to play for, because you only expect a handful of these sub-structures, even if the warm dark matter particle mass is 3.3 keV, the example that I show you here is slightly less than that, it is 2.2 keV in the same kind of notation. So you could see this is very different from cold dark matter. I think once you go to 20 keV or so, then I think from the point of view of astrophysics, we will not be able to tell the difference, but I think there is still that range of values of the mass that can be probed with astrophysics.

H. Murayama I would like to also know how well we might be able to know the self-interaction of dark matter in the future from halo shapes, either galaxies or clusters?

S. White Does anyone want to address interacting dark matter? I can say some things about that. There are already constraints on the interactions of the dark matter with itself. These are collisional interactions which require the mean free path to be long. The cross sections are required to be in the regimes where a typical dark matter particle crossing a halo hits at most one other particle per orbit. In this regime, there are still things allowed. It seems that in order to make data for galaxy clusters, where you do see concentrated dark matter profiles, compatible with dwarf galaxies, where some people would argue that you do not, then you need a velocity-dependent cross section. So this comes back to what Neal was saying about p- and s-waves. You need a model for the dark matter where the s-wave cross section is suppressed, and you get a p-wave cross section with a Sommerfeld enhancement. That means that you can have enhanced effects in the small galaxies compared to the big ones, because of the smaller relative velocities. In those very specific ranges of parameters (Neal you can perhaps address whether it is plausible or not) there are astrophysically interesting possibilities and you could explain why you get different structures in dwarf galaxies and in galaxy clusters, if this turns out to be necessary.

N. Weiner I actually want to comment about this discussion of self-interacting dark matter. If you go back to the original paper of Spergel and Steinhardt, they took some phenomenological models for studying what the self-interaction was, and I think that there is an unfortunate acronym overlap between self-interacting dark matter, SIDM, and strongly interacting dark matter, SIDM, because these things sometimes get conflated, also because the cross section that is often needed to get an interesting effect on halos, at least for sort of GeV to TeV type dark matters, is roughly around a barn. So we look at that and we say that this is a strongly interacting particle, and we then actually start to think about nuclear physics and things like this. But the one thing I just want to point out about this, is that this is not necessarily the case, and it is maybe not even the most natural case. We also know about another large cross section, which is the Thompson cross section, which is actually extremely large, and it is not a particle that is strongly interacting. The reason why you have a large cross section there is because you have a force which is massless. So if the interaction between dark matter is arising from an interaction which is effectively massless compared to the scales of the problem, you can get very large cross sections without any particularly strong interactions. Because of this, you actually end up with this naturally velocity-dependent scattering cross section: at low velocities the mass of the interactions becomes relevant and at high velocities, the mass of the interactions is essentially irrelevant. So, this leads to precisely to the kind of properties that I was talking about, where at low velocities, the scattering rate is small, and at high velocities scattering gets small, but somewhere in between there is some peak. The only reason I bring this up is, not only to say that I think that it is totally reasonable to have dark matter models with this size of a cross section, but that studying the interactions of dark matter at as many different velocity scales as possible is very important, because that is really complementary information.

M. Rees I would like to ask if these simulations place any constraints on the initial fluctuation spectrum on sub-galactic scales, which is clearly relevant for the tilt and curvature on scales far lower than you can observe on the CMB?

S. White Well, I do not know. I can answer that to the extent that the two simulations which Carlos was comparing had effective power spectra which differed in the form of a cut-off on small scales, and that I guess that was also what

Uros was showing. Clearly they do give different results. Anything which affects the power spectrum on those scales or larger would give effects on astrophysical scales. Whether they would be measurable, I guess depends on the model.

M. Rees On scales say, 10^{10} solar masses, can you say anything?

C. Frenk For example, a small tilt or things like that, they could show up as an excess or a variation in the number of lumps that one expects in halos like the ones that I was showing. I think there was a paper when the BICEP2 results were announced. It proposed a solution to the "too big to fail problem" because a small tilt in the power spectrum would then be reflected in the abundance of sub-structures and sub-halos in the Milky Way.

U. Seljak I just wanted to add that it is true you can place constraints on the slope or running from these very small scales but it is not obvious that these constraints are going to be competitive to the constraints that you can get from CMB plus large scale structure, where large scale structure is on somewhat larger scales, at least, in the simplest models where you just have a simple running. If you have a cut-off, then of course these small scale observations become very important, but for simple running cases, I think CMB itself is already very powerful and if you add large scale structure, it probably is even more powerful.

E. van den Heuvel Maybe a naive question. Could someone summarize the reasons why we think that we are dealing here really with matter or with particles and that it is not a modified theory of gravity which could explain the observations?

S. White I have asked Matias Zaldarriaga to do this later in the discussion, but I was leaving this until we got through the discussions of the more conventional things first. So I would like to come back to that, but leave it for now. Does anyone have any more discussion about astrophysical manifestations of dark matter that we heard about from Carlos and Uros?

L. Baudis You said you can get the neutrino mass from CMB alone, but I did not understand to which level you would be able to constrain it. Maybe somebody can comment?

G. Efstathiou From the temperature anisotropies alone, the Planck 2 sigma limit for 2013 is .3 electron volts. There will be an improvement with the 2014 data with polarization, but in principle high resolution polarization experiments would be able to probe to .01 eV. John may say something about that in his rapporteur talk.

J. Carlstrom I will mention that this afternoon. We think that in the next ten years we will get below .02 eV.

D. Spergel I think one does even better if you combine that with gravitational lensing measurements. So the combination of CMB lensing and optical lensing should get below this, so you should have a pretty clear detection of neutrino mass. That should be many sigma for the current neutrino mass differences which are about .06 eV.

U. Seljak If I can just expand a bit more on this. It is true these predictions from the small scale structure are coming at the level of .02 eV for the future. What is interesting is that they are actually limited by the CMB, and not large scale structure. Large scale structure will become so powerful that the limits will be from CMB and in particular from the optical depth τ uncertainty that we discussed yesterday. If that uncertainty could be improved, then combining large scale structure with CMB limits will go below .01 eV, and there is a potential that 21cm experiments for example, could improve the limits on τ and therefore get us another factor of two in the precision.

S. White Can someone remind us of what the current lower limit is from the oscillation measurements on the sum of the neutrino masses?

U. Seljak .06 eV.

S. White So in other words you are saying we get to a place where, if we do not observe it, we will have a serious problem.

U. Seljak Yes.

J. Carlstrom We will have a detection I think that is what that means.

D. Spergel One solution you could imagine is that the neutrino was not stable and had a very long life-time and decayed. That would require adding something. It would be an interesting result but not one that would contradict cosmology.

S. White Is 10^{17} seconds a long time in this context? I mean, for proton decay, it is quite a short time.

D. Spergel I remember back in the days of 17 keV neutrinos, people constructed models where that particle decayed and I am sure those could be revived. People had these familon models.

C. Frenk I have a question for Neal and all the particle physicists here: a growing number astrophysicists are investing quite a lot of time and resources in trying to explore warm dark matter models. In my case, this is because Misha Shaposhnikov bought me dinner once and convinced me that sterile neutrinos were really attractive particles. But I somehow sense amongst particle physicists a not so great enthusiam for particles in this sort of keV mass range. Can one of you, Neal maybe, say what you feel about this sort of mass range for the dark matter?

N. Weiner As I tried to say in my talk, I think that you should be extremely skeptical of determining things on whether or not theorists are enthusiastic about them, because we will become enthusiastic about things for a different set of reasons. The thing that I would say about keV sterile neutrinos, in particular, which I think is very nice, is that if you want to have an indictment of the entire theoretical community's approach to the hierarchy problem, and say that there is going to be nothing at the weak scale, if you want to take that position, then a keV sterile neutrino is in some sense a very minimal set-up. You need something like that to explain neutrino masses, and you need something to be dark matter. And so it is a very economical theory because you already need something like this just to get the neutrino masses we observe. Why not use it to be dark matter as well? I think this is what Shaposhnikov and others have really pointed out. There is just not a tremendous amount to do with it theoretically. There are things to do, but I think that one of the reasons that there is not so much enthusiam is just that.

H. Murayama The short answer is basically just sociological. The particle physicists have been living this culture that if there is no reason to pick up a particular mass scale, then we do not. So far there has not been any other reasons to pick that particular mass scale, that is why it was not sort of enthusiastically embraced, it is purely sociological.

S. White Let us move on to even smaller scales. Tom Abel is going to tell us something about what the dark matter's distribution might look like on even smaller scales. You could ask, for example what it would be like on the scale of our detectors.

Prepared comment

T. Abel Dark Matter Dynamics

Computational Physics allows us to study extremely non-linear systems with fidelity. In astrophysical hydrodynamics and studies of galaxy formation much of the last two decades we have explored various discretization techniques and found subtle differences in some applications. Interestingly numerical studies of collisionless fluids such as e.g. the collapse of cold dark matter to form the large scale structure of the Universe has only been studied meaningfully with one approach; N-body Monte Carlo techniques. We recently introduced a novel approach to analyze N-body simulations[a] and further developed it to write entirely new simulation codes[b] that for the first time can study a collisionless system in the continuum limit in multi-dimensions. This new technique opens a new window in making sense of structure formation as well as many systems in plasma physics. These approaches allow also for much improved predictions for gravitational lensing, putative dark matter annihilation signals, properties of cosmic velocity fields , and a number of other applications.

In its simplest form our approach interprets the particles in an N-body simulations as massless tracers of a fluid that are spread between these marker particles. This is in contrast to historical approaches where the N bodies are viewed as marker of bodies of a some typical shape. In our interpretation the particles form the corners of cells containing the fluid. In the particular application of structure formation in the Universe this corresponds to viewing the initial conditions as a three dimensional manifold in the classical six dimensional

---

[a]Tom Abel, Oliver Hahn & Ralf Kaehler 2012, Tracing the Dark Matter Sheet in Phase Space, MNRAS, 427, 61.
[b]Oliver Hahn, Tom Abel & Ralf Kaehler 2013, A new approach to simulating collisionless dark matter fluids, MNRAS, 434, 1171.

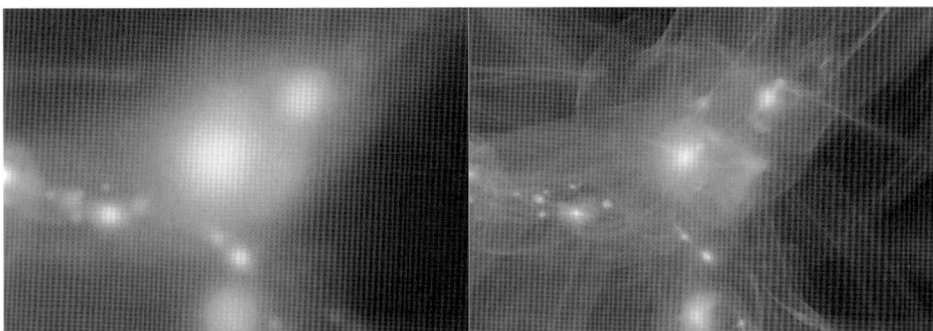

Fig. 1.   A comparison of adaptive kernel smoothing (left) to a new technique discussed here (right) to estimate the dark matter density from an N-body simulation of the formation of structure in the Universe.

phase space formed by position and momentum coordinates. Cold dark matter has negligible initial extent in the momentum directions and hence is described just by a three dimensional manifold indexed by Lagrangian coordinates. Our approach therefore uses not only the information where particles are at a particular time but also uses the information of where they came from to estimate densities, velocities, and other quantities. Consequently it reveals more of the information contained in the simulation. A visual demonstration provided in Figure 1 compares adaptive kernel smoothing and our new approach. It clearly demonstrates that our technique reveals further information in the data as compared to any adaptive smoothing approach which washes details and does not capture the sharp caustics formed in the dynamics of collsionless fluids.

Discussion

S. Tremaine The technique that Tom describes is extremely impressive and a tremendous advance. I just would like to point out that even that does not exploit fully the extra information that is available. You know that the transformation from the initial to the final dark matter sheet has to be symplectic and that is not yet in your calculation, so there is still a lot more that could be done from a theoretical point of view.

T. Abel Yes, in fact on our very first attempts, we just used linear interpolation, and actually piecewise constant in the very first one that we have published. We went on to higher order interpolation exactly for that reason, so this will come out very shortly, actually.

S. White I would like to point out though that it is in fact possible to use the symplectic property in the simulation. What Tom has shown is a way to reconstruct the entire phase sheet of the distribution. It is still a finite element analysis in the sense that you have a finite number of particles, but the elements have switched from being point particles to be an area if you like, or a volume of the initial sheet, which is then followed. This means your resolution is still of the same order as you had in the N-body simulation, in the sense that the finite element has the same mass. But you can actually reconstruct the sheet in a neighborhood of each particle at much higher resolution than that, by following directly the symplectic dynamics in a neighborhood of the particular orbit. This is done effectively by something which the relativists would recognize. You integrate the geodesic deviation equation, which looks at all the orbits of virtual particles neighboring the one you are following. If you do this you can reconstruct the sheet locally with much higher resolution. Much higher means maybe 10^{12} times higher depending on the situation. If you only need to know statistical properties like the densities of the sheets or the caustic passages of individual particles then you can infer these robustly through this technique, right down to a limit which would correspond to, say, a neighborhood in the Milky Way. This has been done by Mark Vogelsberger.

S. Tremaine I agree that that is also a powerful technique, although then of course the standard approximation that all N-body simulations are based on, which is that the forces are dominated by distant material, is no longer strictly valid, and so there is some concern I think, which Mark has addressed to some extent by numerical tests. But the basic approximation of N-body simulations may be not working quite as well as we have come to expect.

S. White Yes, I agree with that. That is the fact that the forces have to be calculated from a stochastic sampling of the field, which is just points for a standard simulation, and is finite-volume elements for what Tom is suggesting.

T. Abel Yes, there is a beautiful thing now in our latest implementation. We carry out refinement of the discretization in phase space. So you actually reconstruct the entire sheet, and then you converge to the answer for that particular set-up that you have. We now can perform convergence tests in a way that we have not been able to when we just had statistical sampling.

R. Blandford Tom, could you say just a little bit about how this is going to work out in practice when you incorporate baryons and the cooling and all the messy things that they do.

T. Abel In principle it is always very simple to combine the two because in all our current computational cosmology codes, the baryons and the dark matter really only communicate via the gravitational potential. So if we have a new technology here to reconstruct the dark matter density in a somewhat different way, it actually comes in very easily by just adding this mass density to the baryonic mass density before you calculate the potential from it. However what is different, when you are trying to get the entire microphysical phase space structure, just typical non-linear physics, things evolve so rapidly in the centers of these halos that within an extremely short time, you actually run out of bits on your computer to store that information.

S. White Let me give an example of a particular case which is significant for other things we have been discussing in this session. If you have completely cold dark matter and you have a three dimensional phase sheet, as it evolves, it makes caustics, which are the kind you could see in Carlos's simulation. So if the matter is completely cold, the density at a caustic in three dimensional space goes to infinity. If you integrate a particle's trajectory through a caustic, then you find that if it has an annihilation cross section, the annihilation probability diverges in the caustic. So in the completely cold case, you would predict that the annihilation radiation is entirely dominated by the caustics. Hence it behooves us, if you want to get the observational result, to calculate this. Now it turns out that actually our cold dark matter particles are not completely cold, they have a small but finite velocity dispersion, and that regularizes the structure of the caustics, so the annihilation rate does not go to infinity. When you put that in, then you find that the total luminosity as the particle goes through the caustic is regularized. In this case it turns out that whether the caustics are important or not just depends on the details of the caustic structure, and for that reason we need to calculate this to predict what the annihilation radiation looks like. Unfortunately for fully three dimensional simulations, or perhaps fortunately, it seems the caustic structure is sufficiently complicated that caustic anihilation is actually sub-dominant to the normal, straightforward one. But this is something one needed to calculate and to do that one had to evaluate the caustic structures.

T. Abel I think it is also interesting, it leads to this surprising aspect that the colder the particle is, the more winding you have very early on and the less density there is in any particular caustic, so you really smear the signal. So for models with a cut-off where you have structure formation somewhat later, you could imagine still getting an appreciable signal in the caustic, it is perhaps counter-intuitive.

M. Kamionkowski So I wanted to comment on the implications of sub-structure for detection, both direct and indirect. So you look at these simulations and most of the space is empty, and so I was worried at some point about the implications for direct detection of dark matter. The simulations do not really address this issue because if you think about a direct detection experiment which lasts a few years, we travel a very small distance in the galactic halo. The sub-structure in the primordial power spectrum in inflationary models can extend as small as 10^{-6} solar masses, and so there is a good likelihood that sub-structures down to the 10^{-6} solar masses be extremely small compared to the distance we have actually travelled during the lifetime of a direct detection experiment. So I was wondering whether it is actually probable that the area that we occupy, the volume that we occupy, is effectively empty of dark matter. It is impossible to do a simulation because there is about 20 orders of magnitude between even the high-tech simulations you showed and the scales relevant for direct detection.

S. White I would like to dispute that and in fact the answer is in Mark Vogelsberger's paper.

M. Kamionkowski Prior to Mark Vogelsberger's paper, Savas Koushiappas and

I looked at this and we developed an analytic approach that allows you to extrapolate the results of simulations. Because a lot of this hierarchical structure formation is a scale invariant process, or roughly scale invariant process. The good news to report is that you can calculate from this analytic approximation a probability distribution: what is the probability, that over the volume probed by a direct detection experiment, the dark matter density is X? That probability distribution is actually very highly peaked around the mean density, about 0.3 (GeV cm^{-3}). There are very few places where the density is much higher, and then many places where it is a little bit smaller. What you can do with this result is take the direct detection experiment, make the measurement what is the limit on the dark matter density locally, evolve that with this probability distribution to actually get the limit to the dark matter density over the entire halo. It turns out that it does not really make a big difference. Now, the implications for annihilation are a little more complicated because that requires an integral over the density squared, but still, my result agrees with what was said, there are enhancements but there are probably not going to be huge enhancements in the annihilation rates.

S. White I agree with that. There is one point though. Because the dark matter is a three dimensional phase space that fills all of the volume of the universe and since it is an Hamiltonian flow, there is formally always dark matter at every point, but of course the point is, the density could be very low, so I think the kind of calculation you are talking about definitely is important.

L. Baudis I have a question to the astrophysicists going back from simulations to observations, namely how well can we actually constrain this density of dark matter at the solar radius in the future, because if I understand now correctly, the numbers that we have, I mean, the precision is limited by the baryons, by how well the surface density of the baryons can be determined. Depending with whom you talk to, you get about a factor of two or so, error in the local density. Can we decrease that in the future and what does it take?

S. Tremaine I think the uncertainty in the dark matter density in the solar neighborhood is a lot smaller than the uncertainty in the cross section of the WIMP's. So I guess we have never been too worried about that.

L. Baudis You get a much worse reconstruction if you put in the current errors, and still some quite optimistic errors on the density, for instance. So I agree that we have many orders of magnitude on cross section, but still.

S. Tremaine I think what it will take is the Gaia spacecraft, which is launched and working and will have its first data release in a couple of years.

C. Frenk A small point. Should you not worry more, rather than about the density, about the velocity structure of the particles, because particularly for the low energy detection threshold, there it really makes a big difference whether it is a Gaussian distribution or not?

L. Baudis With the current set of experiments you cannot really determine any structure that you see in simulations in this velocity distribution. But on the other hand you are right, especially at low threshold, if you could have a better measure of the velocity distribution, there are also other things that I did not mention here, for instance inelastic scattering of dark matter, not inelastic dark matter models, but really inelastic scattering off nuclei where you excite the nucleus, produce a low energy gamma, and then detect the coincidence between the nuclear recoil and the gamma that boosts your energy region of interest to higher energies, but there you have a kinematic lower limit. So, there the velocity distribution actually plays quite some big role, I agree. So what is it?

C. Frenk I do not know, but one thing I do know is that the model that you particle physicists use, the standard halo model, that is really completely unrealistic. It is a spherical symmetric, isothermal distribution, and Gaussian velocities. There are much better things ...

L. Baudis We use a Maxwell-Boltzmann distribution that you have seen also from the Vogelsberger and so paper. It is maybe not reality, but it is not such a bad approximation, at least for the current generation of simulations, and as long as we are setting limits.

N. Weiner I just want to say that, of course there is a scientific concern, but I think that one thing that theorists have made a lot of progress on in the last few years is trying to develop techniques to know exactly when you need to worry about the velocity distribution and when not to, and how to approach these problems. So I think that we are actually in pretty good shape for understanding direct detection experiments if they see something.

S. White I would like to move on to Ed van den Heuvel's question I asked Matias Zaldarriaga to prepare something. He is becoming less and less enthusiastic with time but nevertheless I am going to put him on the spot and ask him whether he thinks the phenomenology we have been discussing could somehow be accommodated by something other than dark matter.

Prepared comment

M. Zaldarriaga: Some Thoughts about Modifying the Laws of Gravity to Replace Dark Matter

Simon asked me to discuss whether in my view there are any interesting and/or potentially viable alternatives to dark matter for explaining the observed phenomenology.

Ever since the discovery of what we now call dark matter there has been an interest in trying to explain the observed phenomena by modifying the laws of gravity. This received a big boost with the discovery of Milgrom's formulas which seemed to encapsulate the observed phenomenology on galactic scales in a simple way. In contrast, in the standard cosmological model the regularities in the data on the scales of galaxies involve the interplay of several complex phenomena we cannot calculate from first principles and can only model crudely.

The simplicity of the observed phenomenology may seem nothing short of an accident.

In the early days it was sensible to explore the possibility that the simple phenomenological relations found were telling us something about the laws of gravity, that the acceleration scale in MOND had some deeper physical meaning. Exploring how the MOND formulas could be extended outside the realm of galaxy rotation curves and scaling relations was an interesting program. Measurements on the scales of clusters seemed inconsistent with MOND predictions. But now the main question has changed, it has become whether one can build a consistent cosmology based on MOND related ideas and use it to understand Large Scale Structure (LSS) in the linear regime. The focus has shifted because in the last decades observations of structure on very large scales have advanced tremendously. Furthermore there is little place to hide for theories in this regime: linear fluctuations around a background solution should be something that we can compute robustly. It is now clear that the observations of the acoustic peaks on degree scales in the Cosmic Microwave Background (CMB) provide very strong evidence for the presence of dark matter completely independent of any evidence on galactic scales. In addition, the observed damping of the CMB fluctuations on arcminute scales would lead to a spectrum of fluctuations in the LSS which is not consistent with the data unless a dark matter component is included. In the linear regime, there is clear evidence of an additional dark component that clusters.

The current MOND related literature has acknowledge this fact and now incorporates additional dark components that can be used to accommodate the linear regime of structure formation. These components do not cluster on galactic scales and then MOND explains the rotation curve phenomenology. In some cases the new component is prosaic, like some sort of hot dark matter, perhaps even neutrinos. Other times they are some of the additional fields that had to be introduced in the relativistic versions of MOND.

The examples where "hot particles" are used have been reasonably well studied. Unfortunately on large scales, there is both dark matter and a larger gravitational force due to MOND so in detail one finds many problems, for example a completely wrong mass function for clusters. Clearly one needs to make the MOND force environment dependent, to change its properties as one goes from the largest scales in cosmology to galactic scales and then also make sure that the theory survives the stringent tests in the solar system. Ideally one would have an additional dark matter component with environmentally dependent sound speed so that it acts like dark matter on large scales but does not cluster past the non-linear scale.

There are several attempts along this line, in the slide I quote from a recent example by Justin Khoury. But there are many similar attempts. One can see the hints of all the ingredients I mentioned in this model. However at the current time many details are still not completely worked out and it is far from clear that such models can be made to agree with observations. The models have a lot of moving parts, basically to hide "Mondian" effects that should have shown up but have not. And it is very much of an engineering challenge to try to put all the pieces together and still satisfy observational constraints.

Finally let me comment on some another issue not related to phenomenology. These attempts to modify gravity are in a sense dangerously playing with fire. For example, the scalar potential responsible for MOND has superluminal excitations. To get lensing to work one needs some sort of field that sets locally a preferred frame. They require fields which in the background are nontrivial but have zero stress energy tensor. At this point one can easily find oneself violating the Null Energy Condition (NEC), spoiling black hole thermodynamics, etc. So not only does one have to try to accommodate the phenomenology but also needs to try to convince oneself that such radical departures are even reasonable to consider.

Most people would just run away from these difficulties choosing to live further away from the abyss. Others view this as an opportunity because once you free yourself from these constraints there are many more things your theory can do. If you can violate the NEC you can use it to create your favorite bouncing cosmology, if black hole thermodynamics is in trouble perhaps black holes need to have hair and thus you can make predictions for future gravitational wave experiments.

In my view the window for doing away with dark matter in favor of a modification of gravity has all but closed. Direct and indirect searches for the dark matter particle are becoming much more sensitive. Astrophysical measurements of structure on small scales might provide additional information about the properties of the dark matter. We might have solved the dark matter puzzle by the time the next Solvay meeting on astrophysics takes places. Even if we have not, I would be very surprised if the field has gravitated towards alternative gravity theories, at least in this context.

Discussion

V. Mukhanov I agree with Matias concerning MOND. In fact what you can do, you can take Einstein equations, get rid of ten equations namely for trace, and then you are getting very simple modification of the Einstein equations which allows you to get so-called dust without any particles as a constant of integration. The same exercise was done by Marc Henneaux and Claudio Bunster when they also got rid of the trace and got dark energy as a constant of integration. In principle, you can operate with slight modifications of the Einstein equations, and it will be indistinguishable for all the substance that is called cold dark matter and even dark energy. In this case of course, it is not a very, perhaps, encouraging thing because you do not expect that all these searches for particles will bring us any success. But if they will not give us anything, then you can refer all this kind of gravitational behavior to a little modification of gravity, not like MOND. It is not ugly, it is actually one third of a line addition to the Einstein Lagrangian.

M. Kamionkowski Show me.

V. Mukhanov I will show you later.

N. Mandolesi I wonder if the detection of the B-modes, the inflationary detection of the B-modes, could have been putting some constraints in these theories.

M. Zaldarriaga I think they are at a much lower level of problems. I mean, in some specific example, probably, you can make a connection, but there is water coming in the ship in every direction, so ...

S. White Ok, since we are rapidly sinking towards lunch here ... The dark matter problem has been with us for a long time, I am not sure if we are making progress, we have lots of ideas and more observations than we used to, but since Jim Peebles has been part of this for a long time, so I thought I would like to finish the session by asking him to give his perspective on where the issues are for dark matter these days.

Prepared comment

J. Peebles Physical Cosmology

The case for the relativistic hot Big Bang cosmology, with its cosmological constant and hypothetical nonbaryonic Cold Dark Matter, is about as good as it gets in natural science. It rests on the rightly celebrated precision cosmological tests — measurements of the supernova redshift-magnitude $(z - m)$ relation and of the effects of acoustic oscillations of the coupled plasma and thermal radiation at redshift $z > 1000$ on the present distributions of galaxies and the thermal cosmic microwave background radiation. But equally important is the considerable variety of less precise but independent checks. Consider, for example, that if we only had the precision $z - m$ measurement people surely would have noticed that it is close to the prediction of the 1948 Steady State cosmology. In this counterfactual situation, having only the $z - m$ measurement, might we have arrived at a cosmology that resembles inflation but with continual creation of matter instead of baby universes? But we have other tests, in a network of cross checks tight enough to make the case for the ΛCDM cosmology. The less precise tests receive less attention now, which is unfortunate, because the universe is large and quite capable of surprising us yet again. Remind your funding agency that serious open issues in cosmology demand looking at the universe from all sides now.

There are three pressing issues. First, what was the universe doing before it could have been described by the Friedman-Lemaître solution? We have an excellent candidate, inflation, an encouraging indication of it, from the tilt from scale-invariant initial conditions, and the chance of another, tensor perturbations. Second, what are we to make of the vacuum energy density that is detected

Fig. 1. The twelve most luminous galaxies closer than 8 Mpc.

in cosmology and so puzzling in quantum physics? None of the ideas under discussion has captured general attention the way inflation did for early universe physics. Achieving that is a great challenge and opportunity. Third, could the dark matter really be as simple as a near collisionless gas of particles, or might it do something more interesting?

If the dark matter is more interesting it will be discovered as anomalies. I offer a possible example. The figure shows the 12 most luminous galaxies within 8 Mpc distance.[a] Two, Maffei 1 and NGC 5128, are ellipticals; two, M 81 and M 31, have elliptical-like classical bulges of stars centered on rotationally-supported disks; and 5 or 6 are pure disk galaxies supported by motions largely in the plane. Our Milky Way galaxy, which is about the 13th brightest, also is pure disk or close to it.[b] These pure disks challenge accepted ideas, as follows.

In ΛCDM-based galaxy formation theory galaxy-size concentrations of baryons and dark matter grew by hierarchical merging, largely at redshifts $z \sim 1$ to 4. At this range of redshifts the global star formation rate per unit baryon mass is estimated to have been an order of magnitude larger than now. This is when most

---

[a]Credits are: NGC 6946: Adam Block, Mt. Lemmon SkyCenter; M 51: NASA/ESA; NGC 253: Star Shadows Remote Observatory and PROMPT/CTIO; M 81: Tony Hallas; M 106: Hubble Heritage Team; M 63: Bill Snyder, Sierra Remote Observatories; NGC 5128: ESO; M 83: ESO Science Archive; M 101: HUBBLESITE; NGC 2903: Tracey and Russ Birch/Flynn Haase/NOAO/AURA/NSF; Maffei 1: Jarrett *et al.* 2003; M 31: Lorenzo Comolli.
[b]The pure disk phenomenon is discussed in Kormendy *et al.*, ApJ, 723, 54 (2010).

of the stars formed. Where are the remnants of these early stars? The natural answer is in ellipticals and the classical bulges of spirals. Hierarchical merging would destroy any early disks and leave these old stars in the smooth concentrations characteristic of ellipticals and classical bulges. A pure disk galaxy, on the other hand, has to have formed by dissipative settling of diffuse baryons to rotational support before formation of appreciable numbers of stars. This is because stars that formed before settling would have ended up in a stellar bulge or halo, which are not prominent features of pure disk galaxies. But how did the pure disks avoid appreciable star formation when the global rate was high, during hierarchical assembly prior to disk formation?

Perhaps pure disk assembly was closer to monolithic than hierarchical, avoiding the dense early clumps where stars are likely to form. This might happen in rare places in ΛCDM, but we are instructed that pure disk galaxies are common nearby, as illustrated in the figure. Surveys at greater distances assign considerable luminosities to bulges in most spirals, but I am instructed that this is not significant because the surveys do not yet have the resolution needed to distinguish classical bulges from high surface brightness flatter pseudobulges in the inner parts of pure disks. The observations are improving; we will see.

We might imagine that conditions at $z \sim 1$ to 4 allowed star formation only in special regions, in protoellipticals and protobulges, where the formation rate would have been prodigious. But what are we to make of the elliptical NGC 5128 and the spiral M 83 in the figure? They are in the same group, and I suppose their ambient conditions were similar at $z \sim 3$, but M 83 managed to keep most of its stars close to the disk while NGC 5128 placed most of its stars in a stellar halo — apart from the dust cloud recently accreted.

The galaxy formation theory community does not express much concern about pure disks, and indeed they may result from the complexities of visible matter physics. But pure disks are a challenge and could be a clue to a still better cosmology. Put more broadly, we have a compelling case for a term in the stress-energy tensor that acts about like CDM, but it is a spectacularly bold extrapolation to suppose that CDM is a full description.

Session 5

Microwave Background

Chair: *George Efstathiou*, Kavli Institute for Cosmology, Cambridge, UK
Rapporteurs: *John Carlstrom*, University of Chicago, USA and *David Spergel*, Princeton University, USA
Scientific secretaries: *Ben Craps*, Vrije Universiteit Brussel, Belgium and *Thomas Hertog*, KULeuven, Belgium

R. Blandford It is the fifth and final session, and a very exciting one. George Efstathiou is going to lead us through recent developments in the microwave background.

G. Efstathiou The session that you have all been waiting for, on the cosmic microwave background. We are so lucky that we have the cosmic microwave background, that it allows us to probe physics that even David Gross would be interested in. We will start off with two rapporteur talks that will take us through the observational data. John Carlstrom will start.

Rapporteur Talk by J. Carlstrom: The Cosmic Microwave Background: Past, Present and Future

Abstract

Since the discovery of the cosmic microwave background (CMB) 50 years ago, increasingly sensitive and accurate measurements of this radiation have provided unique insights into the origin, composition and evolution of the Universe. Measurements of the CMB provide motivation for the theory of cosmic inflation and the quantum mechanical origin of all structure in the Universe, leading to the fertile connection between cosmology and the frontiers of fundamental particle physics, between the largest and smallest size scales, and between the lowest and highest energy scales. The next decade promises an enormous increase in our ability to use the CMB to explore these connections. We will conduct key tests of inflation, corresponding to energy scales a trillion times beyond the reach of the Large Hadron Collider. We will use extremely sensitive temperature and polarization maps of the CMB over a large fraction of the sky and at arcminute resolution to produce maps of the projected mass in the Universe. The CMB and mass anisotropy maps will be used to assemble the story of cosmic evolution. They will allow precision determination of the dynamics in the early Universe, in turn placing stringent constraints on the relativistic contributions from the neutrino sector. Through measurements of the growth of structure over cosmic time, they will lead to a determination of the sum of the neutrino masses. Through measurements of secondary CMB anisotropy, they will provide constraints on dark energy and tests of gravity.

1. Fifty Years of Progress

The CMB was discovered serendipitously 50 years ago by Arnold Penzias and Robert Wilson.[1] Using a sensitive radiometer operating at 5 GHz on Bell Labs' Holmdel Horn Antenna, they discovered a bright signal apparently from all directions on the sky. After a thorough investigation of possible terrestrial contamination that might systematically bias their signal, they concluded it was cosmic in origin and uniform across the sky with an amplitude of 3.5 ± 0.5 K.[1] That they had discovered remnant light from the early Universe consistent with the hot big bang model was readily accepted by cosmologists, as detailed in a paper (in the same issue of *The Astrophysical Journal Letters* as the report of Penzias and Wilson's detection) by Dicke, Peebles, Roll & Wilkinson,[2] which followed the work of Alpher, Gamow, Herman and others in the decades before.[3] This discovery story illustrates the features that have led to the remarkable progress in the field since: (1) large leaps in both raw sensitivity and in the control of possible contaminating and systematic effects, and (2) tight interplay with theory.

Even with continued increases in instrument sensitivity, there was nearly a 30-year gap between the discovery of the CMB and the detection of anisotropy in the CMB temperature. This measurement was so difficult primarily because the CMB anisotropy is tiny, only about one part in 10^5. This high degree of isotropy was difficult for most models of the Universe to explain, because it indicated causal connections in parts of the Universe that were separated by many times the distance light could have traveled in the entire age of the Universe. An explanation for this apparent paradox was provided by the theory of inflation, in which accelerated expansion in the early Universe could move initially casually connected regions

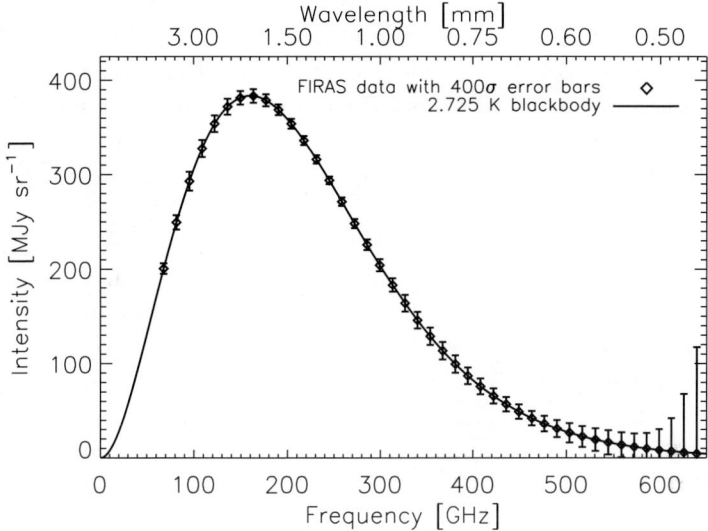

Fig. 1.   The intensity spectrum of the CMB monopole, as measured by the *COBE/FIRAS* instrument. The measured spectrum is consistent with a pure blackbody at a stunning level of precision: the error bars shown here are 400 times larger than the actual 1σ uncertainty on the measured points.

out of causal contact.[1] Inflation also provides a mechanism, namely ground state quantum fluctuations, to provide seeds of all structure in the Universe.[5]

Anisotropy in the CMB was first detected by the Differential Microwave Radiometer (*DMR*) on the COsmic Background Explorer (*COBE*) satellite in 1992.[6] Meanwhile, the Far Infrared Absolute Spectrophotometer (*FIRAS*) experiment, also on the *COBE* satellite, confirmed the black-body spectrum of the CMB with a beautiful measurement that defines the state of the art to this day, as shown in Figure 1. The *COBE/DMR* measurement finally set the level of the anisotropy, allowing theorists to refine their models and experimentalists to design experiments to characterize the anisotropy fully. Within another ten years, CMB measurements[7–10] would show the geometry of the Universe was uncurved within a few percent, that the baryon density accounts for only a few percent of the total universal energy density, consistent with big bang nucleosynthesis, that dark matter was non-baryonic and accounts for only about one quarter of the energy-density of the Universe, and that dark energy accounts for the remaining roughly 70%. The CMB measurements along with the startling discovery from Type 1a supernova that the expansion of the Universe was accelerating,[11,12] led to quick acceptance of the standard ΛCDM cosmological model, a remarkable model which to the present day has survived a battery of increasingly stringent observational tests.

Figure 2 illustrates the history of the Universe and the salient features of the standard ΛCDM cosmological model. Inflation, i.e. the period of extremely rapid

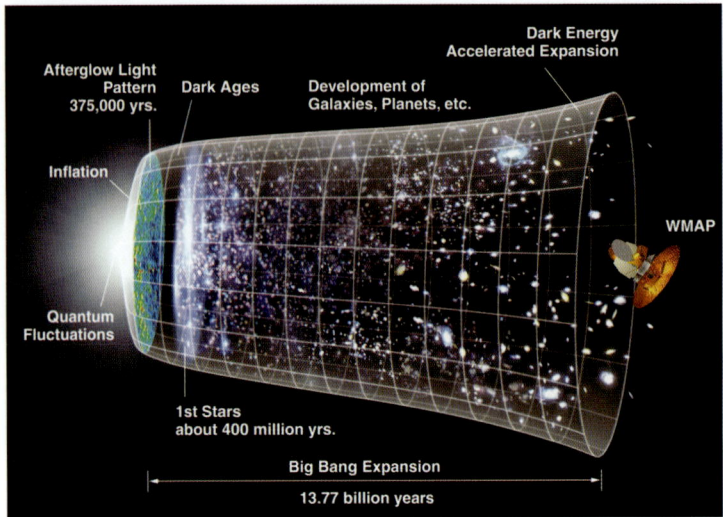

Fig. 2.   The expansion history of the Universe (not to scale). (Image credit: NASA/WMAP Science Team.)

and accelerating expansion in the first fraction of a second of the Universe's existence, is represented on the far left. The era of the last strong electromagnetic interactions between the CMB photons and matter, at $t \simeq 375,000$ years, is shown as a surface; this surface acts as the source for CMB photons we observe. A subsequent period during which very few photons were created (the "Dark Ages") is followed by the ignition of the first generation of stars and the epoch of reionization of the Universe. The era of strong structure growth follows, eventually ending when dark energy comes to dominate the energy budget of the Universe, resulting in another epoch of accelerated expansion.

Measurements of the temperature and polarization anisotropy of the CMB are particularly well suited to provide stringent tests of the cosmological model, including precise and accurate determinations of the model parameters as well as the exploration of extensions to the model. To fully exploit the constraining power of the CMB, what is needed are increased polarization sensitivity on all angular scales and higher angular resolution temperature measurements. Such measurements have been the focus of the CMB experiments over the last 15 years.

Following *COBE*, two more dedicated CMB satellite missions have mapped the entire sky. First, NASA's Wilkinson Microwave Anisotropy Probe (*WMAP*), launched in 2001, mapped the full CMB sky with moderate angular resolution (as fine as 13 arcminutes) in five frequency bands.[13] The European Space Agency's *Planck* mission was launched in 2009 and mapped the full CMB sky with resolution as fine as five arcminutes in nine frequency bands.[14] Figure 3 illustrates the evolution in sensitivity and angular resolution of the satellite missions, showing $\sim$10 deg^2 of the extracted CMB temperature anisotropy over same sky from *COBE/DMR*,

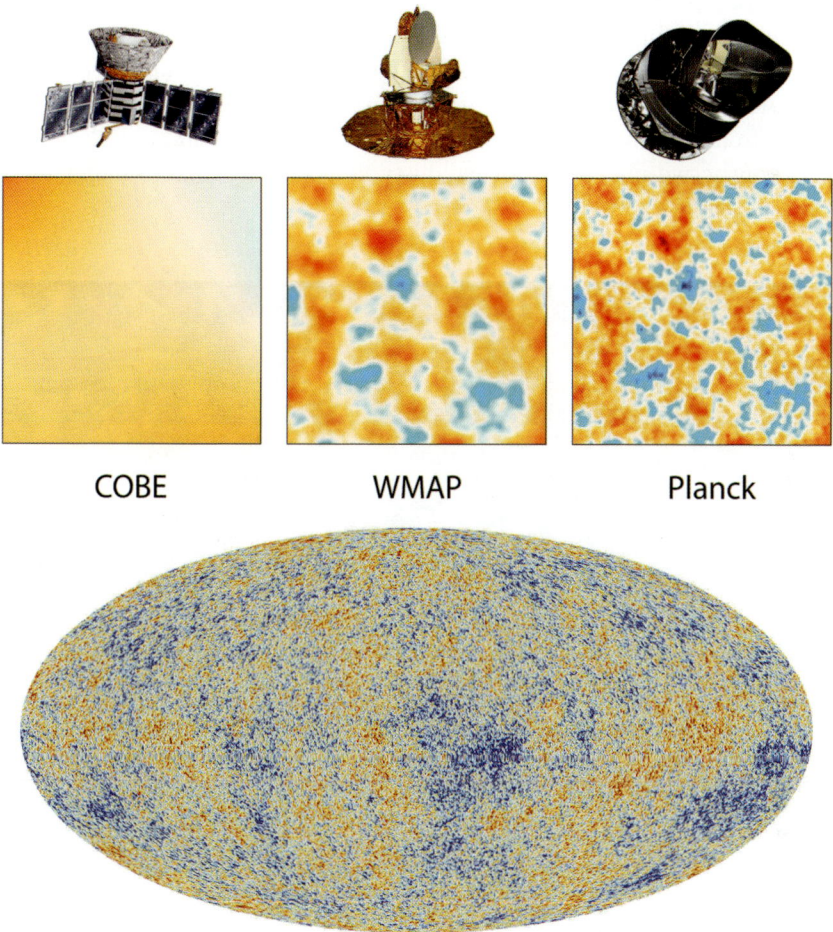

COBE   WMAP   Planck

Fig. 3.   Top Panel: CMB temperature fluctuations over 10 deg^2 of the same sky in *COBE/DMR*, *WMAP* and *Planck* data. (Image credit: NASA/JPL-Caltech/ESA.) Bottom Panel: Full-sky image of the CMB temperature anisotropy from *Planck*. (Image credit: ESA and the *Planck* Collaboration.)

WMAP and *Planck* data. Also shown is the full-sky *Planck* CMB map.

Higher resolution maps of the CMB require larger telescope apertures than are currently practical for space missions. Significant progress in measuring the small-scale temperature anisotropy of the CMB on comparatively small patches of sky (less than 10% of the full sky) has been made recently by ground-based experiments, most notably by the 10m South Pole Telescope (SPT[15]) and the 6m Actacama Cosmology Telescope (ACT[16]). Figure 4 shows 50 deg^2 of CMB sky ($\sim$ 7 degrees on a side), illustrating the improved resolution and sensitivity going from *WMAP* in *W* band to *Planck* at 143 GHz to SPT at 150 GHz. The degree-scale fluctuations, which contain key information about the geometry of the Universe and the relative abundance of baryons, dark matter, and dark energy, are seen clearly in all three data sets. The sub-degree-scale information is captured well in the *Planck* data and completely resolved in the SPT data. Lower-redshift contributions to the millimeter-wave sky also become apparent in the SPT data, in particular the Sunyaev-Zel'dovich effect signatures from massive, high-redshift galaxy clusters (compact dark spots) and emission from active galactic nuclei and dusty, star-forming galaxies (compact bright spots).

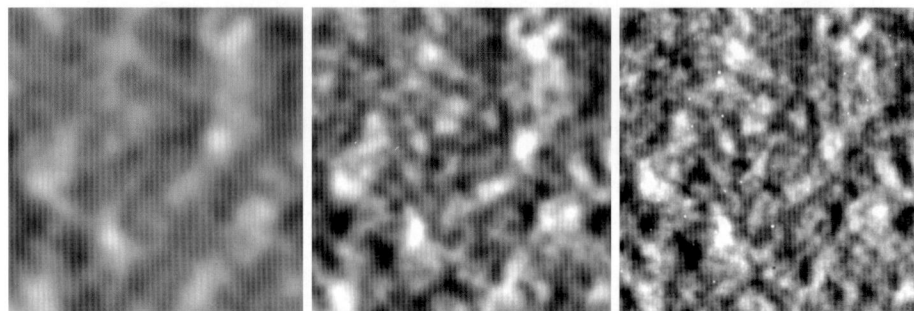

Fig. 4.   50 deg^2 ($\sim$ 7 degrees on a side) of the same sky in (from left to right) *WMAP* W band, *Planck* 143 GHz, and SPTpol 150 GHz data. The degree-scale fluctuations are seen clearly in all three data sets. The sub-degree-scale information is captured well in the *Planck* data and completely resolved in the SPT data. Lower-redshift contributions to the millimeter-wave sky also become apparent in the SPT data, in particular the Sunyaev-Zel'dovich effect signatures from massive, high-redshift galaxy clusters (compact dark spots) and emission from active galactic nuclei and dusty, star-forming galaxies (compact bright spots).

The sensitivity of ground-based CMB measurements is adversely affected by atmospheric emission, primarily from water vapor, and the spatial and temporal fluctuations in this emission. Experiments are therefore located at exceptionally high and dry sites, such the high Antarctic Plateau and the high Atacama Desert in Chile. The sensitivity of the state-of-the-art detectors on the ground based telescopes are background-limited, i.e. limited by the atmospheric noise. Therefore increased sensitivity is being pursued by increasing the detector count. Figure 5 shows the telescopes and current focal planes / detector arrays for the POLARBEAR,[17] ACT, and SPT projects. These experiments and the telescopes on which they are mounted have been designed and built expressly for observations of the CMB and take advantage of the ongoing revolution in the microfabrication of arrays of superconducting detectors—in fact, the experiment teams are largely responsible for driving the scaling of the detector technology. Recently, data from new instruments on these telescopes have led to the first measurements of small-scale polarization anisotropy in the CMB. The advances in detector technology are also being exploited by dedicated small-aperture ground- and balloon-based experiments that are targeting degree scale polarization to search for the inflationary *B*-mode signal, as discussed in Section 2. Figure 6 shows the BICEP telescope[18] at the South Pole

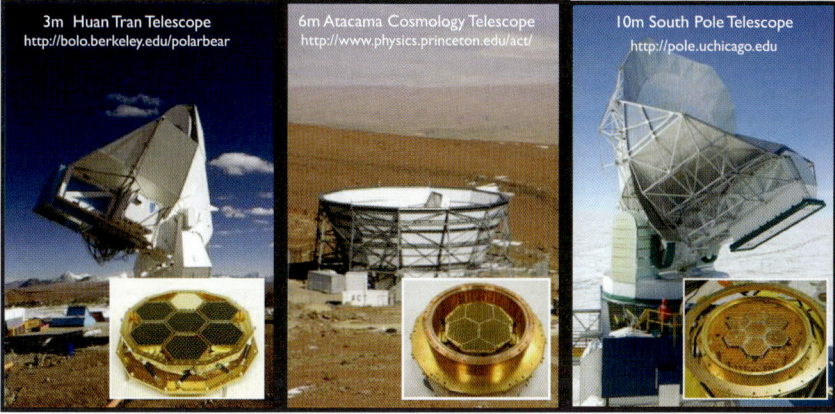

Fig. 5.   The telescopes and current focal plane detector arrays of the POLARBEAR, ACT, and SPT projects.

Fig. 6.   The BICEP telescope with the Keck ground shield in the background, the Spider balloon-born telescope, and an image of the planar array detector modules they employ.

with the ground-shield of the Keck Array[19] in the background, the Spider balloon experiment,[20] and an image of the planar detector technology they both employ.[21]

The fluctuations in the CMB temperature are predicted and measured to be Gaussian-distributed to a high degree of precision,[22] which means that all of the information in the fluctuations are contained in the two-point angular correlation function or its harmonic equivalent, the angular power spectrum. Figure 7 shows the angular power spectrum of the CMB temperature anisotropy from satellite and ground-based data. The spectrum is measured by decomposing a map of the sky into spherical harmonic modes $Y_{\ell m}$ and calculating the variance in the amplitudes of modes with multipole number ℓ. The now-familiar peak-trough structure in the CMB temperature power spectrum is a consequence of acoustic oscillations in the primordial photon-matter plasma and the initial conditions for the phase of the fluctuations, in line with the predictions of inflation.[23]

The primary CMB temperature anisotropy is now well characterized to multi-poles $\ell \sim 3000$, or on angular scales from the full sky to a few arcminutes. The

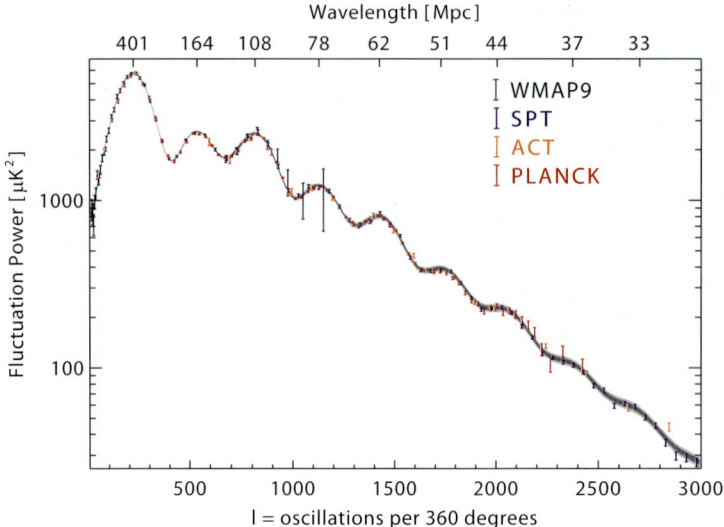

Fig. 7.   The angular power spectrum of the CMB temperature, as measured by the *WMAP*, *Planck*, ACT, and SPT science teams. The best-fit ΛCDM model and 1 and 2σ uncertainties on that model are shown by the light and dark gray bands.

ΛCDM model continues to hold up stunningly well, even as the precision of the CMB-determined parameters has continued to increase. Inflationary constraints include curvature limited to be less than 3% of the energy density, non-Gaussian fluctuations limited to $f_{NL} < 10$, and a small departure from pure scale invariance of the primordial fluctuations detected at 5σ confidence, and the tensor-to-scalar ratio r constrained to be less than 0.11.[22] Also of interest to particle physics, the influence of light relativistic species (i.e. neutrinos and any yet identified "dark radiation") is detected at 10σ, and the effective number of species N_{eff} is shown to be within 1σ of standard cosmology prediction of 3.046; the sum of the masses of the neutrinos is found to be less than 0.66 eV; dark matter is shown to be non-baryonic at $> 40\sigma$, and early dark energy models are highly constrained, as are models of decaying dark matter.[22]

Figure 8 shows the *Planck* constraints on the key inflation parameters n_s and r. Not only are CMB measurements showing strong evidence for the inflationary paradigm, they are also already placing strong constraints on specific models of inflation and ruling out large portions of model space. Constraints on r from CMB temperature fluctuations are now limited not by instrumental sensitivity but by only having one last-scattering surface to measure, i.e. cosmic variance; further progress on constraining this parameter will only come from measurements of CMB polarization.

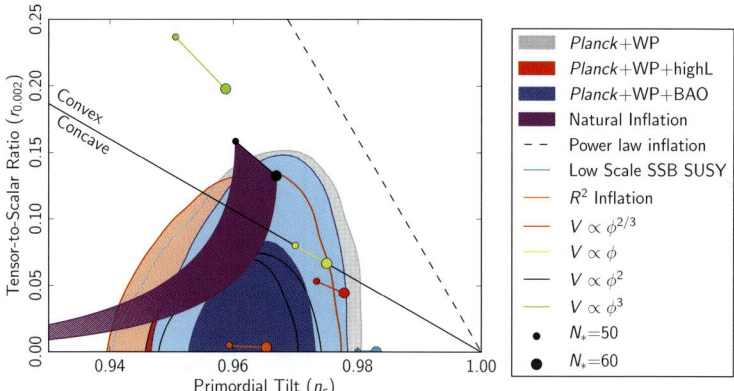

Fig. 8. *Planck* constraints on the inflation parameters n_s and r.[22] A scale-invariant Harrison-Zel'dovich primordial power spectrum ($n_s = 1$) is ruled out at greater than 5σ, providing strong evidence in favor of inflation. The limit on the amplitude of inflationary gravitational waves (tensor-to-scalar ratio $r < 0.11$ at 95% confidence) represents the limit of temperature-only measurements of this parameter; further improvements will only come from CMB polarization measurements.

2. Polarization

The CMB is predicted to be linearly polarized at roughly the 10% level.[24] Initially unpolarized light with a quadrupole moment, scattering off free electrons, will result in a net linear polarization. In the primordial plasma, the largest contribution to the quadrupole results from the Doppler shifts caused by the velocity field of the plasma. In the era of tight coupling between photons and matter, this quadrupole is heavily suppressed; as a result, the polarization of the CMB is primarily generated near the era of photon-matter decoupling, as the photon mean free path increases. This mechanism results in a particular correlation between the direction of polarization and the amplitude of polarized emission, known as E-mode polarization,[25] which exhibits an even-parity, curl-free pattern across the sky. The E-mode polarization is expected to be the dominant component of the CMB polarization, but even this signal is extremely small (another factor of 10 lower in amplitude—100 in power—than the temperature anisotropy), and it was first detected barely a decade ago,[26] roughly ten years after the first detection of temperature anisotropy. Because of the rapid improvements in detector technology, high-significance measurements of the E-mode polarization and temperature-E-mode correlation have now been made over a broad range of angular scales,[27–33] as summarized in Figure 9.

The odd-parity, divergence-free B-mode polarization pattern is not generated to first order in the standard CMB picture with scalar density fluctuations. By contrast, tensor fluctuations, or gravitational waves, present at the time of decoupling lead to roughly equal E- and B-mode CMB polarization.[34] An inescapable prediction of inflation is the production of a background of relic gravitational waves;[35] thus, a detection of B-mode polarization at degree angular scales, corresponding

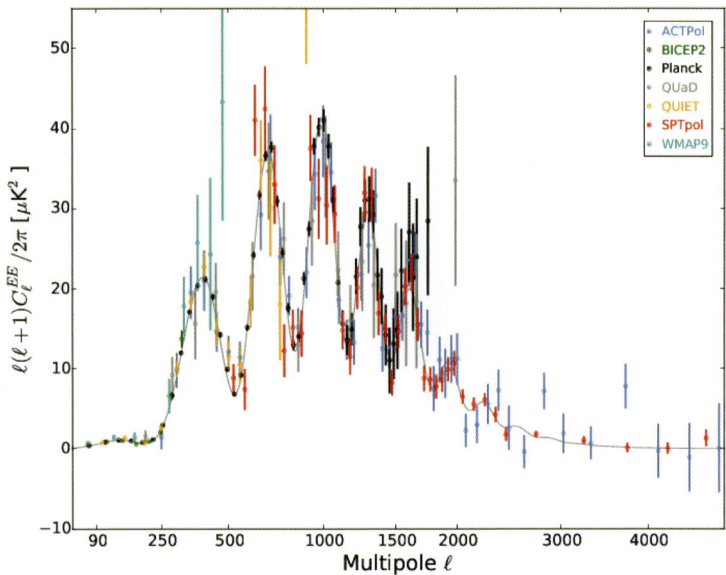

Fig. 9.   A summary of recent high-signal-to-noise measurements of the E-mode power spectrum.

to the angle subtended by the horizon at decoupling, would be a spectacular confirmation of inflation. If inflation happened at the GUT scale ($\sim 10^{16}$ GeV), the B-mode signal would be at level that is within reach of upcoming experiments, and a future generation of CMB experiments would be able to directly map the quantum fluctuations generated by inflation.

The path forward seems simple: make sensitive measurements of the polarization of the cosmic microwave background, make E- and B-mode polarization maps and the corresponding angular power spectra. The B-mode power at degree scales would, in principle, indicate the presence of inflationary gravitational waves.

The apparent simplicity is highly deceptive and in fact it will be extremely challenging to measure the inflationary B-mode signal. The obvious first consideration is that the inflation-induced B-mode polarization signal is extremely weak in even the most optimistic predictions, and, in many inflationary models, the signal may be so weak to be effectively unobservable. Experiments now operating or being built will produce maps with polarization noise levels corresponding to a few tens of nanoKelvin rms over degree scale patches of sky, a level that is already a thousand times lower than the temperature anisotropy and even tens of times lower than the primordial E-mode signal. The experiments therefore require unprecedented control of possible sources of contamination, including the "leakage" of the much stronger CMB temperature and E-mode polarization through a variety of possible mechanisms.

Second, as the recent BICEP2, Keck and *Planck* results have demonstrated,[36] contaminating B-mode polarized components of the Galactic emission may dominate the inflationary B-mode for even the clearest views through our Galaxy. This may be the case even at the current upper limits for the inflationary signal[36]—and most certainly is true at the lowest B-mode levels predicted. Experimenters will need to exploit the fact that frequency dependence of the CMB is distinctly different from that of the contaminating emission components. Precise measurements of the B-mode polarization on the sky in multiple frequency bands will allow for separation of the various components and provide the cleanest interpretation of any detected B-mode signal.

Finally, even the precision measurement and accounting for the foregrounds may not be sufficient. As discussed in Section 3, gravitational lensing of the CMB by the intervening large scale structure in the Universe will lead to the generation of B-mode polarization from the primordial E-mode signal. This will further complicate the extraction of the inflationary B-mode polarization signature, especially if it lies much below the current upper limit. Lensing B modes have the same frequency spectrum as the inflationary signal, and so a different technique, referred to as "de-lensing" must be used to separate them. De-lensing uses correlations within high signal-to-noise CMB temperature and polarization maps to reconstruct the lensing potential and separate the lensing B-mode signal from the inflationary signal.[37]

Figure 10 summarizes the theoretical predictions for the CMB polarization angular power spectra, including the expected level of the B-mode foregrounds and the current B-mode measurements. The measurement of the B-mode power at five orders of magnitude below the temperature anisotropy power is a tour de force. The plot also indicates clearly the formidable challenges, discussed above, that experimenters face in achieving the next orders of magnitude in the search for gravitational waves at levels of $r \sim 0.01$ and lower.

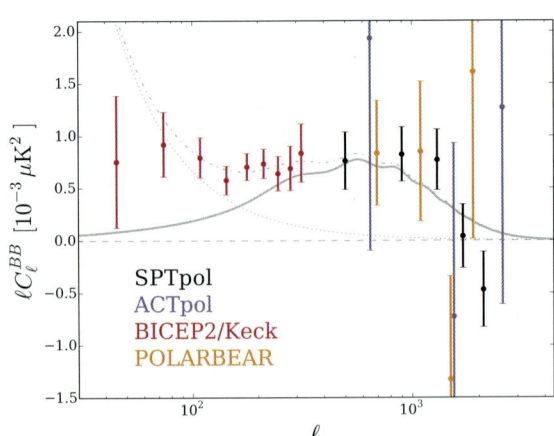

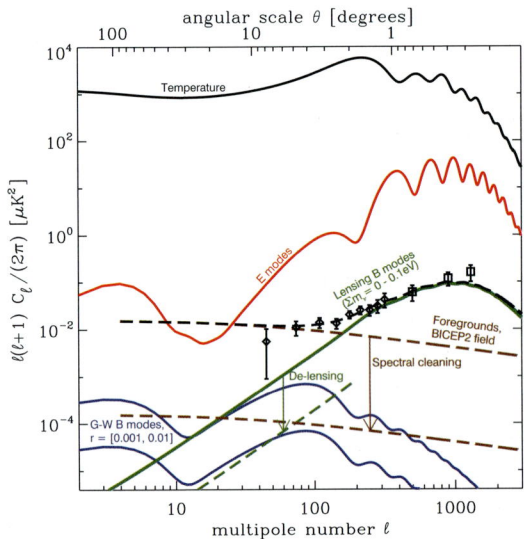

Fig. 10.   Theoretical predictions for and measurements of the angular power spectra of CMB polarization. Top Panel: Compilation of B-mode power spectrum measurements from the literature.[38]
The solid gray line shows the theoretical prediction for the lensing B modes. The dotted gray line
shows the prediction for the B modes from polarized Galactic dust emission.[39] The dot-dashed line
shows the sum of the lensing and dust B-mode predictions. Bottom Panel: Theoretical predictions
for the temperature, E-mode, and B-mode power spectra, with the data points and dust foreground estimation from the top panel included (note the different y-axis scaling between the two
panels). The two contributions to the B-mode power spectrum—inflationary gravitational waves
and gravitational lensing of the E modes—are plotted separately. The expected gravitational-wave
contribution is shown for two values of the tensor-to-scalar ratio r: $r = 0.001$ and $r = 0.01$. The
contribution to gravitational-wave B modes from scattering at recombination peaks at $\ell \sim 80$ and
from reionization at $\ell < 10$. The lensing contribution to the B-mode spectrum can be at least
partially removed from measurements by exploiting the non-Gaussian statistics of the lensing; a
realistic prediction for the level of cleaning ("de-lensing") is represented by the green arrow and
green dashed line. The brown arrow and brown dashed line indicate the expected level of foreground
cleaning that could be achieved with a sufficiently low-noise multi-frequency measurement.

3. Gravitational Lensing of the CMB

Though the last strong electromagnetic interactions between CMB photons and matter occurred many billions of years ago at the last scattering surface (see Figure 2), these photons have undergone small but important gravitational interactions with matter along their entire path from the last scattering surface to CMB telescopes on Earth. The most important of these interactions for cosmology is the gravitational lensing effect. As CMB photons travel to us, their paths are subtly deflected by the gravitational influence of intervening matter. These deflections encode information about the growth of structure. Because massive neutrinos suppress structure growth, the information contained in CMB lensing can provide a valuable constraint on the sum of neutrino masses.[40]

More generally, the effect of gravitational lensing on the CMB can be used to map the projected gravitational potential between the last scattering surface and us—effectively making a map of all the mass in the Universe. This mass map can be combined with tracers of large-scale structure such as galaxies to test cosmological models and constrain properties of the tracer population. These measurements have the potential to test models of structure growth (including tests of the framework of general relativity) and to probe primordial non-Gaussianity.[41,42] As a proof of concept of these types of studies, Figure 11 shows one of the first high-signal-to-noise measurements of the correlation between a mass map from CMB lensing and a tracer of large-scale structure.[43] The contours represent projected mass as measured through the lensing of the CMB temperature in SPT data, and the gray-scale background represents a smoothed version of fluctuations in the cosmic infrared background (CIB) at 500 microns, as measured by the *Herschel-SPIRE* instrument. The correlation between the red contours (indicating increased projected mass) and white regions of the background map (indicating increased 500-micron flux)—and, conversely, between blue contours and black regions—is immediately apparent.

As discussed earlier, the temperature fluctuations in the unlensed CMB are nearly Gaussian-distributed, and the same is true of the polarization fluctuations. The distortions imprinted by lensing break this pure Gaussianity and result in correlations between Fourier or spherical harmonic modes of the CMB that are statistically uncorrelated in a Gaussian field. These distortions also convert a small fraction of a pure E-mode polarization pattern into B modes. As discussed in the previous section, these lensing B modes are a contaminant to the measurement of the primordial gravitational wave signal, but they are crucial to extracting the full power of the CMB to measure projected mass and constrain the properties of the neutrinos. Exploiting the lensing-induced correlations between all combinations of the CMB temperature and polarization fields will both maximize the cosmological power of the lensed CMB and allow us to de-lens the B-mode sky and measure the inflationary gravitational wave signal.

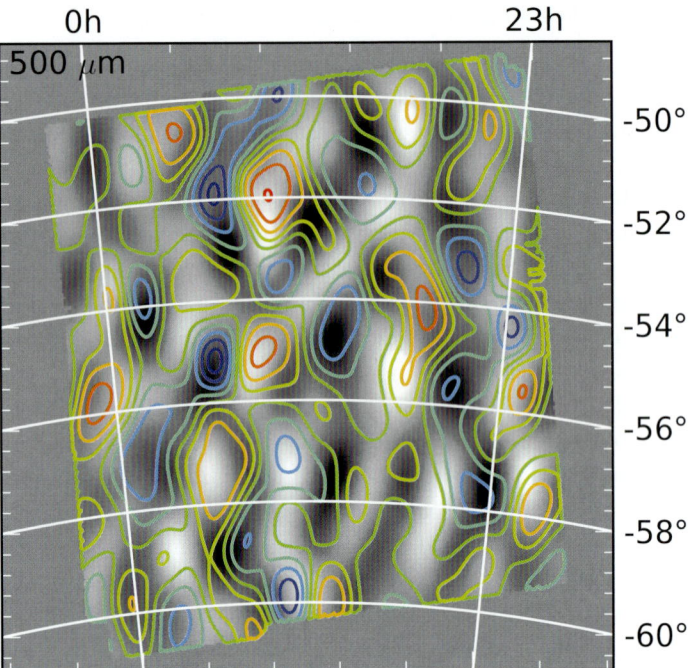

Fig. 11.   The correlation between CMB lensing and *Herschel-SPIRE* 500-micron CIB fluctuations.[43] Colored contours represent projected mass derived from CMB lensing in SPT data; CIB is shown in gray scale. Correlation between the red contours (increased projected mass) and white regions (increased CIB intensity), and between blue contours and black regions, is immediately apparent. The CIB is expected and observed to have roughly 80% correlation with the projected mass measured by CMB lensing.[44,45]

4. A Next Generation CMB Experiment, CMB-S4

There has been phenomenal progress over the last 50 years in the measurement of the CMB and the insights those measurements have provided into the workings of the Universe. Yet, there remains much science to be extracted from the CMB. In fact, we are at the cusp of achieving several major milestones, while at the same time we are saturating the reach of current ground-based experiments. A sea change is required to take the next great leap in experimental CMB research.

Figure 12 shows the "Moore's law"-like increase in sensitivity of CMB experiments, a trend that has continued since Penzias and Wilson's discovery 50 years ago. The trend has recently been sustained by increasing the number of background-limited detectors. The plot defines stages of CMB experiments characterized by their detector counts. Currently, several Stage-2 experiments, each of order 1000 polarization-sensitive detectors are operating.[17,19,46–48] Over the next few years, upgrades to these experiments, as well as new experiments under construction[49] have plans to increase detector counts to order 10,000, offering a significant increase in

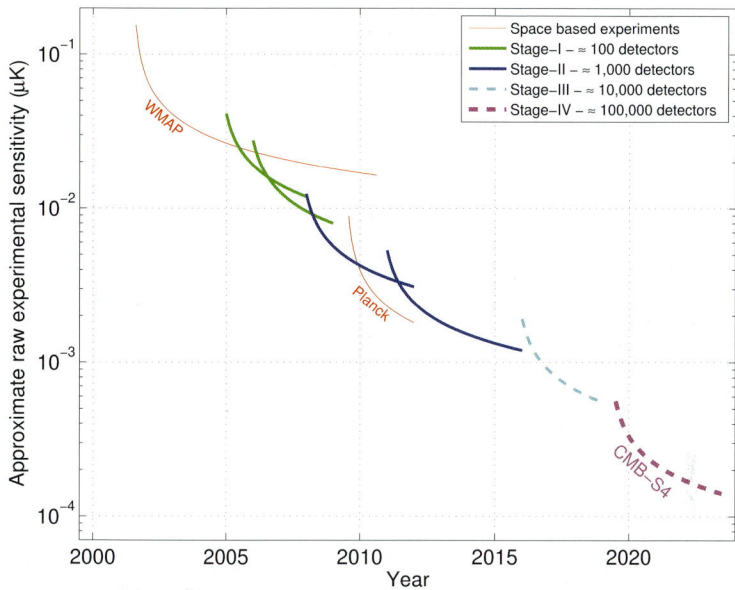

Fig. 12. A "Moore's law"-type plot illustrating the evolution of the raw sensitivity of CMB experiments.[50] The sensitivity is given as the noise in μK that would be achieved if a given instrument concentrated all detectors and observing time on a single resolution element. This total sensitivity scales with number of detectors, and advances in detector technology have enabled a power-law evolution of sensitivity with time.

sensitivity and commensurate improvement in cosmological constraints.

4.1. *Science Reach of CMB-S4*

The projected science reach of the Stage-3 CMB projects is indeed impressive, and data from these projects will provide a dramatic increase over present cosmological constraints in roughly five years. It is at Stage 4, however, with an experiment with order 500,000 detectors, surveying roughly 70% of the sky from degree to arcminute scales with sensitivity of order 1μK arcminute that we will cross critical thresholds for cosmology and physics. Such an experiment, called CMB-S4, was recently endorsed by the Particle Physics Projects Prioritization Panel (P5) in their report *Building for Discovery, Strategic Plan for U.S. Particle Physics in the Global Context.*[51] Here we briefly highlight some of the expected science return from CMB-S4.

As shown in Figure 13, CMB-S4 will be able to detect inflationary B-mode polarization or rule out large field models ($r > 0.01$) with high significance. If no inflationary B modes are detected at $r > 0.01$, a small-area, ultra-deep survey with CMB-S4 could achieve $\sigma(r) < 10^{-4}$, potentially ruling out large classes of inflationary models. If r is detected at the ~ 0.01 level, CMB-S4 would target a robust, cosmic-variance limited measure of r with a large-area survey and set

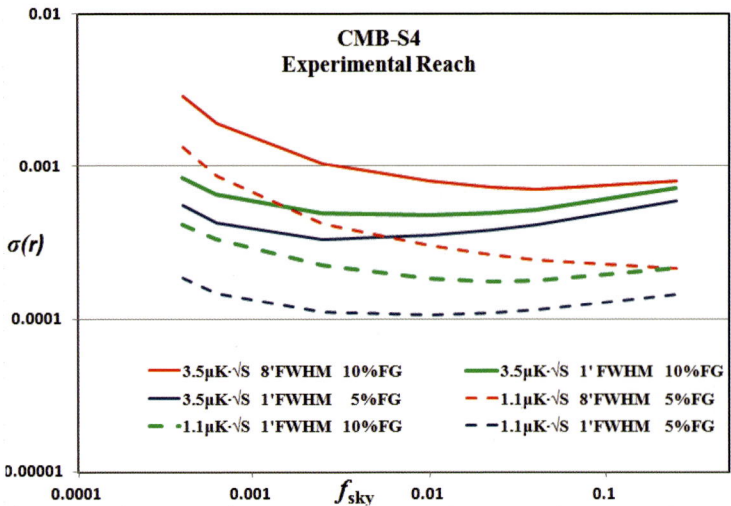

Fig. 13.   Projected CMB-S4 constraints on the tensor-to-scalar ratio r as a function of fraction of sky observed for several possible configurations of instantaneous focal plane sensitivity, beam size, and level of foreground cleaning.[53] The total observing time is kept constant, such that larger sky coverage leads to higher noise per map pixel. For even the most pessimistic assumptions (highest detector noise, largest beam, largest fraction of residual foreground power), CMB-S4 will make a measurement that can distinguish between large- and small-field inflation at greater than 5σ significance.

constraints on the tensor spectral index n_t with an ultra-deep survey.[52]

For neutrino physics, CMB-S4 would exploit the unique ability of the CMB to provide critical constraints on models of the neutrinos and their interactions in the Universe – offering a highly complementary probe to big bang nucleosynthesis studies and to sterile neutrino models.[50] With a projected uncertainty in the effective number of relativistic particle species of only 0.02,[50] finding consistency with the standard cosmology prediction of $N_{\rm eff} = 3.046$ would be a fundamental achievement linking particle physics and our understanding of the evolution of the first seconds of the Universe. Finding a departure from 3.046 would be even more exciting, indicating new physics.

As shown in Figure 14, CMB-4 is highly sensitive to the sum of the neutrino masses, Σm_ν, primarily through its impact on the level of CMB lensing from large scale structure. CMB-S4 will not just set an upper bound on Σm_ν, but by achieving $\sigma(\Sigma m_\nu) = 16$ meV (with a prior from baryon acoustic oscillation measurements) it will provide a detection of neutrino mass, even if at the minimum mass and the normal hierarchy.[50] Figure 15 shows the projected CMB-S4 constraints in the $N_{\rm eff}$ – Σm_ν plane superimposed on the *Planck* 2013 results.[22] The CMB-S4 sensitivity to the sum of the masses is unique and complementary to terrestrial neutrino experiments.

In addition to CMB-S4's sensitivity to CMB lensing, its high resolution and

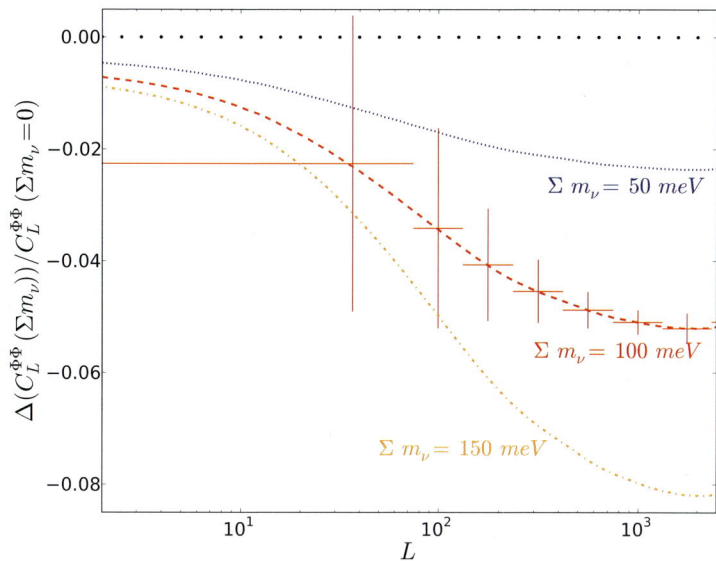

Fig. 14.   An illustration of the projected CMB-S4 constraints on the sum of the neutrino masses $\sum m_\nu$ through measurements of the CMB lensing power spectrum $C_L^{\phi\phi}$.[50] For a given value of $\sum m_\nu$, the fractional change in $C_L^{\phi\phi}$ is plotted relative to the case for massless neutrinos. Projected CMB-S4 constraints on $C_L^{\phi\phi}$ are shown for $\sum m_\nu = 100$ meV.

multiple bands will enable unprecedented measurements of the Sunyaev-Zel'dovich (SZ) effects. The SZ measurements will provide dark energy constraints through the impact of dark energy on the growth of structure. The measurements will also allow tests of gravity on large scales through exploiting the kinematic SZ effect to measure the momentum field and large-scale flows (see contribution in this volume from D. Spergel). The power of these probes will be amplified by combining the CMB lensing and SZ effect data with data from other complementary cosmological surveys, such as those to be conducted by the dark energy Spectroscopic Instrument (DESI), the Large Survey Synoptic Telescope (LSST), *Euclid* and the Wide-field Infrared Survey Telescope (WFIRST).

A noteworthy advance enabled by CMB-S4 will be the ability to use CMB lensing to determine the mass scaling relation of galaxy cluster observables, especially of the CMB-S4 SZ cluster measurements. It is well appreciated that the number density evolution of massive galaxy clusters is highly sensitive to models of the dark energy, but that exploiting this sensitivity is plagued by the large systematic uncertainty in the mass scaling of the cluster observable.[54] Cluster catalogs from SZ surveys are well suited for cosmological studies as they are effectively mass-limited (the SZ brightness does not suffer cosmological dimming) and the SZ flux has low scatter with cluster mass. They remain, however, limited by uncertainties in the mass scaling. CMB-S4 will be revolutionary in that it is expected to be able to

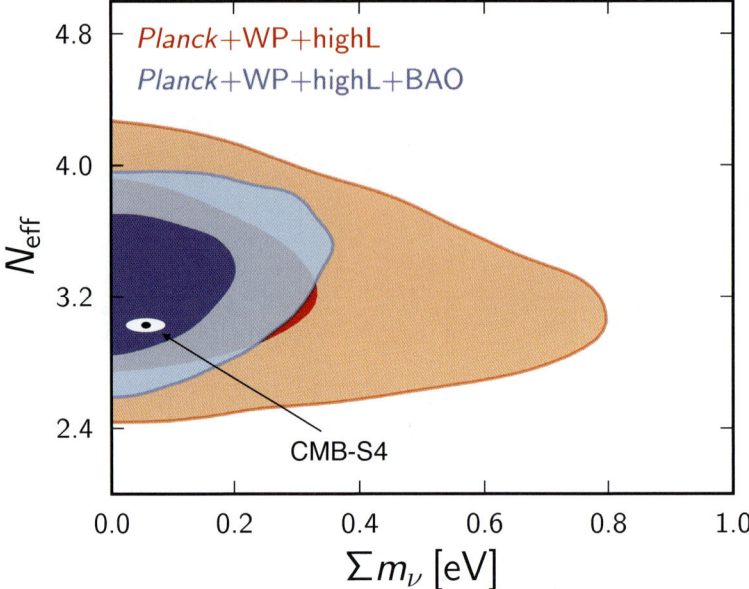

Fig. 15.   An illustration of the dramatic improvement possible with CMB-S4 on the constraints of the effective number of light relativistic species N_{eff} and the sum of the neutrino masses $\sum m_\nu$. The CMB-S4 $\sum m_\nu$ constraints, which assume a DESI BAO prior, are overlaid on a plot of the *Planck* $N_{\text{eff}} - \sum m_\nu$ results.[22]

calibrate the mass scaling to better than 1% through CMB lensing. This coupled with a low mass threshold will enable CMB-S4 to identify of order 100,000 clusters, probe the growth of structure to redshifts beyond $z \sim 2.5$, and thus realize the full potential of galaxy clusters as a probe of dark energy.

4.2. *What is CMB-S4?*

To maintain the Moore's Law-like scaling in detector count and raw CMB sensitivity (see Figure 12), a phase change in the mode of operation of the ground based CMB experiments is required. Two constraints drive the change: (1) CMB detectors are background-limited so we need more on the sky to increase sensitivity; and (2) The pixel count for CMB focal planes are nearing saturation. Even using multi-band pixels and wide-field of view optics, CMB telescopes are able to field only of order 10,000 polarization detectors, far fewer than needed to meet science goals or conduct the definitive ground-based CMB experiment.

CMB-S4 thus requires multiple telescopes, each with a maximally outfitted focal plane of pixels utilizing superconducting, background-limited, CMB detectors. To achieve the large sky coverage and to take advantage of the best atmospheric conditions, the South Pole and the Chilean Atacama sites are baselined, with the possibility of adding a new northern site to increase sky coverage to 100%.

4.2.1. *Inflationary B modes: low-ℓ sensitivity, foregrounds and atmospheric noise mitigation*

At the largest angular scales (lowest values of ℓ), where measurements are critical for pursuing inflationary B modes (as well as testing the ΛCDM model with E-mode polarization), the CMB polarization anisotropy is highly contaminated by foregrounds. Galactic synchrotron dominates at low frequencies and galactic dust at high frequencies, as recently shown by the *Planck* and *Planck*/BICEP/Keck polarization results.[36] Multi-band polarization measurements are required to distinguish the primordial polarized signals from the foregrounds.

Adding to the complexity of ground-based observations of the low-ℓ CMB is the need to reject the considerable atmospheric noise contributions over the large scans needed to extract the low-ℓ polarization. While the spatial and temporal fluctuations of the atmosphere are not expected to be polarized, any mismatches in the polarized beams or detector gains will lead to leakage of temperature signal to apparent polarization. Small-aperture telescopes allow additional modulations, such as bore-sight rotation[55,56] or modulation of the entire optics with a polarization modulation scheme in front of the telescope.[49,57] The cost of a small telescope is dominated by the camera, making it feasible to deploy multiple telescopes each optimized for a single band, or perhaps multiple bands within one of the relatively narrow atmosphere windows.

CMB-S4 is therefore envisioned to include dedicated small-aperture telescopes for pursuing low-ℓ polarization. The default plan for CMB-S4 is to target the recombination bump at $\ell \sim 80$ (see Figure 10), with E-mode and B-mode polarization down to $\ell \sim 20$. If Stage-3 experiments demonstrate that it is feasible to target the reionization bump ($\ell < 10$, see Figure 10) from the ground, those techniques may be incorporated into CMB-S4. More likely, however, this is the ℓ range for which CMB-S4 will be designed to be complementary to balloon- and satellite-based measurements.[20,58,59]

4.2.2. *Neutrinos and dark energy: high-ℓ sensitivity*

At the highest angular resolution (highest ℓ), where measurements are needed for de-lensing the inflationary B modes, constraining $N_{\rm eff}$ and Σm_ν, and investigating dark energy and general relativity with secondary CMB anisotropy, the CMB polarization anisotropy is much less affected by both foregrounds and atmospheric noise. In fact, it should be possible to measure the primary CMB anisotropy in E-mode polarization to multipoles a factor of a few times higher than possible in temperature, thereby extending the lever arm to measure the spectral index and running of the primordial scalar (density) fluctuations. CMB measurements contain useful information about lensing out to ℓ_{max} of order 5000 and secondary CMB measurements are greatly improved with ℓ_{max} of order 10,000 or even higher. Accessing these angular scales requires large-aperture telescopes. Owing to the steep scaling of telescope cost with aperture diameter, it is unlikely to be cost-effective to consider separate large-aperture telescopes each optimized for a single frequency band.

CMB-S4 is therefore envisioned to include dedicated large-aperture, wide-field of view telescopes equipped with multi-chroic detector arrays.

4.3. *Who is Building CMB-S4*

There is an enormous increase in scaling that must be achieved to go from Stage-3 experiments to CMB-S4, especially in detectors, but also in telescopes, data rates, simulations, computing and people. The plan is to build on the Stage-3 experiments by coherently bringing together the current CMB research teams and the broader CMB community, the expertise and facilities at the U.S. Laboratories (Argonne, Fermilab, LBNL and SLAC), as well international partners.

An ongoing task is to refine the rough conceptual design of CMB-S4 outlined above. The first priority is to determine the optimum mix of telescopes sizes and designs, the number of bands needed to mitigate foreground contamination and the required sky coverage. This requires simulations, informed by the best available data and phenomenological models. The entire community is invited to participate in this critical exercise to optimize the science reach of CMB-S4.

5. Final Words

It has been a remarkable 50 years for cosmology. Much of the progress has been driven by the tight interplay of new measurements of the CMB with theoretical and phenomenological advances in cosmology. The next decade holds the potential for major breakthroughs, and likely new discoveries, that can be realized with the next-generation ground-based experiment, CMB-S4. The future is very bright indeed.

Lastly it would be an oversight not to point out the obvious: there is only one CMB sky. It holds a wealth of information on fundamental physics and the origin and evolution of our Universe. While we have learned a great deal from CMB measurements, including discoveries that have pointed the way to new physics, we have only begun to tap the information contained in CMB polarization, CMB lensing and secondary effects. The CMB is the gift that keeps on giving.

Acknowledgments

I thank the organizers for the opportunity to participate in such a fascinating and stimulating meeting and the Solvay family for their generosity and most gracious hospitality. I acknowledge and thank Tom Crawford for his expert and unselfish help in writing this contribution. Partial support to JC and TC was provided by NSF awards PLR-1248097 and PHY-1125897.

References

1. A. A. Penzias and R. W. Wilson, *Astrophys. J.* **142**, 419 (July 1965).

2. R. H. Dicke, P. J. E. Peebles, P. G. Roll and D. T. Wilkinson, *Astrophys. J.* **142**, 414 (July 1965).

3. R. A. Alpher and R. Herman, *Nature* **162**, 774 (November 1948).

4. A. H. Guth, *Phys. Rev. D* **23**, 347 (January 1981).

5. V. F. Mukhanov and G. V. Chibisov, *Soviet Journal of Experimental and Theoretical Physics Letters* **33**, p. 53 2(May 1981).

6. G. F. Smoot *et al.*, *Astrophys. J.* **396**, L1 (1992).

7. A. D. Miller, R. Caldwell, M. J. Devlin, W. B. Dorwart, T. Herbig, M. R. Nolta, L. A. Page, J. Puchalla, E. Torbet and H. T. Tran, *Astrophys. J.* **524**, L1 (October 1999), astro-ph/9906421.

8. P. D. Mauskopf, P. A. R. Ade, P. de Bernardis, J. J. Bock, J. Borrill, A. Boscaleri, B. P. Crill, G. DeGasperis, G. De Troia, P. Farese, P. G. Ferreira, K. Ganga, M. Giacometti, S. Hanany, V. V. Hristov, A. Iacoangeli, A. H. Jaffe, A. E. Lange, A. T. Lee, S. Masi, A. Melchiorri, F. Melchiorri, L. Miglio, T. Montroy, C. B. Netterfield, E. Pascale, F. Piacentini, P. L. Richards, G. Romeo, J. E. Ruhl, E. Scannapieco, F. Scaramuzzi, R. Stompor and N. Vittorio, *Astrophys. J.* **536**, L59 (June 2000).

9. S. Hanany, P. Ade, A. Balbi, J. Bock, J. Borrill, A. Boscaleri, P. de Bernardis, P. G. Ferreira, V. V. Hristov, A. H. Jaffe, A. E. Lange, A. T. Lee, P. D. Mauskopf, C. B. Netterfield, S. Oh, E. Pascale, B. Rabii, P. L. Richards, G. F. Smoot, R. Stompor, C. D. Winant and J. H. P. Wu, *Astrophys. J.* **545**, L5 (December 2000), astro-ph/0005123.

10. N. W. Halverson, E. M. Leitch, C. Pryke, J. Kovac, J. E. Carlstrom, W. L. Holzapfel, M. Dragovan, J. K. Cartwright, B. S. Mason, S. Padin, T. J. Pearson, A. C. S. Readhead and M. C. Shepherd, *Astrophys. J.* **568**, 38 (March 2002), astro-ph/0104489.

11. A. G. Riess, A. V. Filippenko, P. Challis, A. Clocchiattia, A. Diercks, P. M. Garnavich, R. L. Gilliland, C. J. Hogan, S. Jha, R. P. Kirshner, B. Leibundgut, M. M. Phillips, D. Reiss, B. P. Schmidt, R. A. Schommer, R. C. Smith, J. Spyromilio, C. Stubbs, N. B. Suntzeff and J. Tonry, *Astron. J.* **116**, p. 1009 (1998).

12. S. Perlmutter, G. Aldering, G. Goldhaber, R. A. Knop, P. Nugent, P. G. Castro, S. Deustua, S. Fabbro, A. Goobar, D. E. Groom, I. M. Hook, A. G. Kim, M. Y. Kim, J. C. Lee, N. J. Nunes, R. Pain, C. R. Pennypacker, R. Quimby, C. Lidman, R. S. Ellis, M. Irwin, R. G. McMahon, P. Ruiz-Lapuente, N. Walton, B. Schaefer, B. J. Boyle, A. V. Filippenko, T. Matheson, A. S. Fruchter, N. Panagia, H. J. M. Newberg, W. J. Couch and The Supernova Cosmology Project, *Astrophys. J.* **517**, 565 (June 1999).

13. C. L. Bennett, M. Bay, M. Halpern, G. Hinshaw, C. Jackson, N. Jarosik, A. Kogut, M. Limon, S. S. Meyer, L. Page, D. N. Spergel, G. S. Tucker, D. T. Wilkinson, E. Wollack and E. L. Wright, *Astrophys. J.* **583**, 1 (January 2003).

14. Planck Collaboration, P. A. R. Ade, N. Aghanim, M. Arnaud, M. Ashdown, J. Aumont, C. Baccigalupi, M. Baker, A. Balbi, A. J. Banday and et al., *Astron. Astrophys.* **536**, p. A1 (December 2011).

15. J. E. Carlstrom, P. A. R. Ade, K. A. Aird, B. A. Benson, L. E. Bleem, S. Busetti, C. L. Chang, E. Chauvin, H.-M. Cho, T. M. Crawford, A. T. Crites, M. A. Dobbs, N. W. Halverson, S. Heimsath, W. L. Holzapfel, J. D. Hrubes, M. Joy, R. Keisler, T. M. Lanting, A. T. Lee, E. M. Leitch, J. Leong, W. Lu, M. Lueker, D. Luongvan, J. J. McMahon, J. Mehl, S. S. Meyer, J. J. Mohr, T. E. Montroy, S. Padin, T. Plagge, C. Pryke, J. E. Ruhl, K. K. Schaffer, D. Schwan, E. Shirokoff, H. G. Spieler, Z. Staniszewski, A. A. Stark, C. Tucker, K. Vanderlinde, J. D. Vieira and R. Williamson, *PASP* **123**, 568 (May 2011).

16. J. W. Fowler, V. Acquaviva, P. A. R. Ade, P. Aguirre, M. Amiri, J. W. Appel, L. F. Barrientos, E. S. Battistelli, J. R. Bond, B. Brown, B. Burger, J. Chervenak, S. Das,

M. J. Devlin, S. R. Dicker, W. B. Doriese, J. Dunkley, R. Dünner, T. Essinger-Hileman, R. P. Fisher, A. Hajian, M. Halpern, M. Hasselfield, C. Hernández-Monteagudo, G. C. Hilton, M. Hilton, A. D. Hincks, R. Hlozek, K. M. Huffenberger, D. H. Hughes, J. P. Hughes, L. Infante, K. D. Irwin, R. Jimenez, J. B. Juin, M. Kaul, J. Klein, A. Kosowsky, J. M. Lau, M. Limon, Y.-T. Lin, R. H. Lupton, T. A. Marriage, D. Marsden, K. Martocci, P. Mauskopf, F. Menanteau, K. Moodley, H. Moseley, C. B. Netterfield, M. D. Niemack, M. R. Nolta, L. A. Page, L. Parker, B. Partridge, H. Quintana, B. Reid, N. Sehgal, J. Sievers, D. N. Spergel, S. T. Staggs, D. S. Swetz, E. R. Switzer, R. Thornton, H. Trac, C. Tucker, L. Verde, R. Warne, G. Wilson, E. Wollack and Y. Zhao, *Astrophys. J.* **722**, 1148 (October 2010).

17. K. Arnold, P. A. R. Ade, A. E. Anthony, F. Aubin, D. Boettger, J. Borrill, C. Cantalupo, M. A. Dobbs, J. Errard, D. Flanigan, A. Ghribi, N. Halverson, M. Hazumi, W. L. Holzapfel, J. Howard, P. Hyland, A. Jaffe, B. Keating, T. Kisner, Z. Kermish, A. T. Lee, E. Linder, M. Lungu, T. Matsumura, N. Miller, X. Meng, M. Myers, H. Nishino, R. O'Brient, D. O'Dea, C. Reichardt, I. Schanning, A. Shimizu, C. Shimmin, M. Shimon, H. Spieler, B. Steinbach, R. Stompor, A. Suzuki, T. Tomaru, H. T. Tran, C. Tucker, E. Quealy, P. L. Richards and O. Zahn, The POLARBEAR CMB polarization experiment, in *Society of Photo-Optical Instrumentation Engineers (SPIE) Conference Series*, Society of Photo-Optical Instrumentation Engineers (SPIE) Conference Series Vol. 7741, July 2010.

18. P. A. R. Ade, R. W. Aikin, M. Amiri, D. Barkats, S. J. Benton, C. A. Bischoff, J. J. Bock, J. A. Brevik, I. Buder, E. Bullock, G. Davis, P. K. Day, C. D. Dowell, L. Duband, J. P. Filippini, S. Fliescher, S. R. Golwala, M. Halpern, M. Hasselfield, S. R. Hildebrandt, G. C. Hilton, K. D. Irwin, K. S. Karkare, J. P. Kaufman, B. G. Keating, S. A. Kernasovskiy, J. M. Kovac, C. L. Kuo, E. M. Leitch, N. Llombart, M. Lueker, C. B. Netterfield, H. T. Nguyen, R. O'Brient, R. W. Ogburn, IV, A. Orlando, C. Pryke, C. D. Reintsema, S. Richter, R. Schwarz, C. D. Sheehy, Z. K. Staniszewski, K. T. Story, R. V. Sudiwala, G. P. Teply, J. E. Tolan, A. D. Turner, A. G. Vieregg, P. Wilson, C. L. Wong, K. W. Yoon and Bicep2 Collaboration, *Astrophys. J.* **792**, p. 62 (September 2014).

19. C. D. Sheehy, P. A. R. Ade, R. W. Aikin, M. Amiri, S. Benton, C. Bischoff, J. J. Bock, J. A. Bonetti, J. A. Brevik, B. Burger, C. D. Dowell, L. Duband, J. P. Filippini, S. R. Golwala, M. Halpern, M. Hasselfield, G. Hilton, V. V. Hristov, K. Irwin, J. P. Kaufman, B. G. Keating, J. M. Kovac, C. L. Kuo, A. E. Lange, E. M. Leitch, M. Lueker, C. B. Netterfield, H. T. Nguyen, R. W. Ogburn, IV, A. Orlando, C. L. Pryke, C. Reintsema, S. Richter, J. E. Ruhl, M. C. Runyan, Z. Staniszewski, S. Stokes, R. Sudiwala, G. Teply, K. L. Thompson, J. E. Tolan, A. D. Turner, P. Wilson and C. L. Wong, The Keck Array: a pulse tube cooled CMB polarimeter, in *Society of Photo-Optical Instrumentation Engineers (SPIE) Conference Series*, Society of Photo-Optical Instrumentation Engineers (SPIE) Conference Series Vol. 7741, July 2010.

20. J. P. Filippini, P. A. R. Ade, M. Amiri, S. J. Benton, R. Bihary, J. J. Bock, J. R. Bond, J. A. Bonetti, S. A. Bryan, B. Burger, H. C. Chiang, C. R. Contaldi, B. P. Crill, O. Doré, M. Farhang, L. M. Fissel, N. N. Gandilo, S. R. Golwala, J. E. Gudmundsson, M. Halpern, M. Hasselfield, G. Hilton, W. Holmes, V. V. Hristov, K. D. Irwin, W. C. Jones, C. L. Kuo, C. J. MacTavish, P. V. Mason, T. E. Montroy, T. A. Morford, C. B. Netterfield, D. T. O'Dea, A. S. Rahlin, C. D. Reintsema, J. E. Ruhl, M. C. Runyan, M. A. Schenker, J. A. Shariff, J. D. Soler, A. Trangsrud, C. Tucker, R. S. Tucker and A. D. Turner, SPIDER: a balloon-borne CMB polarimeter fkeor large angular scales, in *Millimeter, Submillimeter, and Far-Infrared Detectors and Instrumentation*

for Astronomy V. Edited by Holland, Wayne S.; Zmuidzinas, Jonas. Proceedings of the SPIE, Volume 7741, pp. 77411N-77411N-12 (2010)., July 2010.

21. R. O'Brient, P. A. R. Ade, Z. Ahmed, R. W. Aikin, M. Amiri, S. Benton, C. Bischoff, J. J. Bock, J. A. Bonetti, J. A. Brevik, B. Burger, G. Davis, P. Day, C. D. Dowell, L. Duband, J. P. Filippini, S. Fliescher, S. R. Golwala, J. Grayson, M. Halpern, M. Hasselfield, G. Hilton, V. V. Hristov, H. Hui, K. Irwin, S. Kernasovskiy, J. M. Kovac, C. L. Kuo, E. Leitch, M. Lueker, K. Megerian, L. Moncelsi, C. B. Netterfield, H. T. Nguyen, R. W. Ogburn, C. L. Pryke, C. Reintsema, J. E. Ruhl, M. C. Runyan, R. Schwarz, C. D. Sheehy, Z. Staniszewski, R. Sudiwala, G. Teply, J. E. Tolan, A. D. Turner, R. S. Tucker, A. Vieregg, D. V. Wiebe, P. Wilson, C. L. Wong, W. L. K. Wu and K. W. Yoon, Antenna-coupled TES bolometers for the Keck array, Spider, and Polar-1, in *Society of Photo-Optical Instrumentation Engineers (SPIE) Conference Series*, Society of Photo-Optical Instrumentation Engineers (SPIE) Conference Series Vol. 8452, September 2012.

22. Planck Collaboration, P. A. R. Ade, N. Aghanim, C. Armitage-Caplan, M. Arnaud, M. Ashdown, F. Atrio-Barandela, J. Aumont, C. Baccigalupi, A. J. Banday and et al., *Astron. Astrophys.* **571**, p. A16 (November 2014).

23. W. Hu and M. White, *Astrophys. J.* **471**, p. 30 (1996).

24. J. R. Bond and G. Efstathiou, *Mon. Not. R. Astron. Soc.* **226**, 655 (June 1987).

25. W. Hu and M. White, *Astrophys. J.* **479**, p. 568 (1997).

26. J. M. Kovac, E. M. Leitch, C. Pryke, J. E. Carlstrom, N. W. Halverson and W. L. Holzapfel, *Nature* **420**, p. 772 (December 2002).

27. G. Hinshaw, D. Larson, E. Komatsu, D. N. Spergel, C. L. Bennett, J. Dunkley, M. R. Nolta, M. Halpern, R. S. Hill, N. Odegard, L. Page, K. M. Smith, J. L. Weiland, B. Gold, N. Jarosik, A. Kogut, M. Limon, S. S. Meyer, G. S. Tucker, E. Wollack and E. L. Wright, *Astrophys. J. Supp.* **208**, p. 19 (October 2013).

28. M. L. Brown, P. Ade, J. Bock, M. Bowden, G. Cahill, P. G. Castro, S. Church, T. Culverhouse, R. B. Friedman, K. Ganga, W. K. Gear, S. Gupta, J. Hinderks, J. Kovac, A. E. Lange, E. Leitch, S. J. Melhuish, Y. Memari, J. A. Murphy, A. Orlando, C. O'Sullivan, L. Piccirillo, C. Pryke, N. Rajguru, B. Rusholme, R. Schwarz, A. N. Taylor, K. L. Thompson, A. H. Turner, E. Y. S. Wu, M. Zemcov and The QUa D collaboration, *Astrophys. J.* **705**, 978 (November 2009).

29. QUIET Collaboration, D. Araujo, C. Bischoff, A. Brizius, I. Buder, Y. Chinone, K. Cleary, R. N. Dumoulin, A. Kusaka, R. Monsalve, S. K. Næss, L. B. Newburgh, R. Reeves, I. K. Wehus, J. T. L. Zwart, L. Bronfman, R. Bustos, S. E. Church, C. Dickinson, H. K. Eriksen, T. Gaier, J. O. Gundersen, M. Hasegawa, M. Hazumi, K. M. Huffenberger, K. Ishidoshiro, M. E. Jones, P. Kangaslahti, D. J. Kapner, D. Kubik, C. R. Lawrence, M. Limon, J. J. McMahon, A. D. Miller, M. Nagai, H. Nguyen, G. Nixon, T. J. Pearson, L. Piccirillo, S. J. E. Radford, A. C. S. Readhead, J. L. Richards, D. Samtleben, M. Seiffert, M. C. Shepherd, K. M. Smith, S. T. Staggs, O. Tajima, K. L. Thompson, K. Vanderlinde and R. Williamson, *Astrophys. J.* **760**, p. 145 (December 2012).

30. Keck Array and BICEP2 Collaborations, P. A. R. Ade, Z. Ahmed, R. W. Aikin, K. D. Alexander, D. Barkats, S. J. Benton, C. A. Bischoff, J. J. Bock, J. A. Brevik, I. Buder, E. Bullock, V. Buza, J. Connors, B. P. Crill, C. D. Dowell, C. Dvorkin, L. Duband, J. P. Filippini, S. Fliescher, S. R. Golwala, M. Halpern, M. Hasselfield, S. R. Hildebrandt, G. C. Hilton, V. V. Hristov, H. Hui, K. D. Irwin, K. S. Karkare, J. P. Kaufman, B. G. Keating, S. Kefeli, S. A. Kernasovskiy, J. M. Kovac, C. L. Kuo, E. M. Leitch, M. Lueker, P. Mason, K. G. Megerian, C. B. Netterfield, H. T. Nguyen, R. O'Brient, R. W. Ogburn, IV, A. Orlando, C. Pryke, C. D. Reintsema, S. Richter,

R. Schwarz, C. D. Sheehy, Z. K. Staniszewski, R. V. Sudiwala, G. P. Teply, K. L. Thompson, J. E. Tolan, A. D. Turner, A. G. Vieregg, A. C. Weber, J. Willmert, C. L. Wong and K. W. Yoon, *ArXiv e-prints* (February 2015).

31. S. Naess, M. Hasselfield, J. McMahon, M. D. Niemack, G. E. Addison, P. A. R. Ade, R. Allison, M. Amiri, N. Battaglia, J. A. Beall, F. de Bernardis, J. R. Bond, J. Britton, E. Calabrese, H.-m. Cho, K. Coughlin, D. Crichton, S. Das, R. Datta, M. J. Devlin, S. R. Dicker, J. Dunkley, R. Dünner, J. W. Fowler, A. E. Fox, P. Gallardo, E. Grace, M. Gralla, A. Hajian, M. Halpern, S. Henderson, J. C. Hill, G. C. Hilton, M. Hilton, A. D. Hincks, R. Hlozek, P. Ho, J. Hubmayr, K. M. Huffenberger, J. P. Hughes, L. Infante, K. Irwin, R. Jackson, S. Muya Kasanda, J. Klein, B. Koopman, A. Kosowsky, D. Li, T. Louis, M. Lungu, M. Madhavacheril, T. A. Marriage, L. Maurin, F. Menanteau, K. Moodley, C. Munson, L. Newburgh, J. Nibarger, M. R. Nolta, L. A. Page, C. Pappas, B. Partridge, F. Rojas, B. L. Schmitt, N. Sehgal, B. D. Sherwin, J. Sievers, S. Simon, D. N. Spergel, S. T. Staggs, E. R. Switzer, R. Thornton, H. Trac, C. Tucker, M. Uehara, A. Van Engelen, J. T. Ward and E. J. Wollack, *J. Cos. Astroparticle Phys.* **10**, p. 7 (October 2014).

32. A. T. Crites, J. W. Henning, P. A. R. Ade, K. A. Aird, J. E. Austermann, J. A. Beall, A. N. Bender, B. A. Benson, L. E. Bleem, J. E. Carlstrom, C. L. Chang, H. C. Chiang, H.-M. Cho, R. Citron, T. M. Crawford, T. de Haan, M. A. Dobbs, W. Everett, J. Gallicchio, J. Gao, E. M. George, A. Gilbert, N. W. Halverson, D. Hanson, N. Harrington, G. C. Hilton, G. P. Holder, W. L. Holzapfel, S. Hoover, Z. Hou, J. D. Hrubes, N. Huang, J. Hubmayr, K. D. Irwin, R. Keisler, L. Knox, A. T. Lee, E. M. Leitch, D. Li, C. Liang, D. Luong-Van, J. J. McMahon, J. Mehl, S. S. Meyer, L. Mocanu, T. E. Montroy, T. Natoli, J. P. Nibarger, V. Novosad, S. Padin, C. Pryke, C. L. Reichardt, J. E. Ruhl, B. R. Saliwanchik, J. T. Sayre, K. K. Schaffer, G. Smecher, A. A. Stark, K. T. Story, C. Tucker, K. Vanderlinde, J. D. Vieira, G. Wang, N. Whitehorn, V. Yefremenko and O. Zahn, *Astrophys. J.* **805**, p. 36 (May 2015).

33. Planck Collaboration, N. Aghanim, M. Arnaud, M. Ashdown, J. Aumont, C. Baccigalupi, A. J. Banday, R. B. Barreiro, J. G. Bartlett, N. Bartolo and et al., *ArXiv e-prints* (July 2015).

34. U. Seljak and M. Zaldarriaga, *Physical Review Letters* **78**, 2054 (March 1997).

35. V. A. Rubakov, M. V. Sazhin and A. V. Veryaskin, *Physics Letters B* **115**, 189 (September 1982).

36. BICEP2/Keck and Planck Collaborations, P. A. R. Ade, N. Aghanim, Z. Ahmed, R. W. Aikin, K. D. Alexander, M. Arnaud, J. Aumont, C. Baccigalupi, A. J. Banday and et al., *Physical Review Letters* **114**, p. 101301 (March 2015).

37. U. Seljak and C. M. Hirata, *Phys. Rev. D* **69**, 043005 (February 2004).

38. R. Keisler, S. Hoover, N. Harrington, J. W. Henning, P. A. R. Ade, K. A. Aird, J. E. Austermann, J. A. Beall, A. N. Bender, B. A. Benson, L. E. Bleem, J. E. Carlstrom, C. L. Chang, H. C. Chiang, H.-M. Cho, R. Citron, T. M. Crawford, A. T. Crites, T. de Haan, M. A. Dobbs, W. Everett, J. Gallicchio, J. Gao, E. M. George, A. Gilbert, N. W. Halverson, D. Hanson, G. C. Hilton, G. P. Holder, W. L. Holzapfel, Z. Hou, J. D. Hrubes, N. Huang, J. Hubmayr, K. D. Irwin, L. Knox, A. T. Lee, E. M. Leitch, D. Li, D. Luong-Van, D. P. Marrone, J. J. McMahon, J. Mehl, S. S. Meyer, L. Mocanu, T. Natoli, J. P. Nibarger, V. Novosad, S. Padin, C. Pryke, C. L. Reichardt, J. E. Ruhl, B. R. Saliwanchik, J. T. Sayre, K. K. Schaffer, E. Shirokoff, G. Smecher, A. A. Stark, K. T. Story, C. Tucker, K. Vanderlinde, J. D. Vieira, G. Wang, N. Whitehorn, V. Yefremenko and O. Zahn, *Astrophys. J.* **807**, p. 151 (2015).

39. Planck Collaboration, R. Adam, P. A. R. Ade, N. Aghanim, M. Arnaud, J. Aumont, C. Baccigalupi, A. J. Banday, R. B. Barreiro, J. G. Bartlett and et al., *ArXiv e-prints*

(September 2014).

40. J. Lesgourgues, L. Perotto, S. Pastor and M. Piat, *Phys. Rev. D* **73**, 045021 (February 2006).

41. N. Dalal, O. Doré, D. Huterer and A. Shirokov, *Phys. Rev. D* **77**, 123514 (June 2008).

42. D. Jeong, E. Komatsu and B. Jain, *Phys. Rev. D* **80**, p. 123527 (December 2009).

43. G. P. Holder, M. P. Viero, O. Zahn, K. A. Aird, B. A. Benson, S. Bhattacharya, L. E. Bleem, J. Bock, M. Brodwin, J. E. Carlstrom, C. L. Chang, H.-M. Cho, A. Conley, T. M. Crawford, A. T. Crites, T. de Haan, M. A. Dobbs, J. Dudley, E. M. George, N. W. Halverson, W. L. Holzapfel, S. Hoover, Z. Hou, J. D. Hrubes, R. Keisler, L. Knox, A. T. Lee, E. M. Leitch, M. Lueker, D. Luong-Van, G. Marsden, D. P. Marrone, J. J. McMahon, J. Mehl, S. S. Meyer, M. Millea, J. J. Mohr, T. E. Montroy, S. Padin, T. Plagge, C. Pryke, C. L. Reichardt, J. E. Ruhl, J. T. Sayre, K. K. Schaffer, B. Schulz, L. Shaw, E. Shirokoff, H. G. Spieler, Z. Staniszewski, A. A. Stark, K. T. Story, A. van Engelen, K. Vanderlinde, J. D. Vieira, R. Williamson and M. Zemcov, *Astrophys. J.* **771**, p. L16 (July 2013).

44. Y.-S. Song, A. Cooray, L. Knox and M. Zaldarriaga, *Astrophys. J.* **590**, 664 (June 2003).

45. Planck Collaboration, P. A. R. Ade, N. Aghanim, C. Armitage-Caplan, M. Arnaud, M. Ashdown, F. Atrio-Barandela, J. Aumont, C. Baccigalupi, A. J. Banday and et al., *Astron. Astrophys.* **571**, p. A18 (November 2014).

46. J. E. Austermann, K. A. Aird, J. A. Beall, D. Becker, A. Bender, B. A. Benson, L. E. Bleem, J. Britton, J. E. Carlstrom, C. L. Chang, H. C. Chiang, H.-M. Cho, T. M. Crawford, A. T. Crites, A. Datesman, T. de Haan, M. A. Dobbs, E. M. George, N. W. Halverson, N. Harrington, J. W. Henning, G. C. Hilton, G. P. Holder, W. L. Holzapfel, S. Hoover, N. Huang, J. Hubmayr, K. D. Irwin, R. Keisler, J. Kennedy, L. Knox, A. T. Lee, E. Leitch, D. Li, M. Lueker, D. P. Marrone, J. J. McMahon, J. Mehl, S. S. Meyer, T. E. Montroy, T. Natoli, J. P. Nibarger, M. D. Niemack, V. Novosad, S. Padin, C. Pryke, C. L. Reichardt, J. E. Ruhl, B. R. Saliwanchik, J. T. Sayre, K. K. Schaffer, E. Shirokoff, A. A. Stark, K. Story, K. Vanderlinde, J. D. Vieira, G. Wang, R. Williamson, V. Yefremenko, K. W. Yoon and O. Zahn, SPTpol: an instrument for CMB polarization measurements with the South Pole Telescope, in *Society of Photo-Optical Instrumentation Engineers (SPIE) Conference Series*, Society of Photo-Optical Instrumentation Engineers (SPIE) Conference Series Vol. 8452, September 2012.

47. M. D. Niemack, P. A. R. Ade, J. Aguirre, F. Barrientos, J. A. Beall, J. R. Bond, J. Britton, H. M. Cho, S. Das, M. J. Devlin, S. Dicker, J. Dunkley, R. Dünner, J. W. Fowler, A. Hajian, M. Halpern, M. Hasselfield, G. C. Hilton, M. Hilton, J. Hubmayr, J. P. Hughes, L. Infante, K. D. Irwin, N. Jarosik, J. Klein, A. Kosowsky, T. A. Marriage, J. McMahon, F. Menanteau, K. Moodley, J. P. Nibarger, M. R. Nolta, L. A. Page, B. Partridge, E. D. Reese, J. Sievers, D. N. Spergel, S. T. Staggs, R. Thornton, C. Tucker, E. Wollack and K. W. Yoon, ACTPol: a polarization-sensitive receiver for the Atacama Cosmology Telescope, in *Society of Photo-Optical Instrumentation Engineers (SPIE) Conference Series*, Society of Photo-Optical Instrumentation Engineers (SPIE) Conference Series Vol. 7741, July 2010.

48. Z. Ahmed, M. Amiri, S. J. Benton, J. J. Bock, R. Bowens-Rubin, I. Buder, E. Bullock, J. Connors, J. P. Filippini, J. A. Grayson, M. Halpern, G. C. Hilton, V. V. Hristov, H. Hui, K. D. Irwin, J. Kang, K. S. Karkare, E. Karpel, J. M. Kovac, C. L. Kuo, C. B. Netterfield, H. T. Nguyen, R. O'Brient, R. W. Ogburn, C. Pryke, C. D. Reintsema, S. Richter, K. L. Thompson, A. D. Turner, A. G. Vieregg, W. L. K. Wu and K. W.

Yoon, BICEP3: a 95GHz refracting telescope for degree-scale CMB polarization, in *Society of Photo-Optical Instrumentation Engineers (SPIE) Conference Series*, Society of Photo-Optical Instrumentation Engineers (SPIE) Conference Series Vol. 9153, August 2014.

49. T. Essinger-Hileman, A. Ali, M. Amiri, J. W. Appel, D. Araujo, C. L. Bennett, F. Boone, M. Chan, H.-M. Cho, D. T. Chuss, F. Colazo, E. Crowe, K. Denis, R. Dünner, J. Eimer, D. Gothe, M. Halpern, K. Harrington, G. C. Hilton, G. F. Hinshaw, C. Huang, K. Irwin, G. Jones, J. Karakla, A. J. Kogut, D. Larson, M. Limon, L. Lowry, T. Marriage, N. Mehrle, A. D. Miller, N. Miller, S. H. Moseley, G. Novak, C. Reintsema, K. Rostem, T. Stevenson, D. Towner, K. U-Yen, E. Wagner, D. Watts, E. J. Wollack, Z. Xu and L. Zeng, CLASS: the cosmology large angular scale surveyor, in *Society of Photo-Optical Instrumentation Engineers (SPIE) Conference Series*, Society of Photo-Optical Instrumentation Engineers (SPIE) Conference Series Vol. 9153, July 2014.

50. K. N. Abazajian, K. Arnold, J. Austermann, B. A. Benson, C. Bischoff, J. Bock, J. R. Bond, J. Borrill, I. Buder, D. L. Burke, E. Calabrese, J. E. Carlstrom, C. S. Carvalho, C. L. Chang, H. C. Chiang, S. Church, A. Cooray, T. M. Crawford, B. P. Crill, K. S. Dawson, S. Das, M. J. Devlin, M. Dobbs, S. Dodelson, O. Doré, J. Dunkley, J. L. Feng, A. Fraisse, J. Gallicchio, S. B. Giddings, D. Green, N. W. Halverson, S. Hanany, D. Hanson, S. R. Hildebrandt, A. Hincks, R. Hlozek, G. Holder, W. L. Holzapfel, K. Honscheid, G. Horowitz, W. Hu, J. Hubmayr, K. Irwin, M. Jackson, W. C. Jones, R. Kallosh, M. Kamionkowski, B. Keating, R. Keisler, W. Kinney, L. Knox, E. Komatsu, J. Kovac, C.-L. Kuo, A. Kusaka, C. Lawrence, A. T. Lee, E. Leitch, A. Linde, E. Linder, P. Lubin, J. Maldacena, E. Martinec, J. McMahon, A. Miller, V. Mukhanov, L. Newburgh, M. D. Niemack, H. Nguyen, H. T. Nguyen, L. Page, C. Pryke, C. L. Reichardt, J. E. Ruhl, N. Sehgal, U. Seljak, L. Senatore, J. Sievers, E. Silverstein, A. Slosar, K. M. Smith, D. Spergel, S. T. Staggs, A. Stark, R. Stompor, A. G. Vieregg, G. Wang, S. Watson, E. J. Wollack, W. L. K. Wu, K. W. Yoon, O. Zahn and M. Zaldarriaga, *Astroparticle Physics* **63**, 55 (March 2015).

51. S. Ritz *et al.* (2014).

52. S. Dodelson, *Physical Review Letters* **112**, p. 191301 (May 2014).

53. K. N. Abazajian, K. Arnold, J. Austermann, B. A. Benson, C. Bischoff, J. Bock, J. R. Bond, J. Borrill, E. Calabrese, J. E. Carlstrom, C. S. Carvalho, C. L. Chang, H. C. Chiang, S. Church, A. Cooray, T. M. Crawford, K. S. Dawson, S. Das, M. J. Devlin, M. Dobbs, S. Dodelson, O. Doré, J. Dunkley, J. Errard, A. Fraisse, J. Gallicchio, N. W. Halverson, S. Hanany, S. R. Hildebrandt, A. Hincks, R. Hlozek, G. Holder, W. L. Holzapfel, K. Honscheid, W. Hu, J. Hubmayr, K. Irwin, W. C. Jones, M. Kamionkowski, B. Keating, R. Keisler, L. Knox, E. Komatsu, J. Kovac, C.-L. Kuo, C. Lawrence, A. T. Lee, E. Leitch, E. Linder, P. Lubin, J. McMahon, A. Miller, L. Newburgh, M. D. Niemack, H. Nguyen, H. T. Nguyen, L. Page, C. Pryke, C. L. Reichardt, J. E. Ruhl, N. Sehgal, U. Seljak, J. Sievers, E. Silverstein, A. Slosar, K. M. Smith, D. Spergel, S. T. Staggs, A. Stark, R. Stompor, A. G. Vieregg, G. Wang, S. Watson, E. J. Wollack, W. L. K. Wu, K. W. Yoon and O. Zahn, *Astroparticle Physics* **63**, 66 (March 2015).

54. A. Albrecht, G. Bernstein, R. Cahn, W. L. Freedman, J. Hewitt, W. Hu, J. Huth, M. Kamionkowski, E. W. Kolb, L. Knox, J. C. Mather, S. Staggs and N. B. Suntzeff, *ArXiv Astrophysics e-prints* (September 2006).

55. E. M. Leitch, C. Pryke, N. W. Halverson, J. E. Carlstrom, J. Kovac, W. L. Holzapfel, M. Dragovan, J. K. Cartwright, B. M. Mason, S. Padin, T. J. Pearson, A. C. S. Readhead and M. C. Shepherd, *Astrophys. J.* **568**, 28 (2002), astro-ph/0104488.

56. K. W. Yoon, P. A. R. Ade, D. Barkats, J. O. Battle, E. M. Bierman, J. J. Bock, J. A. Brevik, H. C. Chiang, A. Crites, C. D. Dowell, L. Duband, G. S. Griffin, E. F. Hivon, W. L. Holzapfel, V. V. Hristov, B. G. Keating, J. M. Kovac, C. L. Kuo, A. E. Lange, E. M. Leitch, P. V. Mason, H. T. Nguyen, N. Ponthieu, Y. D. Takahashi, T. Renbarger, L. C. Weintraub and D. Woolsey, The Robinson Gravitational Wave Background Telescope (BICEP): a bolometric large angular scale CMB polarimeter, in *Proc. SPIE, Vol. 6275, Millimeter and Submillimeter Detectors for Astronomy III*, eds. J. Zmuidzinas, W. S. Holland, S. Withington and W. D. Duncan (SPIE Optical Engineering Press, Bellingham, July 2006).

57. A. Kusaka, T. Essinger-Hileman, J. W. Appel, P. Gallardo, K. D. Irwin, N. Jarosik, M. R. Nolta, L. A. Page, L. P. Parker, S. Raghunathan, J. L. Sievers, S. M. Simon, S. T. Staggs and K. Visnjic, *Review of Scientific Instruments* **85** (2014).

58. M. Hazumi, J. Borrill, Y. Chinone, M. A. Dobbs, H. Fuke, A. Ghribi, M. Hasegawa, K. Hattori, M. Hattori, W. L. Holzapfel, Y. Inoue, K. Ishidoshiro, H. Ishino, K. Karatsu, N. Katayama, I. Kawano, A. Kibayashi, Y. Kibe, N. Kimura, K. Koga, E. Komatsu, A. T. Lee, H. Matsuhara, T. Matsumura, S. Mima, K. Mitsuda, H. Morii, S. Murayama, M. Nagai, R. Nagata, S. Nakamura, K. Natsume, H. Nishino, A. Noda, T. Noguchi, I. Ohta, C. Otani, P. L. Richards, S. Sakai, N. Sato, Y. Sato, Y. Sekimoto, A. Shimizu, K. Shinozaki, H. Sugita, A. Suzuki, T. Suzuki, O. Tajima, S. Takada, Y. Takagi, Y. Takei, T. Tomaru, Y. Uzawa, H. Watanabe, N. Yamasaki, M. Yoshida, T. Yoshida and K. Yotsumoto, LiteBIRD: a small satellite for the study of B-mode polarization and inflation from cosmic background radiation detection, in *Society of Photo-Optical Instrumentation Engineers (SPIE) Conference Series*, Society of Photo-Optical Instrumentation Engineers (SPIE) Conference Series Vol. 8442, September 2012.

59. A. Kogut, D. J. Fixsen, D. T. Chuss, J. Dotson, E. Dwek, M. Halpern, G. F. Hinshaw, S. M. Meyer, S. H. Moseley, M. D. Seiffert, D. N. Spergel and E. J. Wollack, *J. Cos. Astroparticle Phys.* **7**, p. 25 (July 2011).

Rapporteur Talk by D. N. Spergel: Microwave Background Observations as Probe of Cosmology: Past and Future

Abstract

Our universe is both remarkably simple and remarkably strange. The statistical properties of the initial fluctuations appear to be nearly as minimal as possible: the fluctuations are adiabatic, Gaussian random-phase, with a power spectrum that is described by only two numbers, an overall amplitude and a slope. The existence of these super-horizon scale fluctuations requires either that the universe underwent an extended period of early accelerated expansion, inflation, or that these fluctuations were already present at the start of the universe's expansion. On its large-scales, the universe is nearly uniform, close to geometrically flat with no observable signatures of topology. However, the composition of the universe is strange: atoms comprise less than 5% of the universe with the bulk in the form of dark matter and dark energy. Our measurements from the Wilkinson Microwave Anisotropy Probe (WMAP) help establish this standard model and determine its basic parameters. The recent measurements from the Planck satellite has reinforced the model and determined its basic parameters to even higher precision. This now standard model fits not only our microwave background measurements but also observations of large-scale structure, measurements of the Hubble constant, supernovae observations, measurements of element abundances and gravitational lensing measurements. While we have a consistent cosmology that can be described by a handful of basic parameters, we are still left with a number of profound questions whose resolutions will require physics beyond the standard model of particle physics: What is the dark matter that dominates on galaxy scales? What is the nature of the dark energy that is driving the accelerating expansion of the universe? Is inflation the source of the primordial fluctuations? I will conclude by discussing how upcoming microwave background measurements can help address these questions.

1. Introduction

Over a decade ago in our analysis of the first year of WMAP data,[1] we concluded:

> Cosmology is now in a similar stage in its intellectual development to particle physics three decades ago when particle physicists converged on the current standard model. The standard model of particle physics fits a wide range of data but does not answer many fundamental questions: What is the origin of mass? Why is there more than one family? etc. Similarly, the standard cosmological model has many deep open questions: What is the dark energy? What is the dark matter? What is the physical model behind inflation (or something like inflation)? Over the past three decades, precision tests have confirmed the standard model of particle physics and

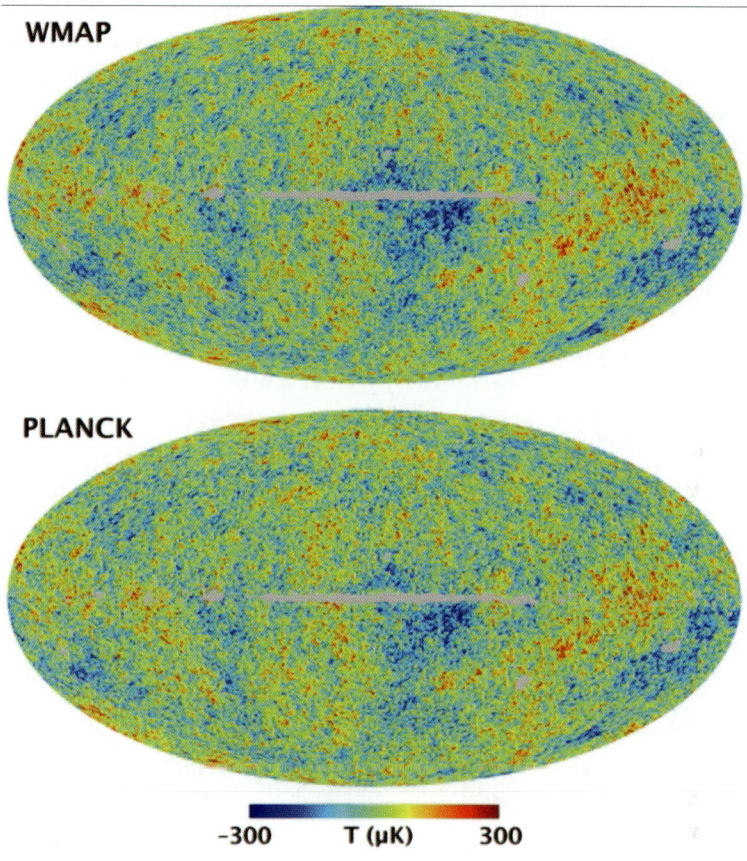

Fig. 1. This figure compares the WMAP ILC map with the Planck SMICA map smoothed to the WMAP resolution. The Planck data has additional small scale structure that is lost in the smoothing. The striking visual agreement between the two maps has been checked by detailed quantitative comparisons.

searched for distinctive signatures of the natural extension of the standard model: supersymmetry. Over the coming years, improving CMB, large-scale structure, lensing, and supernova data will provide ever more rigorous tests of the cosmological standard model and search for new physics beyond the standard model

In this paper, I will summarize the progress over the subsequent years with a focus on the recent results from Planck and ground-based experiments. Section 2 will show that the improving data has provided ever more rigorous tests of our standard cosmological model; however, it has not yet revealed any new physics nor provided powerful new hints about dark matter, dark energy or the physics of inflation. Section 3 will review the statistical properties of the fluctuations. Section 4 looks forward to future measurements and their potential to address the nature of dark energy, the connection between dark matter and galaxies and the physics that generated the primordial fluctuations.

2. Consistent Cosmology

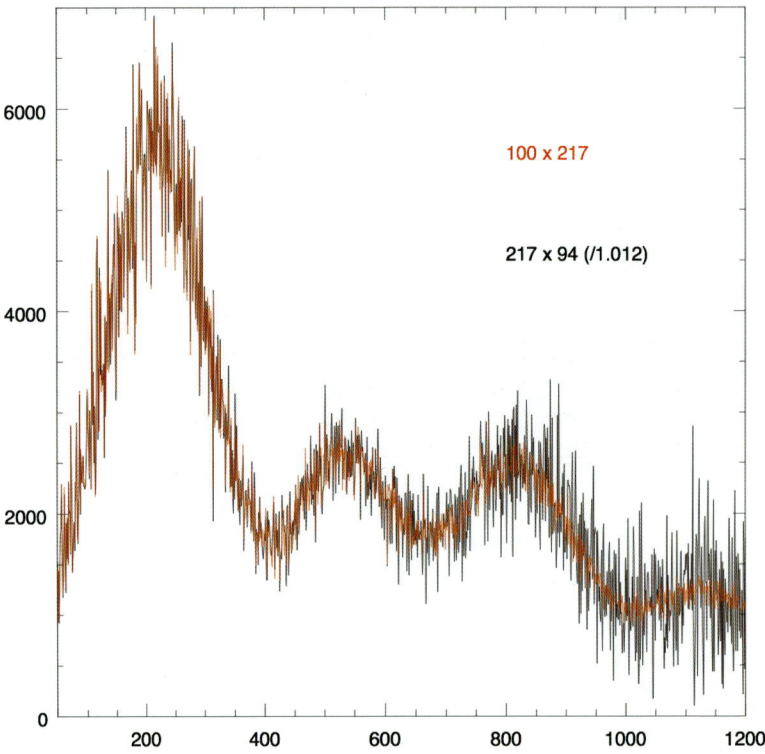

Fig. 2.   This figure compares the cross-correlation between the WMAP 94 GHz map and the Planck 145 GHz maps with the cross-correlation between the Planck 100 GHz and 145 GHz maps. The data used in the figure is from the 2013 Planck release but has been rescaled with the Planck 2015 normalization.

The remarkable agreement between different microwave background experiments is a tremendous accomplishment for our experimental colleagues. As Figure 1 shows, the WMAP and Planck experiments agree in detail in their measurements of microwave background fluctuations across the sky. With the new Planck calibration,[2] the detailed quantitative agreement between the two experiments is extremely impressive.[3] Figure 2 shows a comparison between the cross-correlation between Planck 145 GHz and two nearly overlap WMAP and Planck maps showing the excellent agreement outside the galactic plane. Near the galactic plane, there are modest differences between the two maps due to molecular lines that fall within one or the other of the WMAP and Planck bandpasses.[4] Not only is these excellent detailed agreement between Planck and WMAP, but also between the space-based and the ground-based experiments.[5]

As many talks in this conference have emphasized, the same basic model with the now accurately determined parameters (see e.g. Table 1) that fits the cosmic microwave background data fits a host of astronomical observations: measurements of large-scale structure, gravitational lensing observations, determinations of the Hubble constant, globular cluster ages, supernova distance measurements, and detailed studies of galaxy clusters. While there is a modest tension between the amplitude of fluctuations determined from the most recent Planck results and the amplitude of matter fluctuations inferred by cluster observations,[6] I suspect that careful analyses of systematics and more data will most likely resolve this potential discrepancy.

Table 1. Best Fit LCDM Parameters: The first column shows the best fit parameters determined from a combination of WMAP9 and ACT data[7] and the second column shows the best fit parameters from the Planck 2013 temperature data.[8]

		WMAP9+ACT	Planck
Spectral Index	n_s	0.973 ± 0.011	0.9603 ± 0.0073
Matter Density	$10\Omega_m h^2$	1.146 ± 0.044	1.199 ± 0.027
Baryon Density	$100\Omega_b h^2$	2.260 ± 0.040	2.205 ± 0.028
Hubble Constant	H_0	69.7 ± 2.0	67.3 ± 1.2

Our best current data imply that we live in a geometrically and topologically simple universe. The universe is very close to being geometrically flat with the combination of Planck data and measurements of the baryon acoustic oscillations implying $\Omega_K = 0.000 \pm 0.005$,[6] thus the curvature scale must be at least 10 times larger than the horizon size. Searches for "matched circles" in the cosmic microwave background[9] implies that the universe is either infinite or the size of the fundamental domain is very large. The limits from the most recent general analyses of the WMAP7[10] increases the lower limit on the length of the shortest closed null geodesic that self-intersects at our location in the Universe (equivalently the injectivity radius at our location) to 98.5% of the diameter of the last scattering surface or approximately 26 Gpc. A more recent search for topology used both the Planck polarization and temperature data[11] but restrict the search to topologies with antipodal symmetries.

3. Simple Initial Conditions

Measurements of the microwave background fluctuations reveal that the primordial fluctuations are remarkably simple and with properties consistent with the simplest versions of the theory of inflation. The data are very well fit by a model where the fluctuations are purely adiabatic with Gaussian random phase fluctuations. In this model, the statistical properties of the primordial fluctuations are described by two numbers, an overall amplitude and a scale dependence,

$$< \Phi(\vec{k})\Phi(\vec{k}') >= P(k)\delta^{(3)}(\vec{k} - \vec{k}') \tag{1}$$

where $P(k) = Ak^{n_s}$. Thus, all of the statistical properties of the primordial fluctuations seen in WMAP and Planck are fit by just two numbers with four additional numbers (the density of baryons, the density of matter, the age of the universe and its optical depth) needed to characterize all of the properties of the temperature and polarization fluctuations at the surface of last scatter. While there are a number of important open issues that theorists need to confront with our current understanding of inflation (see[12]), the observational successes of the generic predictions of the model are striking.

The data show no evidence for primordial non-Gaussianity.[13–15] These constraints imply that the physics that generated the fluctuations was very weakly self-coupled and rules out many theoretical ideas for generating these initial fluctuations. The data also show that the primordial fluctuations were adiabatic with no evidence for any form of isocurvature fluctuations with the most recent Planck constraints that utilize both precision temperature and polarization measurements[6] improving significantly on earlier constraints on non-adiabatic fluctuations from WMAP.

The temperature and polarization fluctuation seen by WMAP and Planck are that the fluctuations must be super-horizon scale fluctuations: an observation with profound implications for our understanding of the early universe. While temperature fluctuations can be generated either at the surface of last scatter or along the line-of-sight, polarization fluctuations are only generated through electron-photon scattering transforming a local quadrupole in the photon distribution function into a polarized fluctuation. Thus, fluctuations in polarization must come from the surface of last scatter. "... Because polarization is generated at the last scattering surface, models in which fluctuations are causally produced on subhorizon scales cannot generate correlations on scales larger than $\sim 2°$. Inflationary models, on the other hand, predict a peak in the correlation functions at these scales: its detection would be definitive evidence in favor of a period of inflation."[16] WMAP[1,17] and Planck[2] clearly see the predicted temperature and polarization cross-correlation and the large-angle polarization auto-correlation that requires either the universe underwent a period of superluminal expansion or that the fluctuations were there on super horizon scales as the universe exited the initial singularity. Thus, the fluctuations we see in the microwave sky today were generated by physics that occurred either during the first 10^{-30} seconds of the universe's expansion or before the big bang.

4. CMB Observations: Future Directions

If inflation generated the superhorizon fluctuations in the density of the universe, then the same expansion should also generate superhorizon scale gravitational waves. While the amplitude of this gravitational wave signal depends on the expansion rate of the universe during inflation, the existence of these large-scale gravitational waves are an inevitable prediction of inflation.

Last year's excitement over the initial claimed detection of gravitational waves[18] emphasized their potential importance. The detection of gravitational waves would be a glimpse into the first moment of our universe, a direct probe of physics near the Planck scale, an energy scale a trillion times larger than that explored by the LHC, the direct evidence for gravitons and the quantum nature of the gravitational field. Unfortunately, the detection claims have turned to dust as galactic emission appears to account for most, if not all of the reported signal.[19–21] Fortunately, upcoming experiments will be even more sensitive than the BICEP2 experiment and should soon be able to detect gravitational waves as long as the amplitude of the tensor mode signal is greater than 1% of the scalar mode amplitude. Proposed space-based polarization measurements (e.g., PIXIE,[22] LITEBIRD[23] and CORE[a]) hold the promise of detecting even weaker gravitational wave signals.

In this final section of the talk, I want to focus on some of the science goals of the Advanced Atacama Cosmology Telescope, AdvACT, a recently selected MSIP project led by my colleague, Suzanne Staggs. While the text of this talk focuses on the potential science return from AdvACT, other next generation CMB experiments such as SPT and Polar Bear will also make very similar measurements. The proposed CMB-S4 program[24] will be an even more powerful instrument for addressing these questions.

While the search for gravitational waves will be an important goal for upcoming polarization experiments, the more sensitive temperature and polarization measurements will enable cosmologists to make progress on a number of important open questions in cosmology. Microwave background observations are a powerful probe of the integrated properties of the universe at $z < 1100$: the gravitational lensing signal is proportional to the integrated mass along the line-of-sight, the thermal Sunyaev-Zel'dovich effect is proportional to the integrated pressure and the kinematic Sunyaev-Zel'dovich effect is proportional to the integrated momentum. By combining CMB observations with measurements of large-scale structure, we can use these effects to trace the evolution of structure and the large-scale motions of baryons.

The upcoming CMB experiments will contribute in several distinct ways to advancing our understanding of cosmic acceleration:

- AdvACT's measurement of the EE polarization will halve the present uncertainty on many of the basic cosmological parameters including the baryon and matter densities and the spectral index, key inputs to any calculation of the evolution of the universe.[25]
- AdvACT will probe the growth rate of structure, $f(z)$, with measurements of

---

[a]http://www.core-mission.org/documents/CoreProposal_Final.pdf

CMB lensing and its cross-correlation with optical lensing surveys. Since lensing of E modes is the dominant mechanism for producing B modes on scales smaller than 1°, AdvACT's lensing map will be very sensitive, with S/N~200 and an effective resolution of 0.1°.

While the LSST will use weak lensing to measure the projected matter distribution to $z \sim 2$, AdvACT will measure it to $z \sim 1100$, making the two methods highly complementary. AdvACT lensing will increase the DE figure of merit (FOM) from LSST by a factor of 3.5, and with AdvACT's clusters the factor is 10. A measurement of the cross-correlation of these two lens sheets with each other and with a spectroscopic sample (e.g. SDSS-III), could determine the multiplicative bias in the LSST optical lensing to 0.3% for sources at $1 < z < 1.5$, a factor of 10 improvement over LSST self-calibration. This alone doubles the LSST DE FOM.

– AdvACT will detect $\sim 27,000$ clusters with $M > 2 \times 10^{14} M_\odot$ with a sample purity of 99.95%, and ~50,000 to 90% purity), through their thermal SZ (tSZ) signatures. To first order, the tSZ effect depends on mass and not redshift, making the half-sky AdvACT catalog an excellent complement to the full-sky eROSITA X-ray survey, SDSS-III and LSST. The cluster counts, when calibrated by weak lensing data from SDSS-III, HS, DES and LSST, serve as a direct probe of the growth rate of structure. An even more powerful combination is to combine the cluster counts with measurements of weak lensing power spectrum.[26]

– The final way AdvACT will probe dark energy is through measurement of the kSZ effect with a technique pioneered by the ACT team.[27] By comparing the galaxy momentum field traced by kSZ to the density field, including the effects of redshift space distortions, the combination of AdvACT and spectroscopic surveys will make determinations of $f(z)$ and of the Alcock-Paczynski effect that are *not limited by cosmic variance*.

Planck has already produced a striking map of the large-scale distribution of matter through its lensing observations.[28] The upcoming polarization experiments will improve the fidelity of these maps significantly. These will be useful not only for studying dark energy but as overall tracers of the evolution of matter fluctuations and their connection to luminous matter. By cross-correlating lensing maps with galaxy and quasars surveys, we can use the measurements of bias to determine the properties of the host halos.[29]

The upcoming CMB polarization observations have the potential to produce even more profound insights into the physics of the early universe than our temperature measurements. I look forward to these developments with excitement.

References

1. D. N. Spergel, L. Verde, H. V. Peiris et al., *Astrophys. J. Supp.* **148**, 175 (September 2003).

2. Planck Collaboration, R. Adam, P. A. R. Ade, N. Aghanim, et al., *ArXiv:1502.01582* (February 2015).

3. D. Larson, J. L. Weiland, G. Hinshaw and C. L. Bennett, *Astrophys. J.* **801**, p. 9(March 2015).

4. Planck Collaboration, R. Adam, P. A. R. Ade, N. Aghanim, et al., *ArXiv:1502.01588* (February 2015).

5. T. Louis, G. E. Addison et al., *J. Cos. Astroparticle Phys.* **7**, p. 16(July 2014).

6. Planck Collaboration, P. A. R. Ade, N. Aghanim, et al., *ArXiv:1502.01589* (February 2015).

7. E. Calabrese, R. A. Hlozek, N. Battaglia et al., *Phys. Rev. D* **87**, p. 103012 (May 2013).

8. Planck Collaboration, P. A. R. Ade, N. Aghanim et al., *Astron. Astrophys.* **571**, p. A16 (November 2014).

9. N. J. Cornish, D. N. Spergel, G. D. Starkman and E. Komatsu, *Physical Review Letters* **92**, p. 201302 (May 2004).

10. P. M. Vaudrevange, G. D. Starkman, N. J. Cornish and D. N. Spergel, *Phys. Rev. D* **86**, p. 083526 (October 2012).

11. Planck Collaboration, P. A. R. Ade, N. Aghanim et al., *ArXiv:1502.01593* (February 2015).

12. A. Ijjas, P. J. Steinhardt and A. Loeb, *ArXiv: 1402.1980* (February 2014).

13. E. Komatsu, A. Kogut, M. R. Nolta et al., *Astrophys. J. Supp.* **148**, 119 (September 2003).

14. G. Hinshaw, D. Larson, E. Komatsu et al., *Astrophys. J. Supp.* **208**, p. 19 (October 2013).

15. Planck Collaboration, P. A. R. Ade, N. Aghanim, et al., *ArXiv:1502.01592* (February 2015).

16. D. N. Spergel and M. Zaldarriaga, *Physical Review Letters* **79**, 2180 (September 1997).

17. C. L. Bennett, D. Larson, J. L. Weiland et al., *Astrophys. J. Supp.* **208**, p. 20 (October 2013).

18. P. A. R. Ade, R. W. Aikin, D. Barkats et al., and Bicep2 Collaboration, *Physical Review Letters* **112**, p. 241101 (June 2014).

19. R. Flauger, J. C. Hill and D. N. Spergel, *J. Cos. Astroparticle Phys.* **8**, p. 39 (August 2014).

20. M. J. Mortonson and U. Seljak, *J. Cos. Astroparticle Phys.* **10**, p. 35 (October 2014).

21. BICEP2/Keck and Planck Collaborations, P. A. R. Ade, N. Aghanim, Z. Ahmed, R. W. Aikin, K. D. Alexander, M. Arnaud, J. Aumont, C. Baccigalupi, A. J. Banday and et al., *Physical Review Letters* **114**, p. 101301 (March 2015).

22. A. Kogut, D. J. Fixsen, D. T. Chuss et al., *J. Cos. Astroparticle Phys.* **7**, p. 25 (July 2011).

23. T. Matsumura, Y. Akiba, J. Borrill et al., *Journal of Low Temperature Physics* **176**, 733 (September 2014).

24. K. N. Abazajian, K. Arnold, J. Austermann et al., *Astroparticle Physics* **63**, 66 (March 2015).

25. E. Calabrese, R. Hložek, N. Battaglia et al., *J. Cos. Astroparticle Phys.* **8**, p. 10 (August 2014).

26. M. Takada and D. N. Spergel, *Mon. Not. R. Astron. Soc.* **441**, 2456(July 2014).

27. N. Hand, G. E. Addison, E. Aubourg et al., *Physical Review Letters* **109**, p. 041101 (July 2012).

28. Planck Collaboration, P. A. R. Ade, N. Aghanim et al., *ArXiv:1502.01591* (February 2015).
29. R. Allison, S. N. Lindsay, B. D. Sherwin et al., *ArXiv:1502.06456* (February 2015).

Discussion

G. Efstathiou Thank you. I did ask Spergel to tweak the audience, but he went a bit further than I had anticipated. I think that we can take some questions that people have on the rapporteur talks before we start a set of discussions.

M. Begelman What are the prospects for detecting a nonthermal SZ effect from maybe some relativistic particles accelerated by feedback or something like that, or shocks?

D. Spergel I think the challenge will be modeling dust in the cluster and understanding its spectrum. There is a well-defined frequency dependence you expect from the relativistic contribution, but you need to separate that – because it is a small signal – from dust in the individual cluster. We need to make sure that we understand that component – understand it well. It is in principle doable, but for this and for many things, understanding dust well will be quite important.

J. Ostriker You observed the secondary effect of gravitational lensing. What did you learn from it? And will future experiments teach you more than you know now?

D. Spergel I think the precision measurements that have so far been published – and there is new stuff coming soon from SPT – have come from Planck, and that has given us an independent measurement of the amplitude of fluctuations around redshift 1 to 2 consistent with ΛCDM at about 20σ – it is about 5% measurement. 40σ was promised for 2014.

J. Ostriker That is just a consistency check. Did it tell you anything new?

D. Spergel I am getting there. So I think at this point it is a consistency check, which is important, but I think soon it will become a way of measuring neutrino mass. We have talked about looking forward to one that is 200σ. At CMB-S4 you can look to making a measurement of neutrino mass with a precision better than 0.02 eV. So we will look forward to a detection of neutrino mass. And I think when we look at cross-correlating that very good lensing map with the optical lensing map, that will turn out to be very good tracers of the mass distribution. And to provoke people who do optic lensing maps, they have been so far systematics limited. The precision that is anticipated for things like LSST is incredible if you believe that it is statistics limited. I think at the end of the day what we are going to end up doing is taking the CMB lensing map, correlating with the different optical lensing maps, and that will turn out to be perhaps the cleanest way of measuring the evolution of the matter fluctuations and achieving a lot of the goals for constraining dark energy, that are part of projects like LSST.

G. Efstathiou So Ostriker, you are a very hard person to please, right? The lensing verifies the ΛCDM model purely from the CMB, all the way from redshifts of 1000 to redshifts of 2.

J. Peebles Your S4 project brings together powerful groups and promises great things, I am all for it. But recent events teach us that one does well to be cautioned against the unconscious bias that affects any project in science. That is why particle physicists have independent teams. Do you contemplate any arrangement for blind analysis of the data, particularly if you have – what did you say – Greenland combined with South Pole?

J. Carlstrom We are doing a few things. I think blind in particle physics and blind in things we do are a little bit different, because we cannot go in and make a cut and change the results or find that there is one more dark matter candidate or not. So we need to look at the data pretty intensely, but we for instance are doing a BP ?? autocorrelation analysis and we are doing that blind. We look at difference maps but we do not look at the result. We are also doing experiments in which the different teams promised to combine but in fact everyone is doing their own analysis independently. So I do not think we will give up that. In the project that we are talking about – S4 – I rushed through that – is not a single telescope either. It is going to be of order ten platforms just to get that many detectors on the sky.

J. Peebles North and South?

J. Carlstrom Chile and South Pole.

J. Peebles That is not really very North and South.

J. Carlstrom Chile is a lot further north than the South Pole. There is a group pushing for Greenland as well.

A. Guth As you might guess I would like to take some issue with the last remark that Spergel made, which I am sure he expected me to take issue to, concerning the use of the anthropic principle as a necessity for inflation. As far as I know at least, the anthropic principle is often invoked, but not at the level of understanding the mechanics of inflation, but rather at the level of trying to make plausible the potential energy function that one starts with. In the standard model of particle physics we just make up the potential energy function and do not ask where it came from. So at the same level, inflation just works, like the standard model works, and anthropic principles only come in at a higher level of analysis, where one tries to guess why the potential looks the way it apparently looks. I would add further that as far as not needing inflation because you are using the anthropic principle anyway, when inflation uses the anthropic principle it is to fix a few parameters in the potential. If you were to just try to explain the homogeneity and isotropy of the Universe by the anthropic principle by itself, first of all you cannot really do it, because you do not need as large a homogeneous region for anthropic as we see, so inflation gives you a good justification for why the Universe is so much larger than it needs to be anthropically. Furthermore I point out that if you are just trying to fix the homogeneity and isotropy of the Universe you are talking about fixing maybe a million different Fourier modes independently if you do not have some theory that connects them. With inflation, however, instead of fixing a million different Fourier modes, you are just fixing a few parameters of the potential. So I would say that inflation is still in very good shape and not really in danger of suffering from an anthropic sickness.

D. Spergel What about the multiverse? I like Sasha's recent paper that said we can avoid the multiverse in certain models, but in most models you end up finding yourself with a multiverse, anything can happen and then you are forced to have anthropic selection of what universe we are in.

A. Guth The statement that anything can happen does not mean that anything can happen with equal probability. That is very important to keep in mind. We have always since the advent of quantum mechanics been dealing with probabilities rather than certainties and all you are saying here is that now we have probabilities for what the part of the landscape that we are living in might look like. But it all just comes down to fixing the potential. Once you fix the potential, the implications for inflation and our part of the multiverse are pretty much fixed by that. It is true that the multiverse, which I do consider to be a plausible outcome of inflation, does open this measure problem of how one defines probabilities and I agree completely with people who say that that is not solved yet. We do not really know for sure how to define probabilities in the multiverse. But still I think it makes good sense to count on what has always happened in physics, that when you go from one theory to a more sophisticated theory, from classical mechanics to quantum mechanics, from classical mechanics to relativity, even though you are opening up a new door, where until you have understood that door, anything might happen, it is always safe ground to believe that where you have already verified things they will continue to work and what you will be finding is some extension. So we do need to understand multiversal statistics for example in principle be able to predict how likely our vacuum is compared to other vacua that exist in the landscape of string theory, but as far as predicting things like what the cosmic microwave background should look like, I think we are perfectly safe in assuming that the naive predictions previously made will still hold.

H. Murayama But what of the questions like the initial condition of the inflaton has to be smooth enough to cause inflation to get started to begin with, and the fact that you also have to refer to a horizon size smaller than Planck scale at the beginning?

A. Guth There are two separate questions. As far as initial conditions, I do not worry about initial conditions, largely because I do think the multiverse is plausible and I think initial condition problems disappear in the multiverse, that is, as long as there is *some* probability of the Universe starting, it approaches some steady state and that is what we now see. As far as the – I think you are talking basically about the trans-Planckian problem, that the fluctuations we are looking at now started out sub-Planckian – I think that there are still things to be learned there. I think that is really a string theory question, which I do not consider myself an expert on. But various people have studied possible scenarios of what could happen at sub-Planckian distances and by and large the opinion of those papers has been that any complicated sub-Planckian physics gets washed out as you go to more macroscopic wavelengths. The point is that the Bunch-Davies vacuum that we take as our starting point in inflation is really the only locally Lorentz invariant state, so any perturbations that preserve the idea of local Lorentz invariance will get us what we are expecting to get.

G. Efstathiou Thank you. I think we should cut off this discussion now, because we have got a special slot after coffee to discuss theory and I think we can pick up on it then. One of the quotes from this meeting was a remark that Mukhanov made to me, which was that Planck, BICEP and theory cannot coexist. That is the topic for the first discussion and to lead us through to introduce the topic Puget will discuss some of the Planck results and we are fortunate that Pryke, one of the co-leaders of the BICEP team, is here at this meeting.

Prepared comment

J.L. Puget: Relevant Observations for Primordial Universe Models

The inflation paradigm is an excellent solution for the description of the universe before it can be described by a Friedman-Lemaitre solution. An obvious question is then: what are the observable predictions and their status? There are four testable generic predictions of inflation, one associated with the initial exponential expansion leading to a flat space universe and three associated with the quantum origin of the initial perturbations giving rise to all structures we see. The quantum origin is the only model we have so far for the origin of the structures. This should lead to an almost scale invariant power spectrum of gaussian, adiabatic perturbations. The necessity to end inflation leads to a small deviation from scale invariance and a strong upper limit of 0.97 for the index of the perturbations power spectrum n_s.

The flatness of space on large scale is very well established by the combination of Planck Cosmic Microwave Background (CMB) and Baryonic acoustic oscillations observations ($\Omega_K = 0.000 \pm 0.005$).[a] The tilted CMB anisotropy spectrum index has been measured by Planck CMB observations ($n_s = 0.9653 \pm 0.0048$) away from scale invariance by more than 7σ. The Gaussianity is tested by comparing the three point correlations with the predicted one from the two point correlation if the perturbations are Gaussian. The Planck result for temperature and polarization combined is, for example $f_{NL} = 0.8 \pm 5.0$ for the local configurations. The adiabaticity is demonstraed by the very small allowed amplitude of the isocurvature modes ($\alpha = 0.0003 \pm 0.0016$).

The CMB observations have thus verified with high accuracy the four generic inflation and quantum origin of fluctuations predictions. My comments are related to the last observable associated with them. Inflation models also predict tensor modes in addition to the scalar ones. This is very interesting as it could provide

---

[a]The Planck with external data results in this paragraph are from Planck 2015 Results.XIII. Cosmological parameters, arXiv150201590.

information about the energy scale of inflation but there is no generic lower limit of the ratio r of tensor to scalar modes from inflation models which can verified. Specific models like the Starobinsky model have definite predictions of relations between n_s and r: $n_s \approx 1 - \frac{2}{N} \in (0.960, 0.967)$, and $r \approx \frac{12}{N^2} \in (0.003, 0.005)$. Planck using temperature only together with external data give an upper limit obtained on the tensor to scalar ratio $r < 0.11$.

A more direct measurement of the primordial gravitational waves can be performed using polarization data. The tensor modes imprint specific polarization patterns, the B modes on the CMB (not invariant under parity), both during the recombination and reionization periods. Thus, searching for these primordial CMB B modes is now probably the most important observational goal for primordial cosmology. In March 2015 the BICEP2 team announced the detection of B modes at 150 GHz from their south pole experiment on about 1 % of the sky in a region of low galactic emission at high latitude. The data showed an excess in the power spectrum around multipole 80 where the primordial B modes generated at recombination should peak. This is an impressive observational result. Nevertheless the interpretation requires that we estimate the galactic dust contribution and they argued initially, on the basis of too simple models, that the dust emission was negligible and that this excess could be a detection of primordial B modes with an r value of 0.2. This was a surprisingly high value both from a theory point of view and observationally as it is significantly higher than the previously published Planck upper limit and incompatible with it.

A statistical analysis of the polarized galactic dust emission at high latitude has been performed with the Planck high frequency channel (353 GHz).[b] This analysis shows that there is a general correlation of the E and B modes power spectra with dust emission intensity, scaling nearly like the square of the intensity as expected. Nevertheless the dispersion in polarization fraction for a given intensity spans a range of 2 to 20 percent. It is thus not possible to evaluate or remove the galactic dust B modes on the basis of intensity measurements only. The Planck analysis showed that the most probable value of the dust B modes amplitude for the intensity observed in the BICEP2 field was very close to the measured value by the BICEP2 team.

At the time of the Solvay meeting, a collaboration between the Planck and BICEP teams was doing the correlation analysis of the Planck dust and the BICEP2 150 GHz maps and finally demonstrated that a large fraction of the B modes in the BICEP2 field is due to galactic dust emission with no significant detection of the primordial B modes.[c]

---

[b]The related figures shown at the meeting are taken from: Planck collaboration Intermediate papers arXiv 14095738.

[c]The joint analysis was not yet available at the time of the meeting. It gives an upper limit on r very close to the Planck one, in a paper to be published in Physical Review Letters: BICEP2/Keck Array and Planck collaboration, A joint analysis of BICEP2/Keck Array and Planck collaboration data, 2015.

The point made at the meeting is that ground based experiments reach now an excellent sensitivity on the B modes at 100 or 150 GHz which could give interesting constraints on inflation models but do not have high frequency channels to clean their observations from the galactic dust component. The present upper limit is limited by the Planck 353 GHz data on 1 % of the sky. Improvements to get to higher sensitivities will require an improvement of the Planck data at 353 GHz (ongoing) and larger areas of the sky. The south galactic pole B modes emission shown at the meeting from the Planck paper cited in the footnote show a large area with a slightly lower level of contamination than the BICEP2 field. There is also a similar area in the North. The combination of ground based measurement of the B modes around the peak of the CMB (150 GHz) on larger areas in the best regions shown by the Planck dust B modes maps or balloon borne experiments both measuring the dust at higher frequencies will give much better constraints in the next 10 years before the next generation space CMB experiment flies.

Discussion

G. Efstathiou Thank you. Any comments?

N. Mandolesi Just for completeness let me just add that on the left side of the frequency spectrum – on the 30 GHz side ?? – we understand quite well the synchrotron, so from that point of view I think there is no problem in understanding the BICEP2 results.

J. Dunkley This is just a comment, really. I agree that there are these unexpected features in the dust map that we do not fully understand and that will maybe tell interesting things about the dust physics, but the overall level of dust we are seeing from Planck is not unexpected, right? We were saying years ago we are not sure if it is 1% polarized, 10% polarized, perhaps a little more. And so in planning for this current generation of the next five years of CMB experiments, we have had that in mind. We know we have to do multifrequency, we know we have to use the Planck 353 map, in combination with our lower frequency measurements. I think this is just a reminder that we have to keep going with many frequencies. We have had in our minds that the dust might be this high and we might have to deal with this for a while now.

G. Efstathiou Any other remarks?

V. Mukhanov It is a question. Combining BICEP and Planck data, Seljak recently did it already and a paper appeared in which he got a bound on the gravity waves that r should be smaller than 0.09 instead of 0.11 at 95% level. Correct? So how much better you can go compared to this?

G. Efstathiou I do not think you can do this type of thing without directly cross-correlating the maps.

V. Mukhanov I did not do it. There was a paper on the net as you know.

G. Efstathiou It does not matter. Even my illustrious colleague Spergel cannot do that.

J. Carlstrom You need to know all the modes that are measured and the weights etc. to do it accurately.

D. Spergel You need to know the number of modes in common which you can measure off the reported results.

M. Zaldarriaga Even though the Flauger et al result did not differ too much from the paper that Planck put out, so it is not that you get nonsense when you scan these maps.

R. Blandford Taking this discussion in a slightly different direction, the other thing one could measure is circular polarization which can, in principle, be produced by dust. I think the general expectations are that it should be very very small. And if that is a true statement, then trying to measure circular polarization provides some sort of null calibration. I want just to ask John Carlstrom and others what plans are there for trying to do state-of-the-art circular polarization observations in the future if any.

J. Carlstrom It is a good question. Right now I do not believe that any of the CMB experiments are set up to measure circular polarization at all. There are of course interferometers and stuff out there that can measure small scale circular polarization – even in the dust. ALMA could do it eventually – but at these scales there is nothing planned down, that is something we should look into.

M. Kamionkowski So along those lines but very slightly more sober, theorists have thought about what might circularly polarize the CMB, and it is extremely difficult to come up with anything, even in the minds of the craziest of our theoretical colleagues.

J. Carlstrom That is saying a lot.

M. Kamionkowski Along these lines, though, there is this idea called cosmic birefringence, which is a physical phenomenon that arises in some theories for dark energy. Some of the first theories for dark energy after the discovery of cosmic acceleration were quintessence, which involved a single scalar field rolling slowly down its potential. That potential had to be very flat, so Carroll wrote a paper pointing out that an axion-like field could have a flatness that is preserved by a shift symmetry. Axion-like fields tend to have a coupling to the electromagnetic field and the slow rolling of that electromagnetic field tends to cause a rotation of the window of polarization as the photon propagates from the surface of last scattering. So if you have only E-modes at the surface of last scatter, some fraction of those E-modes are converted into B-modes. Those induced B-modes are correlated with the E-modes, so you would expect to measure an EB cross-correlation, which is parity violating, which is a consequence of the fact that you are rotating either to the right or to the left. The l-dependence of the EB power spectrum is the same as the EE power spectrum, which we have now measured very precisely and know well from theory. So you can look for this and people have looked for this and there is a result from a variety of ground-based experiments and WMAP about 1 degree. The experimentalists tell me that the big problem is systematic effects: How do we actually calibrate what the orientation of the detectors is? I think it will be very very very interesting, especially in the spirit of the last Solvay Conference of 1973 and having Peebles and Sunyaev in the room, both of whom wrote very very prescient papers in 1969, whose predictions took over 30 years to actually be detected and which we are now measuring precisely. I think it will be interesting to think about how it is that we can make measurements of crazy effects like this more precise as we invest in the next generation of experiments.

J. Carlstrom You are right and that would be great. The EB correlation of course we look for that now and use that as a way to see if our calibration makes sense. If we get something that is far away from zero, that would make us pretty nervous. But I think we need to build in the ability to independently measure that and not use that as a calibration.

G. de Bruyn The synchrotron component of course is known to be circularly polarized, although it has not been measured. Could that neglect be important in the decomposition of the synchrotron and the dust polarization components? It should be at the level of 0.1%, depending on frequency.

D. Spergel At this point we are not at the level of removing foregrounds to 0.1% level. Right now we are at the level where the foregrounds are comparable to the signal. I think one of the challenges for the next few years and the work Puget described on measuring whether there is variation in dust polarization properties across the sky – and I think we will be learning more about synchrotron across the sky – will determine whether we can get down to removing foregrounds so the residuals are 10% or 1% or 0.1% and I think if we get to the 0.1% we will have to start worrying about these issues.

J.-L. Puget Just a short comment on that. We will need anyway to know these numbers and these properties to this kind of accuracy to take advantage of the 200,000 detectors.

G. Efstathiou I would encourage people who are not CMB experts to participate in the discussions. You are amongst friends. So maybe you can think about that over coffee.

G. Efstathiou Before moving on to the next discussion topic, I thought it would be useful if Pryke said a few words about BICEP.

C. Pryke I would like to push back just slightly on the idea that the pre-Planck dust models easily accommodated power at the level we detected. I certainly know that most of my colleagues' funding proposals did not say so. There was a wide spread, perhaps group delusional, belief that dust power was considered to be lower at 150 GHz than what we have detected. The other thing that I would like to emphasize is that we are now actively working with the Planck group on a joint analysis doing cross-spectrum analysis between all of the frequencies of Planck and BICEP2. We are actively working on that but I cannot tell you what the results may or may not be showing - but the hope is to have a result from that at or before the main Planck release towards the end of this year. So that is active and ongoing work and we will get back to you with the results as soon as possible. I would like to emphasize that the joint Planck/BICEP2 data is currently the best data available in the world to set limits on r, on this incredibly important physics. It is definitely the right thing to do and we are actively working on it. I do want to say that the Planck 353 GHz map for the kind of cross-correlation with small patch, high sensitivity, lower frequency experiments like ours - it is no panacea - that map has limited signal to noise. For the future CMB experiments, in particular small patch experiments such as ours, are going to have to limit and control foregrounds using other frequencies of their own. So it is not like once the Planck 353 GHz map is available that this will be a dust solution for the future. It is the best thing that exists right now but it is temporary and in the future we will have to do better.

G. Efstathiou Let us move on to our next discussion. Silverstein will introduce the discussion, which is on the theoretical implications of the CMB results.

Prepared comment

E. Silverstein: r and Quantum Gravity

The relation (sometimes called the 'Lyth bound') between the inflaton field range and the tensor to scalar ratio indicates that for r larger than about .01, the inflationary dynamics is strongly sensitive to quantum gravity. Using the standard Wilsonian method of effective field theory to parameterize our ignorance of high energy physics, the system is sensitive to an infinite sequence of Planck-suppressed higher dimension operators, which could introduce steep features in the potential, violating the inflationary slow roll conditions. However, again in the Wilsonian framework, a shift symmetry may protect against such corrections. Nonetheless, it is important to understand the origin of such a symmetry in a ultraviolet completion of gravity.

In string theory, the parameterized-ignorance (effective field theory) picture of random quantum-gravity induced features within each range of field is replaced, in axion directions in field space, by a much more regular structure. Axions descend from higher-dimensional analogues of electromagnetic potential fields, and couple to quantized extra-dimensional magnetic fluxes via gauge-invariant couplings which depend directly on the axions. There is an underlying periodicity in the theory, leading to a branched structure in the potential. On each branch, there is a large-field potential, flattened at large field values relative to $m^2\phi^2$, and with residual oscillatory features. The earliest example gives a $\phi^{2/3}$ potential with $r \approx .04$; other powers ϕ^p arise as well and multiple field versions spread out the tilt of the spectrum toward the red relative to single-field models. This mechanism is among those currently allowed by CMB data.

In general, the theory gives a very uninformative, broad prior on r – but it is very different mechanisms which produce visible r as compared to those which predict undetectably small r. It will be very interesting to see where the upcoming data settle down in terms of r and n_s, and how they constrain the possibility of residual structure in the power spectum and non-Gaussianities.

Discussion

G. Efstathiou Thank you. Any comments?

H. Murayama A question rather. Can you comment on the curvature, which could be a potential signal?

E. Silverstein Right, people like to think about the possibility of an origin in terms of a bubble nucleation process, which puts you in a negatively curved cosmology that then redshifts away. If we have only a minimal number of e-foldings, then it would be logically possible to see the remnant of that curvature. So it is a very interesting possibility to look for and it would be great to bring down the limits on spatial curvature to the level of the underlying perturbations if that is possible. I think it requires a lot more luck than the usual luck we require for theoretical models just because it does depend on having a minimal number of e-foldings and in my view there is not a good argument for that. Sometimes people try and invoke sort of a tuning argument, but inflationary theory, especially in these simplest cases, is Wilsonian natural, it is controlled by symmetries, there is no real reason to have something at the 50th or 60th e-fold. That is my opinion.

M. Zaldarriaga Of course a related signature to this curvature is maybe a deficit of power at low l, which was advocated by previous measurements, so maybe somebody wants to comment on whether those measurements will improve or what will happen.

G. Efstathiou That is the subject of the next little discussion slot. Surely there are questions that people would ask about the theoretical implications of the CMB.

R. Blandford Can I try the question I would ask on behalf of many other people who do not work on the CMB and in particular on inflation? Do you see it as at all falsifiable?

E. Silverstein I do not know a way of practically falsifying it, but you can certainly falsify the simplest versions of it.

V. Mukhanov It would have been better if you had asked this question 15 years ago, because now practically everything is confirmed. What were, 15 years ago, all the predictions that we made? $\Omega = 1$, adiabatic perturbations, Gaussianity, logarithmic spectra. They were in no way in agreement with observations as you know very well. Now these four predictions have been confirmed perfectly well. There is one more prediction, about gravity waves, but for gravity waves there is a lower bound. And I tell you that, for instance, this lower bound, right now, after measuring of the spectral index, if it will be improved, is precisely at the level of $10-3$. If people will not find gravity waves at the level of $10-3$, 50 times less or maybe 100 times less than BICEP, then in this sense inflation will be dead, precisely as it would have been dead as a physical theory if none of the first four predictions had been confirmed. After that you ask the question: Could you save inflation? Of course you can save inflation. As Landau was saying: Give me two free parameters and I will fit you any experimental curve. But the problem is to have a theory with number of parameters less than the number of predictions. And what people call simple inflation is the only inflation which makes sense as a physical theory. And it was confirmed, in fact. In this sense, the question is a little bit too late. You should have asked it 10 years ago.

A. Guth I agree that inflation is essentially confirmed already, but I would also say that I think the idea of falsifying something like inflation is kind of looking at it the wrong way even maybe 15 years ago. You can always falsify any definite theory. But inflation is not a definite theory, it is really a vague idea that can be incorporated into many different possible theories and it is much harder to ever falsify something like that. It is maybe an idea similar to gauge theories and I would say it is as well-established as gauge theories, but you never tried to do an experiment to falsify gauge theories, rather you built gauge theories, specific gauge theories, and you tested those, the standard model in particular, which in the end worked. So I think the real question is not how do we falsify inflation, but how do we build models of inflation that fit what we observe and how do we narrow down those models to ultimately discover either one model that works or maybe no models that work. But it is a model building question, I think, and not a question of falsification.

N. Weiner So sort of building on that, a question to Silverstein would be: We have heard what Mukhanov thinks about r and inflation, and I wonder if you could comment on that generally, but also more specifically in the context of your models, whether there is any particular measurement that... You said that there is some range of rs and that different models have it, but is there some lower bound where if they pushed r below that, then you would say that it is probably not going to be these monodromy models?

E. Silverstein These are falsifiable based on $r > 0.01$. The lowest particular example we happen to have found so far – two ways now actually – is a potential $\phi^{2/3}$, which gives $r \approx 0.04$ or something. I do not know that to be the lower bound, but the whole mechanism is in this large field range category, so I regard it as falsifiable based on r. Of course if r is discovered then we are in the large field range category and then there is this argument to do with understanding of quantum gravity and we try to make the most of that with additional signatures or more work on the theory to be as systematic about exactly what the possibilities are. So far we have a discrete set of possibilities that are understood and... That is the short answer. We would like to work on being more systematic about the whole thing in terms of – just theoretically – what the range of predictions could be, but I think it is a large field mechanism, so...

G. Efstathiou I will take one more question on this topic before moving on.

J. Ostriker Blandford asked the question that this non-specialist was interested in and I listened to all the answers and it seemed to be that the answer was no. Then I have another question for the experts, and that is: Is there an alternative, does anybody have an alternative to inflation? Not having one does not say there is not one, it just means that nobody in this room knows one. But is there an alternative that gives homogeneity, isotropy, the spectrum and these other things that we love?

G. Efstathiou There are some in the perimeter of the Universe.

D. Gross One of your colleagues...

R. Blandford One of your colleagues at the perimeter of Princeton University?

J. Ostriker I am interested in people's views on that, because it is an interesting question.

G. Efstathiou Does anybody want to tackle that?

D. Spergel I just want to propose a hypothetical test of inflation. If we knew that there was a stable gravitino that was going to be produced copiously when the temperature of the Universe was 10 TeV, and we had pretty convincing evidence of that from experiment, and would predict Ω in gravitino that was so large that it would overclose the Universe and we did not see it. That would tell us that the temperature of the Universe never got to 10 TeV. A measurement like that would seem to be something that would be problematic for any idea like inflation, where the Universe was once much hotter. We do not have evidence for this, this is not an experiment we can do today, but I think it is important to think about whether there are tests we could hypothetically do for the model.

A. Guth I guess I do not know all the details of your hypothetical model, but if you can arrange for inflation to happen at a lower energy than 10 TeV, which maybe you could, maybe there would still be a chance of making things work. I do not know. It seems to me that if there is no good other model it is really just a question of what is the most plausible model for our Universe and I think inflation has that by far.

G. Efstathiou I think we should move on to the next topic, because Zaldarriaga already referred to anomalies, and there are some peculiar features about the large scale temperature pattern that were first seen by WMAP and that have been seen by Planck as well. Martin Rees has followed the path of his illustrious predecessor Sir Harold Jeffreys and invented a scale for assessing statistical significance. On the Rees scale, you ask whether you would bet your goldfish, or your dog, or your life. I personally think that the anomalies are worth about your goldfish. Bond, who will introduce this topic, I think may be at the dog level, but we will find out.

Prepared comment

R. Bond: Anomalies: Beyond the Standard Model of Cosmology?

The tilted ΛCDM model emerged in the post-COBE era as the standard model of cosmology (SMc) even before the Supernova Ia observations demonstrated that the Universe is accelerating on large scales. Apart from the dark energy, dark matter and baryonic components, the model is defined by an extremely simple form for the primordial curvature, a nearly statistically isotropic and homogeneous Gaussian random field $\zeta(\mathbf{x})$ fully determined by a primordial scalar power spectrum $< |\zeta(k)|^2 >$ which has just two parameters, an amplitude and a uniform tilt in the comoving wavenumber k, $n_s(k) = constant$. The model may or may not have a measurable tensor component induced by primordial gravity waves, whose amplitude depends upon the specifics of the inflation (or alternative) model which generated $\zeta(\mathbf{x})$.

The sequence of Boomerang plus COBE, then WMAP then Planck, augmented by many ground-based higher resolution CMB experiments, especially in conjunction with redshift and other surveys, added ever increasing precision to the case for the SMc. After SMc emergence a major goal has been to find subdominant elements which take us "Beyond the Standard Model". Our quest is much the same as in particle physics, in which a sequence of accelerator experiments have continued to strengthen the case for its SMpp, but new physics which is Beyond the SMpp is the hope. BSM emerges first as anomalies relative to the SM, until the statistics improve to allow one to hone in on a theoretically-motivated physical explanation of the anomalies, where few sigma indications lock into > 5 sigma *bona fide* detections in the particle physics convention. Alas in cosmology we are stuck with the sample variance of having just one projected sky from which to infer statistically-averaged quantities, which is all that the SMc and its variants predict for us to test: the ergodic theorem relating area/volume averages to that for the ensemble averages is fine at moderate to high resolution, but it does not work at low resolution / large scales with our one sky.

So far the CMB data has not revealed anomalies for multipoles $\ell > 50$, nor any indication of theoretically-well-motivated subdominant phenomena such as primordial isocurvature fluctuations, with the best limits again coming from the Planck data. When reconstruction of the primordial ζ power spectrum, $< |\zeta(k)|^2 >$, is done in multiple wavenumber bands, it fits the uniform tilt of the SMc over comoving wavenumbers $(0.003 - 0.3)/Mpc$ superbly well. This number of temperature modes between ℓ_{min} and ℓ_{max} for a fraction of sky f_{sky} is $\sim f_{sky}(\ell_{max}^2 - \ell_{min}^2)$. This amounts to $\sim 5,000,000$ modes at Planck resolution, all pointing so far to the SMc, albeit with varying statistical precision for each. Further, reconstruction of the primordial power using the Planck 2015 data on the E mode of polarization alone also fits the uniform tilt of the SMc over a similar comoving wavenumber range, adding another few million modes to the SMc evidence. The Planck data also has improved significantly the WMAP constraints on primordial non-Gaussianity, with no indication of non-Gaussianity in $\zeta(\mathbf{x})$ at high k. This result, and less form-specific Planck 2013 constraints on deviations from Gaussianity, also give no indication of anomalous localized ζ-structures at high k.

A few anomalies that have been with us since the earliest WMAP release, and verified with Planck, are concentrated at lower multipoles, and thus at low k, and may point to more early Universe complexity in a full BSMc theory. Anomalies are explored as a phenomenology of extremum events in the data as well as with distributed statistics, e.g. as linear, quadratic, trilinear combinations of temperature and polarization modes. When we address whether an anomaly is primordial, the first thing is to demonstrate it is flat in frequency for $T, Q, U(?)$. With its nine frequency bands, Planck has been quite good for this but one always worries whether a combination of effects could mimic frequency-flatness at the precision obtainable, even with Planck.

With everything looking SMc-normal for $\ell > 50$, there are $\lesssim 500$ modes to store anomalies in. But since the first release of WMAP we have seen spatial anomalies in component-separated CMB maps, e.g. the low quadrupole and the alignment of axes defining the octupole and quadrupole orientation; these have been verified using the Planck data with its broader frequency coverage and very different observing strategy. There are a few elongated phase-coherent hot and cold structures that seem to extend through the Galactic plane that underly these quite low ℓ phenomena, hence are defined by rather few modes, and so sample variance is large. This $\ell \lesssim 10$ regime is still of great interest, housing as it does the entire reionization-history signal, within $\lesssim 100$ modes, and any phase-locking accompanying non-standard cosmic topology. In Planck, WMAP, and COBE data no (lasting) evidence for topological signatures has been found.

These anomaly indicators are linear-map-derived. A quadratic map is a compression of $(T, Q, U) \times (T, Q, U)$ data into a parameterized ensemble of correlated trajectories, $< |\zeta(k)|^2 >$ as a function of k being our prime example. These ζ-power trajectories reveal a robust anomaly relative to the SMc uniform-$n_s(k)$ case for $k \lesssim 0.003$: as k gets smaller a dip followed by a less significant rise then an even lesser significant dip driven by the low quadrupole. This is largely due to the deficit in the CMB TT spectrum over $\ell \sim 20 - 30$ that has been stable since WMAP, and presumably will always be with us. The derived ζ-power shape is not fit by running of the tilt. The deviation from the SMc is only a few sigma effect, tantalizing but far from definitive as a BSMc indicator. The TT spectrum has another anomaly: a hemispherical difference at the $\sim 7\%$ level at $\ell \lesssim 50$ (though with no statistically significant difference at high ℓ). Another quadratic anomaly is the TT correlation function at angles $> 60°$ being compatible with zero, another low ℓ result hard to get by statistical chance within the SMc.

The CMB hot and cold spot distributions are mostly compatible with those expected for a Gaussian random field. An exception in the Galactic Southern sky is the famous WMAP cold spot, a $> 4.5\sigma$ rare excursion on multipole scales $\ell \sim 20$, which would happen $\lesssim 1\%$ of the time for Gaussian random skies. Attempts have been made to relate the cold spot to the 3D distribution of galaxies, most recently to a large scale void found at redshift about ~ 0.2 in a combined galaxy survey that could provide at least some of the signal.

A vast number of modes in the 3D field $\zeta_{\ell m}(\chi)$, now expressed in multipoles at a comoving distance χ away from us, are not "illuminated" by the 2D CMB field. We get just one mode per ℓm from temperature observations, and another mode from polarization. We have great hopes for "illuminating" $\zeta_{\ell m}(\chi)$ with large scale structure surveys probing greater and greater volumes of space. Mining the χ-direction is termed tomography, invariably as a redshift-space map with complex statistics. In LSS tomography the 2D mode number can be augmented by a large multiplicative factor, $\sim f_{sky}\ell_{max}^2 k_{max}\chi_{max}$, involving an effective survey depth χ_{max}, and a corresponding radial k_{max}. How large the effective mode number can be is the question of the day, but there are high hopes, for LSST, Euclid, and other large volume experiments. A cautionary note with LSS is an unfortunate history of the SMc being declared to be at odds with various rare-event structures, followed by more refined statistical determinations reinstating the SMc. Still, large volume surveys with good systematics control will undoubtedly illuminate our future path to *bona fide* 3D BSMc anomalies. Till that happy day we are stuck with the various few-sigma low-ℓ CMB anomalies relative to the SMc that have proven to be robust. A subject cosmologists differ on is how to treat the "look elsewhere effect": if you dig in the data with myriad statistical tests, you are bound to unearth anomalies, but what prior probability should you assign for such searches? We need a grand unified BSMc picture tying the elements together, and thereby increasing the significance of the combined anomaly. An example is reconstructing allowed inflaton-potential shapes that may point us to a BSMc; these shapes follow from the ζ-power and tensor-power (to be greatly improved with future B-mode experiments). Inflation theory (often with extra light non-inflaton fields) can also produce anomalies, but it is not easy to protect the SMc agreement at moderate to high k from BSMc modifications designed for low k in a natural way.

Discussion

N. Mandolesi Suppose that the anomalies we see at low ℓ in temperature, with Planck, should happen as well in polarisation, after we have looked at the possibility of systematics, and knowing that at low ℓ what dominates is cosmic variance. What should we do?

R. Bond Of course this is one of the things that we hope to have revealed in the Planck 2015 release, which is the relationship of the power spectrum in EE and the cross spectrum in TE to these power spectrum anomalies in C_ℓ^{TT}. So as fascinating as the pulling power of the EE and the TE relative to the TT is, it is not so great as to change the basic physics story – at least that is the forecast. Of course we will be talking about this in the Planck papers.

G. Efstathiou But any signal at very low wave numbers produces a very tiny signal in polarisation.

R. Bond That is right. It is not pulling that solution very much.

T. Piran The usual pattern when you prove experiments is that a one or two sigma result in a lower resolution experiment becomes more and more significant My naive question is whether any of the anomalies in the WMAP data were confirmed with higher significance in the Planck data, and so should we begin to be suspicious. Or did all of them disappear and we now have new anomalies which may or may not be confirmed in the next generation experiments.

R. Bond The WMAP and Planck anomalies are essentially the same. That is nontrivial because when Planck does its cleaning it has many more frequencies drawn on to try to do the cleaning and yet the basic anomalies have survived. One point that I think is also relevant here is that people are talking anomaly by anomaly, but there is really one region of the sky that is driving a few of the anomalies. So there is a little bit of a mini-unification of anomalies that has occurred. Unfortunately there is still not a grand unification of anomalies so that we understand anomalies in a fundamental way theoretically. But at least some of the anomalies are all associated with the same patterns on the sky and those patterns are stable. In fact you can even see them in COBE. Our view now is that with Planck we can go closer to the plane of the Galaxy because of the high frequencies covered by Planck.

G. Efstathiou To follow up on that, these anomalies are on scales where the WMAP data and the Planck data are signal dominated. So you should see exactly the same sky and you do see exactly the same sky. The reason I said I would only bet a goldfish is that where people claim a high significance level I think this is based on very naive frequency statistics, which I think can lead to an exaggerated statistical significance level.

R. Wijers Bond was the first one to even mention non-Gaussianities. Are there no new developments in that area? Maybe as an outsider I implicitly missed the point about them. Are they no longer interesting?

G. Efstathiou Planck set very tight limits on non-Gaussianity and it is really very difficult to improve on those limits from CMB observations.

R. Bond I would point out that you do see some non-Gaussianity in these maps, but it is at low multipoles. Planck increased the ℓ range considerably, enhancing constraints on short-scale non-Gaussianity. These Planck results took some of the wind out of the sails of the theoretical community who was actively researching this non-Gaussian game, because now the limits are so good.

M. Kamionkowski I want to comment on this anomaly business and in particular on the hemispherical power asymmetry. Here's my entry into the most legible slide competition: This circle here is the observable universe. This is what we see, and we learn in cosmology classes that we cannot ever learn anything about everything over here which is the rest of the universe. But it is impossible as human beings not to wonder what is going on outside. Now we know that inflation produces fluctuations on all distance scales. And here is my entry into the best animation competition: It is not unreasonable to wonder whether there might be some long-wavelength fluctuation in some field that crosses our observable horizon, and the hemispherical power asymmetry is therefore something you have to look at and wonder about. Is it inconsistent with the standard model? No it is not inconsistent with the standard model, because the first theorem of statistics says that a result is firmly established when everybody agrees it is firmly established, and therefore this is not firmly established. Does that mean it is inconsistent with everything else? And again the answer is no. What some people, my colleagues and I, have been doing is trying to construct models that can explain the hemispherical power asymmetry and also make other observable predictions. There are a number of things we can look for, and maybe this is the tip of the iceberg. It is probably just a statistical fluke but it could be folly not to follow up on it. Some of the things we can do in the near future, and I hope Planck will do, is to measure the power asymmetry in the polarisation power spectrum as well. There are also models that make predictions for non-Gaussianities on smaller scales that can also be sought and I would say that the theoretical models at this point are extremely baroque and not very clever, but hopefully we will continue to think and do better in the future along these lines.

G. Efstathiou Lets move on. The next topic which are two rapporteurs touched upon is consistency with other data. Francis Crick once said that any theory that agrees with all data is bound to be wrong because at any one time some of the data is wrong. Eisenstein will guide us through this

Prepared comment

D. Eisenstein: Consistency of CMB with Low-Redshift Data

I was asked to introduce the topic of the level of consistency of results from the cosmic microwave background anisotropies with the many useful data sets at low redshift. As a caveat to what follows, I note that the comparison of CMB results to those of low-redshift data is model-dependent; I will focus on the usual classes of ΛCDM and varients.

Within this set of models, the consistency is extremely good. We see excellent agreement with Type Ia supernovae, galaxy clustering, intergalactic medium clustering, galaxy cluster abundances and baryon fractions, weak lensing, measurements of H_0, and so on. I'll discuss a couple of mild issues next, but the main message is one of exceptional success. This is particularly impressive in light of the orders of magnitude improvement of the data in the last 15 years.

Given the context of a CMB session, I want to stress the very close connection between CMB and large-scale structure. The observational methods differ, of course, but there is a direct physical link. Most dramatically, we have the detection of the baryon acoustic oscillations (BAO) both in the CMB and in low-redshift data. The SDSS-III BOSS measurement of the acoustic peak in the galaxy clustering data is now about 10σ. BOSS also detects the acoustic peak at 5σ in the correlation of the Lyman α forest between different quasar lines of sight.

These measurements tie together the cosmic distance scale from redshifts 0.5 to 2.5 to 1000, using a single physical mechanism. The acoustic peak itself relies on simple physics and can be simulated in great detail, showing excellent robustness due to the large scale of the effect.

There is good consistency between the BAO distance scale and the results from Type Ia supernovae. The supernovae data is more constraining at low redshifts and the BAO at high redshift; the two together are very powerful for cosmological constraints.

The distance-redshift relation is a strong method for measuring the expansion history of the Universe and hence the cosmic composition. For example, the combination of Planck and low-redshift BAO data offers strong constraints on models that allow for spatial curvature and a power-law evolution of the dark energy density (i.e. constant equation of state w). Current data shows excellent consistency with a flat model with a cosmological constant, with sub-percent precision on curvature and 6–10% precision on w (depending on exact choices of data sets). This is remarkable, in that we have opened two degrees of freedom relative to the flat ΛCDM model and found that the data drives us back to the phenomenologically simpler model with tight errors.

I'll turn now to a couple of areas of imperfect agreement. I begin by stressing that neither of these should be considered as strong conflicts. They may well go away as the data and analyses improve. But these are items that stick out now and therefore are getting lots of attention. They're also notable because the possible discrepencies can easily connect to new physics.

First, the Hubble constant H_0. If one interprets the CMB in the usual ΛCDM model, then one gets 67–68 km/s/Mpc. This remains true in much more general models if one combines CMB, BAO, and supernovae. The physics is simple: the CMB determines the acoustic scale, the galaxy surveys uses the BAO to measure the distance to $z = 0.5$ and calibrate the supernova luminosity, which then implies the distance scale at $z < 0.1$. We call this the inverse distance ladder, since it relies on a calibration at $z = 1000$ instead of $z \sim 0$.

The issue is that analyses that local measurements, particularly those using calibrations of Cepheid variables to calibrate the luminosity of supernovae tend to give answers in the low 70's. This is only a 2σ difference, but it repeats in more than one analysis.

If this difference is borne out in future data, then could be resolved by adding new relativistic energy density at $z > 1000$, some new light relic particle. This shrinks the sound horizon and the BAO-based distances, thereby increasing H_0. Second, the amplitude of clustering, is often denoted as σ_8. Here, we have a $z = 1000$ prediction from the CMB, using the measured optical depth, and a measurement from lensing of the CMB. These amplitudes are a little higher than what is inferred from a variety of low-redshift measurements, such as from the counts of massive galaxy clusters and some weak lensing analyses.

Much of the disagreement comes down to the calibration of the mass of clusters. The number of clusters is a very sensitive function of mass, but this also means that the predicted abundances are exceptionally sensitive to this calibration. The calibration shifts required to match the CMB ΛCDM results are substantial, 20–30% in the mass. There have been a couple of analyses in this last year that shift this calibration in the direction of reconciliation, but the issue remains unsettled.

If the amplitude of low-redshift clustering is in fact lower than predicted from the CMB in the standard ΛCDM model, then there are at least two potential ways to alter the model. First, and most famously, one could add a small amount of hot dark matter, such as giving neutrinos a mass of a few tenths of an eV or introducing some new massive relic particle. This suppresses the growth of structure on small scales. Second, one can consider alterations to general relativity that might decrease the growth of structure, particularly in a manner connected to the emergence of cosmic acceleration. In considering these extensions to the model, it is worth remembering that these different measurements of clustering occur on different scales and at different epochs, which should offer the opportunity to separate a relic population effect, which should tax the formation of structure smoothly from $z = 1000$ to today, from a dark-energy connected effect, which might only arise recently.

In closing, the development of such a diverse set of strong cosmological tools is a great accomplishment of the last few decades. The CMB data is beautiful and highly effective in its own right, but the combination with low-redshift data sets provides enormous opportunity for the cosmological tests, particularly those as regard the cosmic expansion history and the gravitational growth of structure. The ΛCDM model was essentially in place in 2000, with the early supernova cosmological results and the detection of the first acoustic peak in the CMB. It is astonishing to consider the weight of tests to which this model has been the subject in the last 15 years, with WMAP, Planck, SDSS, and many other data sets. The next 15 years will see another factor of ~ 100 improvement in the core data sets, continuing our opportunities to explore cosmology over a wide range of length and time scales.

Discussion

J. Ostriker I am wondering whether the difference in the two Hubble constants is greater than you would expect from cosmic variance, because the ones that measure it out to a redshift a half or two are measuring a big patch of the universe whereas Cepheids are measuring a rather small patch. Any small patch is either expanding faster or slower than the average of the universe and you can compute what you would expect for cosmic variance.

D. Eisenstein Let me say a few words about that. That is a good question. What is really happening in the local measurements is that you are using the Cepheids to calibrate nearby supernovae hosts. Then you are going out to the supernovae at redshifts .5 to 1 which are tied to the Hubble flow.

J. Ostriker At .5 or .1 there is still cosmic variance.

D. Eisenstein That is correct, I agree. So there could be an issue with cosmic variance but it is not an issue with how far out we can measure Cepheids, that would be at even smaller scales. We are tying to the Hubble flow at 200 or 300 megaparsec. But I agree with you, you don't need very much difference here. It is not a very big change in density to cause this. It is about a 10% change, it goes as the cube of the change in the Hubble constant. The Hubble constant is measuring the radius of the sphere and the density then goes as the cube of that, so I think we are talking about a 10% change in the fluctuation and that seems to resolve this.

U. Seljak To address this point, people have looked at this and it seems that at .1, which is where they matched up with the Hubble flow, the effect of fluctuations is too small, it is of order 1%.

S. Zaroubi There was a recent measurement using X-ray clusters, with this kind of full sky X-ray thing, and they discovered there is a tiny local bubble. It is not very big, within one or two sigma, but if they are correct it is enough to reduce this difference so that it becomes very small. Recently also there was a paper about the local supercluster and void. Clearly these are on the scale of the discrepancy. On a scale of 100 megaparsec you have these fluctuations, which you can actually reconstruct from the data we have now, and I think that reduces the discrepancy.

G. Efstathiou Could I comment on this low H_0 problem, because the low value of the Hubble constant from the Planck data was controversial within the Planck project and afterwards when we published the results. I got a lovely email from Gustav Tamnann who said it is a shame that Allan Sandage is not alive to see these results, but your value for H_0 is a bit high. After those papers came out I spend some time looking at the Cepheid data, and I have a different perspective on it. The difficulty, if you forget about photometric difficulties with Cepheids, is getting an accurate distance anchor. The most reliable distance anchor is the NGC4258 megamaser galaxy. After a long campaign of VLBI spectroscopy the megamaser group revised the distance of NGC4258 galaxy upwards. That decreased the Hubble constant by three kilometers per second per megaparsec. So I think the best Cepheid based value of the Hubble constant at present is 70.6±3.3 which is not discrepant with the CMB. You can also think about using the LMC as a distance anchor, and there is a very small sample of Milky Way Cepheids with parallaxes so you can try using those. The difficulty with the LMC is that the results are very sensitive to any metallicity dependence of the period luminosity relation. The difficulty with the Milky Way Cepheids is that the sample is very small. Basically only one of those ten has overlap with the period range used to calibrate the supernovae host galaxies through Cepheid measurements. So the results are very unstable. And I can show you cuts where you count the distant Cepheid period luminosity data and get values of H_0 of 80 ± 3, so the results are very unstable and I think you have to discount those.

P. Podsiadlowski I also have some questions about how to fit in supernovae cosmology with CMB cosmology. The previous discussion has somewhat addressed this. But my more general question is that if there are these discrepancies what can one learn by trying to resolve them with CMB cosmology. Or is the CMB cosmology so advanced that it does not need the supernovae anymore?

G. Efstathiou You definitely need the supernovae to test the equation of state.

D. Eisenstein I hope I did not leave a contrary impression. The supernovae are really important. The BAO and supernovae distance scales work wonderfully together. They kind of fix each other's problems. The supernovae give very accurate relative measurements at lower redshifts. The BAO gives more accurate measurements at higher redshifts. We are fortunate that we can get a good overlap at a redshift of about one half to tie them together and get some cross check. But the two work very well together. The results are highly consistent.

P. Podsiadlowski You are aware that there are predicted evolutionary effects that may mean that is not the case.

D. Eisenstein Yes. We are not in a position now, at least on the BAO side, to test this. I don't think our error bars are small enough to test the kind of evolutionary issues that the supernovae folks are worrying about.

G. Efstathiou I think we should move on, because I like de Bernardis to say something about the future of CMB research and in particular future satellite missions.

Prepared comment

P. de Bernardis: The Future of CMB Measurements

Despite the incredible progress of CMB measurements in the past 10 years, we are not done yet. Important progress is expected in CMB polarization, anisotropy and spectrum measurements.

• CMB polarization : E-modes measurements are not cosmic variance-limited yet. They probe reionization and allow to measure its optical depth τ which is very important for understanding the process. The nature of the detected B-modes has to be confirmed and analyzed, extending frequency and ℓ-coverage to understand foreground contributions and possibly to detect the reionization peak in the BB spectrum. The contribution from lensing of E-modes is now starting to be detected, and measurements should be refined to constrain better the deflection spectrum; these measurements represent a necessary high-redshift complement of those of EUCLID and similar optical surveys. Accurate cross-spectra like TB and BE will provide sensitive tests of fundamental symmetry violations.

• CMB anisotropy : Accurate measurements of CMB lensing - and deflection spectrum probe dark matter and neutrino masses and hierarchy. High S/N and resolution maps probe CMB non-gaussianity and inflation physics. The nature of cold spot(s) and low-ℓ anomalies is still to be understood. The color/spectrum of CMB anisotropy has very-high discovery potential: space-based full sky multi-band surveys of the SZ effect have the potential to detect/catalogue $> 10^6$ clusters with $M > 10^{14} M_\odot$; Rayleigh scattering of the CMB can be detected, as well as absorption from primordial molecules and atoms; accurate spectral measurements of the anisotropy of the CIB can provide its redshift tomography .. and much more !

• CMB spectrum : Low-level spectral distortions probe inflation, dark particles decays, the redshift range z=10^3-10^6, recombination lines, ^{4}He recombination, reionization. These should be carried out by a small space mission and will face formidable challenges: in particular the internal reference blackbody has to be precise to 10^{-7}-10^{-8}, galactic and extragalactic foregrounds must be recognized and removed to exquisite precision to extract cosmological distortions.

• Instrumentation : We have seen a tremendous progress in detector array technology and readout/multiplexing techniques. Large format arrays of bolometers are already taking data. Three new and effective technologies are avaliable: TESs-KIDs-CEBs. And new devices are being developed: multichroic pixels, polarization modulators, spectrometers (absolute and differential FTS, on-chip) multimode bolometers .. Exploiting the best ground-based sites (Atacama, South Pole, Dome-C ..) will provide complete surveys at $\nu < 200$ GHz; long-duration stratospheric balloons will provide coverage of higher frequencies and the necessary monitoring of dust emission; these will also represent an important test bed for the final space mission devoted to a wide frequency coverage polarization sensitive survey of the entire sky at mm/submm wavelengths.

• Space : one detector at 150 GHz (220 GHz) in space is worth >30 (>300) detectors on the ground in the best sites. Higher frequencies are necessary for effective foreground monitoring, and cannot be exploited efficiently from the ground, due to the atmospheric background and its fluctuations. The coverage of large angular scales, where most of the inflationary B-modes signal is, is very difficult from the ground, due to strong ground spillover. Finally, space-based observations take advantage of extreme environment and instrument stability, as demonstrated by the successful COBE, WMAP and Planck missions.

• Roadmap : Ground-based measurements (small/medium scales, $\nu < 200$ GHz) complemented by balloon missions (monitoring higher ν and large scales) represent the current tools of CMB research. They are producing very exciting results and provide very efficient testbeds of new dedicated detectors, devices and analysis methods. In the long run these ground/balloon measurements can be competitive and overtake what can be achieved with a small space mission.

The next step is a medium-size space mission with 1.5m-class telescope, full sky coverage, >15 spectral bands covering at least the interval 60 to 600 GHz, and 1-3 μK arcmin survey sensitivity. This will allow clean measurement of B-modes if the tensor to scalar ratio is r >0.001; detection of $> 10^5$ Sunyaev-Zeldovich clusters; measurements of m_ν and indications of the neutrino mass hierarchy. See EPIC[a], COrE[b] etc. studies for details. A large-size mission can include a large telescope (3.5m), and can accommodate an absolute spectrometer. In addition, an ancillary satellite can provide ultra-accurate polarimetric calibration. This has been thoroughly discussed in the EPIC and PRISM[c] studies and represents our wildest dream for the future, with the potential to map accurately the faintest details of the CMB and the distribution of dark matter and baryons in the entire observable universe.

Discussion

G. Efstathiou We have just gone over our allotted hour. We did invite a few important contributions.

E. Komatsu These future missions are wonderful. Although Guth and Mukhanov said inflation has been proven already, I just cannot help but look forward to using these future missions to help test inflation even further. Any one glitch now will rule out inflation. My question is the following. Everyone wants to find B-modes. Mukhanov said there is a lower bound on B-modes of 10^{-3}. This view is not shared by everyone. But let us assume for a moment we detect B-modes. If General Relativity holds and if the null energy condition is not violated then the tensor tilt has to be negative. What if we detect B-modes at an amplitude greater than 10^{-3} but convincingly see that the tensor tilt is positive? Do you still think that inflation will be proven, or why not? I would like to ask that question to both Mukhanov and Guth.

V. Mukhanov It is a very good question, but first detect it and then I will think about it

E. Komatsu The next Solvay conference on Cosmology may only be in another 20 years...

V. Mukhanov You observe gravity waves in a very small range of (large) l where cosmic variance is still large. What accuracy do you expect for the tensor tilt?

E. Komatsu .02 or .03. What if we say that n_T is .06 at 3σ.

V. Mukhanov Then n should not be very different from one. There will be no significant running. In the $m^2\phi^2$ theory there was essentially running because it was a square root of a logarithm. When I speak of a lower bound on gravity waves it assumes that the spectral index is practically indistinguishable from

---

[a] see e.g. arXiv:0906.1188.

[b] www.core-mission.org, arXiv:1102.2181

[c] http://www.prism-mission.org/

zero. If you find a spectral index of .98 ± .01 at the level of 3σ I would be in trouble. Also, r is then about 10^{-3}. There are a lot of options, but first detect it.

E. Komatsu What about Guth?

A. Guth What we are looking for is the best model we have for the early universe. So far the simplest inflationary models work beautifully and that is great. If we discover something very peculiar like what you are saying and if that is really convincing - which is your hypothesis which, I point out, has not happened - then we would be forced to some really weird looking model of inflation. Then the question would be whether anyone can come up with a more plausible model, and that would remain to be seen. But so far the simplest inflationary models work incredibly well.

M. Zaldarriaga I want to make a comment about the lower limit on r. There are two slow roll parameters in inflation, ϵ and η. I think what Mukhanov has in mind is that if gravity waves are small this means ϵ is small, and therefore n_s is given by the other slow roll parameter, η. Having measured the tilt of the spectrum you then know, with the tilt we have seen, that r is of order $1/N^2$. So it is very important for this upper limit to measure the tilt better. If the tilt moves in one direction, determining the upper limit goes down. It is very sensitive because it is the difference between n_s and one that fixes ϵ and determines whether the tensor to scalar ratio goes as $1/N^2$ or $1/N^3$ where $N = 50$. If we go from N^2 to N^3 we are not going to see it. We are looking forward to Planck, because this is a very important number for these considerations.

V. Mukhanov To N^3 it will probably not go because then the tilt would be .94 which is already out by 3σ. And I hope it will soon be out by 6σ.

G. Efstathiou If you have a hybrid end to inflation you could get whatever r you want.

M. Zaldarriaga Even in hybrid models of inflation the slow roll parameters decay as one over N to some power. That power for small ϵ is fixed by η. So there is an argument.

G. Efstathiou I think we should end the session there. Let us thank all of the contributors.

Closing Session

Souvenirs

R. Blandford Last night, fortified by Belgian beer, I sat down to write a summary. My list of topics was very long and rambling. So, I thought it might be more fun and at least be an interesting experiment if, instead, we went around the room and we each had one 15 seconds sentence in which you are allowed to say one thing that is new to you, in an area outside your domain. It is either an idea, an observation, a measurement or a prospect. Something that is exciting to you. There will be some repetitions but that does not really matter. I am looking for one souvenir of the Solvay conference to remind ourselves of the exciting discussions we have had over the last three and a half days. This is not a popularity contest. George you are first.

G. Efstathiou I have been working for the past seven years exclusively on Planck and I do not know whether I will be able to think about something else ever again. But at this meeting the talks by Pierre Madau and Richard Ellis have excited a dormant part of my brain, I think there is a real possibility of a resolution of this entire problem of the reionization of the universe.

D. Spergel I have been tweeting for the last few months so I can now do this very easily. Star formation history traced by gamma ray bursts differs from star formation history traced by optically selected galaxies.

J. Carlstrom Dust is important for the CMB, but I did not know that dust is no longer important for reionization studies.

P. Madau The universe gets reionized a little bit later.

J. Peebles I regret the reluctance to think outside the box.

J. Ostriker I suspect that in 20 years the plain, vanilla LCDM model will still look better and better.

W. Hoffman In 10 years we may know whether WIMPS exist or not.

R. Romani There were many impressive science results but I would like to comment here on the impressive hospitality of our Belgian hosts.

M. Kramer I think the lack of a theory does not mean there is no better theory

T. Abel I look forward to come back to the next Solvay conference to discuss the agreement of precision measurements and simulation results of the microphysical phase space structure of dark matter all around us.

P. Podsiadlowski I was amazed by the amazing phenomenology of the cosmic dawn which was completely new to me

G. Raffelt Working in neutrino physics part time I was not expecting a systematically significant measurement of the neutrino mass from cosmology in the next few years, but I think this conference has proved me wrong. At least I am looking forward to see that.

M. Zaldarriaga I continue to be amazed by the developments from my experimental colleagues but I am worried we are not in for any new surprises.

J. Dunkley I am going to take things like axions and sterile neutrinos more seriously and try to lower my theoretical prior on WIMPS being the right answer.

N. Weiner I was surprised to learn that apparently I do not know what τ is and that Planck is just going to provide me with another upper limit, and that from conversation, I apparently care a great deal.

S. Furlanetto I was excited to hear about the great advances in the pulsar timing and that we can look forward to a detection of stochastic gravitational waves in the near future.

G. de Bruyn To make progress there must be interaction I learnt. To learn about dark matter it must interact with the standard model. To learn about the universe observers must interact with theorists.

M. Begelman I learnt that Rees was being too conservative at the Solvay meeting in 1973 when he warned the community about being too speculative, and I also learnt the term dark photons.

E. Komatsu I did not know that in 10 years from now a direct detection experiment may hit the neutrino background. That was completely new to me. It is good to know that maybe there will be a stopping point.

A. Fabian I suggest that serendipitous discoveries made by observatories of all type will continue to play a major role in our work.

R. Ellis I think we are very lucky to have a project like Planck alongside the results of WMAP and I look forward to disagreements between them.

S. White I was scared to discover that we might be fried by a flare from a magnetar but then more relieved when I realised that perhaps this may only make the ozone layer go away.

J.L. Puget The problem of reionization has despite an impressive amount of work not been resolved at all.

T. Damour I am thrilled by the prospect of observing to a few Schwarzschild radii of the black hole at the center of the Galaxy in a few years.

F. Englert I learnt a lot from many talks at the meeting.

T. Piran I was surprised to learn how important our ignorance is about the initial mass function in star formation processes, and how deeply it is relevant to many of the issues that were discussed here even during the last CMB session.

E. Silverstein I am intrigued by the prospects of imaging black hole horizons and I look forward to that.

N. Mandolesi On top of sophisticated and expensive experiments I wish that more would be invested in theoretical groups for the quest of new ideas.

P. De Bernardis I was relieved to hear that MOND is dead, and I was interested to understand that the only potential direct detection is not to be believed, and so we better hurry up on that.

M. Rees I was interested to note that the properties of dark energy did not seem to play any role in any of the topics we discussed.

L. Brink I am just an observer to the meeting. I did observe that you make very nice observations and I learnt a lot from that.

S. Tremaine Given the long history of the Solvay conferences and the conscientiousness with which the remarks are recorded, it is worth asking which of our discussions will sound most naive at the Solvay conference a hundred years from now. My vote is for our discussions about the nature of dark matter.

E. van den Heuvel I was happy to learn from Matias Zaldarriaga that dark matter must really be a kind of matter since you cannot talk it away by modifying gravity theories, since these would just require epicycles on epicycles on epicycles.

S. Kulkarni I learnt there are a diversity of scientific styles. A few cheetahs make big kills...

C. Aerts I learnt that simulations of galaxy evolution do not yet use the latest results from single and binary massive star evolution models and I look forward to see cross-fertilisation of the fields of stellar physics and extra-galactic studies to make sure the simulations have more diagnostic power.

P. Ramond I am just an observer, a particle physicist. It reminds me of Proust's a la recherche du temps perdu, but I found a Madeleine which is the ability to measure neutrino masses to such an accuracy which totally blew me away.

J. Hawley I would be curious what the participants of the 1973 Solvay Conference would think about our discussions, in particular the level of detail, the fact that we are arguing about plus or minus three kilometers per second for the Hubble constant. I wonder which of the anomalies we talked about will turn out to be the fruitful ones.

C. Kouveliotou I found the dark matter session very illuminating. I was surprised because I realised that the progress in that field is not necessarily commensurate with the significance of the potential results.

V. Mukhanov I was impressed with the tremendous progress not only in CMB physics and I suggest not to postpone the next Solvay conference on Astrophysics and Cosmology by another 40 years.

R. Wijers I am very optimistic about galaxy evolution. It used to be completely murky, it now seems we have made a lot of progress and, as Aerts already noted, I think there is an enormous step forward that is completely for free by simply applying what we already now about stellar evolution better into galaxy evolution.

S. Phinney I am looking forward to some redshift twenty gamma bursts and fast radio bursts making irrelevant all the planned cosmic dawn experiments.

R. Sunyaev I was very impressed today by the great perspectives of ground based CMB experiments, but for me it was very important to hear the remark yesterday that we need a small mission to measure the spectral distortions in the CMB spectrum, and that this is possible only to do as a low ℓ physics experiment form space. Therefore we should all remember PIXY [??], because this is a proposal that can possibly solve these two great problems which are really very important for CMB physics.

D. Eisenstein I am just glad to see the maturity of the 21 cm techniques and I am looking forward to seeing results from LOFAR and other experiments on that power spectrum.

M. Kamionkowski I am going to take out a life insurance policy on George Efstathiou's goldfish and I still look forward to a sub 15 second contribution from Bond.

R. Genzel I found it astounding that our luck has held up now to the fourth generation of CMB experiments but I am a pessimist and I think your luck is running out now because you are running into serious gastrophysics.

L. Baudis I did not know that you can actually measure the neutrino mass from the CMB and I think that is very exciting, and I look forward to the number.

H. Murayama Personally I got a lot more excited about the cosmic dawn business, and I also would love to see the rest of the universe one day.

C. Frenk I was really shocked to find out that there are people who love neutron stars particularly when some of them form black widow sisters which are going to eat their lovers. I find them fascinating but I hate them because they get in the way of finding out what that gamma radiation from the center of the galaxy really is.

A. Guth I was rather shocked to learn that the sensitivity of dark matter detection is improving by a factor of ten every two years.

U. Seljak I was impressed by the cosmic dawn session and in particular I look forward to the experimental cosmology in 21 cm as a way to eliminate the dark ages.

V. Kaspi I learnt a huge amount at the reionization cosmic dawn session and I even have a new idea I want to think about based on something Jerry Ostriker and Ed van den Heuvel [?] said.

P. Goldreich I was surprised there was no discussion of the detectability of cosmic magnetic fields.

C. Pryke I think the simulations of the first stars are really beautiful but I worry if we ever know whether they reflect what really happened.

J. Lattimer I had not realised the difficulty of making very low metallicity stars.

R. Bond I was very pleased with Richard Ellis's "bright cosmic morning" discussion. It was a revelation that all that activity had occurred relative to the rather "dim cosmic dawn" that I thought was our state of uncertainty (especially in theory), a state we were going to be in for a long time.

S. Zaroubi I am very pleased that so many of you are enthusiastic about cosmic dawn and 21 cm. That will keep us busy for a while. In a broader sense, I thought there was much less emphasis on controversy than it should have been. I would have liked to see more of that rather than the things we agree on.

U. Pen I was most impressed by what I learnt about the phase space simulations that Tom Abel presented and I am not the one who actually did the simulations...

M. Henneaux I am an outsider, but I learnt a lot. I think it was a great conference, and it is clear we should not wait another 41 years before organising the next Solvay conference on Cosmology and Astrophysics.

D. Gross After watching all your amazing contributions I was blown away by the fact that you all managed to stick to 15 seconds.

R. Blandford This is obviously much better than the list I would have read, but let me throw in my souvenir. I look forward to the prospect of Advanced LIGO measuring the coalescence of high mass black holes at high redshift.

Address by the Director of the International Solvay Institutes M. Henneaux

I am very pleased that this conference has been a great meeting. The program was wonderful. Our scientific committee was right to choose "Astrophysics and Cosmology" as the subject of the 26th Solvay conference and we should not wait another 41 years before organizing the next Solvay conference on this topic!

I would like to repeat once more the gratitude of the Solvay Institutes to Roger, the conference chair, for all the careful work that went into the scientific organization of the conference. All scientific sessions, but also extra conference activities, were meticulously thought through. The preparation started almost two years ago. This tells a lot!

We are also grateful to the session chairs and the rapporteurs, who were actively involved in this careful preparation. Before the conference started, we got all the rapporteurs' talks. These could be put online on the conference web page. This is a 'première' that should be warmly praised!

I would also like to thank you, all the participants, who made the scientific discussions proceed vividly in the spirit of the Solvay conferences. Without your effective participation, the conference would not have been a success.

As you know, we will publish proceedings. Since discussions are central, they will be reproduced in the proceedings. Transcribing heated discussions into a good written text requires a lot of editorial work. We are grateful to all the auditors who accepted to help us in this important enterprise, and to Alexander Sevrin, the secretary of the scientific committee, who is organizing the editorial work. You will hear from him in the coming days!

The success of the 26th conference makes us look forward with confidence to the next Solvay Conference on Physics. The scientific committee already started to discuss its theme, but it is too early to make any revelation.

With these optimistic closing words about the future of the conferences, I wish you a very nice trip back home.

Address by the Chair of the Conference
R. Blandford

Epilogue

In this concluding summary, I would like to add some personal reflections, having had a few months to try to digest all that I heard at the meeting - the established, the novel and the expectant. Anyone reading my colleague's "souvenirs" must surely be struck by the ongoing vitality of astrophysics and cosmology. It is not just that the subject as described in Solvay XI, XIII and XVI is "a foreign country" but that there has been significant observational progress in each of our five themes in the few months that have elapsed since the conference.

Most attention has been devoted to the new Planck results.[1] These validate and detail more accurately the cosmological model discussed at the conference essentially that was described mathematically long ago by Lemaître and Bondi. The universe is essentially flat and dominated by what, at least behaves like a cosmological constant and the remaining ingredients comprise 26 percent cold dark matter and 5 percent baryons, determined with roughly one percent accuracy (not precision!). The Hubble constant is measured to be 67.7 ± 0.5 km s^{-1} Mpc^{-1}, in some tension with earlier determinations. There seem to be three neutrinos with combined mass less than ~ 190 meV and the initial fluctuations are adiabatic and Gaussian with a spectrum that appears to be a power law with power spectrum $P(k) \propto k^{-3.03}$, consistent with the simplest versions of inflation. The results – described by a six parameter fit – are generally within the limits established by earlier measurements, including, mostly, those of WMAP. One small shift is the reduction in the measured Thomson optical depth through recombination to $\tau_T \sim 0.07$ which has the encouraging implication, discussed at the meeting, that the epoch of reionization becomes more accessible to 21 cm searches.

The B-modes reported by the BICEP-2 collaboration are still present but they are now agreed to be consistent with a Galactic dust origin and do not require tensor modes derived from the epoch of inflation as erroneously claimed.[2] While this is undoubtedly disappointing to some and was forecast by others, we should not lose sight of the fact that is a good example of the way science is supposed to arrive at objective conclusions (in contrast to relying upon the ebb and flow of fashion or the loudness of opinion). Furthermore, it has focussed attention on what should be measurable by the next generation of ground-based CMB telescopes.

Although, significant departures could appear at any time and there are limits to its scope, the standard cosmological model is holding up extremely well (just like the standard model of particle physics). We have no need, as yet, to invoke baroque elaborations or empirical corrections to fit the data. The fifty year program involving CMB, optical, radio and X-ray telescopes that has brought us to this point is an extraordinary scientific, technical and managerial success.

Much else has happened in the areas we covered since our meeting. Let me give a few idiosyncratic examples. A highly luminous quasar, SDSS J010013.02 with a redshift $z = 6.3$ and an estimated black hole mass of ~ 12 billion solar masses has been found, impacting our understanding of how large galaxies are assembled and how they ionize their surroundings.[3] The active galactic nucleus PDS 456 has been shown to drive powerful winds over a large solid angle that will impose an effective stellar birth control in the host galaxy.[4] The recently released ALMA images of galactic gravitational lenses promise to reinvigorate their use in studies of cosmography, large scale structure and massive black holes.[5] A supernova has been multiple-imaged by a gravitational lens.[6] An intermediate mass back hole has just been found in the galaxy NGC 2276[7].[8] The NuSTAR X-ray telescope has found a neutron star radiating at ~ 100 times the Eddington limit![9] The pulsar PSR J1906+0746 has been observed to disappear through geodetic precession of its radio beam out of our line of sight.[10] IceCube is detecting more cosmic neutrinos with PeV energy and seeking temporal associations with other transients.[11] Fast radio bursts are reported to be up to 42 percent circularly polarized.[12] DES and Fermi set new limits on dark matter annihilation in dwarf galaxies.[13] There are few fields of physical science that can boast such a high rate of durable and incipient discovery as astrophysics and cosmology.

I was struck by the strong interconnects between the different topics discussed at the meeting. As we saw, the quite messy business of discovery constantly thwarts efforts to stovepipe the management of scientific programs. Most of the major advances that have transformed our view of the universe and in many cases, basic physics, were "unscripted" and not the result of careful experiments to detect or measure what was actually found. The first neutron star was observed using equipment designed for lunar X-rays. Pulsars were recognized using a telescope built to study interplanetary scintillation. Evidence for the first black holes came from surveys of cosmic X-ray sources. Gamma ray bursts were found by military satellites seeking terrestrial nuclear explosions. Magnetars showed up in searches for gamma ray bursts. Short period exoplanets emerged from a search for distant brown dwarf and giant planet companions. The study of cosmological dark matter began long ago with investigations of the dynamics of rich clusters of galaxies and nearby galaxies like Andromeda and the Milky Way. More recently, the measurement of the acceleration of the universe was enabled by detailed observational studies of supernovae that revealed that their luminosities could be deduced with surprising precision. Perhaps the most famous example was the discovery of the microwave background itself where Penzias and Wilson were trying to remove the effects of interference and noise in a communications antenna although, in fairness, a "scripted" search would have got there soon after, as may also be true for the discovery of neutron stars and black holes.

The other side of this particular coin is that confident advances in one area can build a secure foundation in another. The accuracy with which the standard cosmological model has been specified has transformed general extragalactic astronomy, enabling accurate measurements of the sizes, powers, masses ... of distant objects. There are now very few investigations where the remaining uncertainty in the cosmological parameters is limiting progress. Calculations will no longer be cluttered by scaling everything to the Hubble constant, etc. The exquisite affirmation of weak field general relativity in so many different ways has emboldened the application of this theory to strong field environments such as neutron stars and disks orbiting black holes and the calculation of gravitational waveforms from merging cosmic objects. We now calculate boldly using GR rather than tentatively adopting some multi-parameter theory of gravity. Our new description of massive and mutable neutrinos, that may soon be completed experimentally, informs our modelling of core collapse supernovae, neutron star cooling and the sun as well as cosmology. (It is impressive how well cosmologists are already constraining the neutrino mass sum and number and how much better they are doing and will likely do in the future relative to the best laboratory experiments.)

In practical terms, there is a third way in which this mingling occurs and this is in the relationship between signal and noise. Understanding interstellar dust is proving to be critical to measuring cosmological parameters with the CMB and probing the epoch of inflation. A sophisticated appreciation of how Galactic cosmic rays are accelerated and propagate is the key to validating any claimed indirect dark matter detection. The subtleties of galactic evolution are important in maximising the accuracy with which the kinematics of the expanding universe can be described using supernovae. While no serious person would claim that the detection of signal from the $\sim 10^{-35}$ s young universe would be an annoying intrusion in our efforts to characterize the dielectric properties of interstellar grains, cosmologists are both grateful for all the hard observational effort that has been carried out over the years in the less frenetic study of the interstellar medium and are bringing new insights, observations and ideas to the subject while they strive to purge its influence from their data.

There has been a tension for many years between astronomers using observatories and physicists performing experiments, between enabling fresh discovery and measuring, (or, more often than not, in limiting), parameters that allow calculations to be completed and a qualitative description to become quantitative. This tension is probably healthy and creative because it forces both cultures to re-examine their arguments and perhaps to find common ground. It has also been very fruitful at the technical and organizational level in big projects. A very good example, out of many, is provided by the Large Synoptic Survey Telescope, LSST, which is under construction by astronomers and physicists working together. Although many physicists see this as measuring the equation of state of dark energy when the observations are completed and the data analyzed in the mid 2030's astronomers are highly confident that it will be a discovery engine from first light in six years time and that most of what it will find has not yet been imagined. It has appeal in both communities and the data archive that will be essential to both lines of inquiry will be more robust and more broadly accessible though having to satisfy many different needs. However, this is not always possible. It is hard to imagine a gravitational radiation or an underground dark matter detector discovering much outside the science for which it was designed (although Planck scale fluctuations and proton decay are being sought). Here, as in these examples, one must demonstrate that there is a reasonable expectation of making a discovery or a novel measurement and that, in the case of large experiments, the goal will not be achieved by other means sooner and or better. Expensive affirmations and upper limits do not and should not carry the appeal of discoveries and fresh measurements. With the long lead time necessary for large projects, this is becoming a serious issue.

Simulation is playing a central role in each of the five themes. For example, the shape of a pulsar magnetosphere, the gravitational wave templates produced by merging black holes, the subtle radiative transfer through the epoch of recombination, the descriptions of the formation of the first stars, the subsequent growth of structure in an expanding universe dominated by dark matter and a cosmological constant are each computational triumphs which have enabled or are enabling major observational studies. Furthermore, observational data are, increasingly, being presented in the format of simulations and vice versa. The benefit is obvious; the concern is that too much superstructure may discourage radical improvements in observation and changes in the interpretation at a time when this may be most needed.

Astrophysics and cosmology are very exciting right now because there are several open questions that could be resolved within a decade or so. Dark matter is the topic where there may be most engagement. Although there are no guarantees, complementary approaches are being pursued above, on and below ground and the race to "elucidate" it fosters optimism. The upper limits are solid and already valuable to particle physicists and if and when a positive identification happens, the next steps are relatively easy to imagine. However, we must contemplate the possibility that the Large Hadron Collider, the experimental physics marvel of our generation, performs flawlessly, makes even more exquisite and detailed demonstrations of the validity of the standard model Lagrangian and yet fails to find any qualitatively new physics. In addition, the only signal seen by the direct detection experiments may be the solar neutrino background while Fermi, AMS et al. set ever tighter annihilation and decay upper limits from dwarf galaxies and identify nearby pulsars through their positrons. Of course, no one wishes these outcomes. It is far more fun to have new physics and cosmology to explore. However, if they do happen, we should be neither apologetic nor dismayed. Exquisite scientific technique will have revealed the world as it actually is and this is progress of which we should be collectively proud. Many possible avenues of inquiry will have been sealed off and new ones, such as more extensive searches in the axiverse and elsewhere will ensue.

Even stronger statements can be made concerning dark energy. The present measurements from Planck strongly support earlier claims that it is well-approximated, at the five percent level, by Einstein's cosmological constant, Λ. This ethereal entity is ubiquitous, eternal and relativistically compliant, unlike its nineteenth century antecedent. It is described by just one number, is devoid of dynamics and condemns us to a depressing future of alienation and senescence! (A more optimistic viewpoint is that we got out of this predicament in the past during the reheating epoch at the end of inflation and we will do it again!) Either way, if we fail to measure any difference from Λ, field theorists will have to overcome their repugnance and perhaps take seriously the notion that it has nothing whatsoever to do with the quantum mechanical vacuum; that, instead, it is a manifestation of physics on the largest scale, not the smallest. The square peg of quantum field theory need not be forced into the round hole of Riemannian geometry. This brings up the multiverse and the anthropic principle. Causality, "Things are as they are because they were as they were" is replaced by "Things were as they were because we are as we are". I am glad that we did not get sidetracked into neo-scholastic disputation but there is assuredly a world outside our cosmological horizon. A sailor in the seventeenth century knew that there was ocean and land beyond what he could see from the top of a mast and this larger domain was the source of local phenomena, such as gravity, wind and rain. Metaphor can be misleading in science but it may now be time to set aside a strictly reductionist explanation of Λ, as well as possible interpretations in which it "emerges" from the cooperation of microphysical elements and, instead, to explore the possibility that it is "situational" not "fundamental" and is simply associated with the larger neighbourhood within which our private universe is located, just as the acceleration of a falling cannonball or apple is a consequence of the larger Earth beyond our view.

As we saw in the discussion of the Cosmic Dawn, there should be great progress over the next few years in writing the narrative history of the birth of the first galaxies, stars and planets. LOFAR, MWA, LWA and so on are taking data at low radio frequencies and learning how to cope with the ionosphere, man-made and natural interference. The observations of galaxies and quasars are pushing out to greater and greater redshifts and the James Webb Space Telescope will carry them even further, while our understanding of the density, composition and ionization state of the intergalactic medium is improving apace. The puzzles are fascinating. "Which came first, the galactic chicken or the black hole egg?". "Can we really account for the burst of ionizing photons using young hot stars alone or must we invoke additional, possibly exotic, sources?". "How are the first metal-starved stars born and how do they die?". *Ab initio* simulations are starting to connect with the most detailed observations and should provide answers before too long.

To me one of the most exciting prospects is that we will soon be able to make a three dimensional map of the actual universe that we inhabit interior to the sphere associated with the epoch of recombination. The goal is to take the gravitational potential, as sampled on the recombination sphere by the low spherical harmonics of the CMB and combine this with local surveys of galaxies and quasars, together with future structure measurements from epoch of reionization studies to determine with the coarsest resolution how the three dimensional equipotential surfaces are nested. (This nesting can be described by a tree which will "grow" as we increase the resolution of the map or, equivalently include modes that left the horizon later during inflation.) This is a quite different enterprise from the one that currently dominates our thinking which is to overcome cosmic variance and infer the rules that govern the initial conditions and the development of structure in an ensemble of universes. Instead of studying physics we would be plying the traditional trade of the astronomer and describing the one universe we can observe. In order to aid us in this task, we need to assume that the modes which we are trying to measure evolve linearly according to the standard model and that their initial amplitudes are drawn from a gaussian distribution with random phase and measured variance. The goal is to understand where the major 1-5 Gpc scale under- and over-densities are to be found without being too particular about their precise form. Although this is primarily a "geographical" exercise it could assist our specification of standard cosmology by adding important new priors to the great Bayesian analyses that underlie the measurement of fundamental parameters.

Perhaps the surest bet, though, is that we will learn much more about compact objects specifically neutron stars and black holes. Sub-millimeter interferometry such as those made using the Event Horizon Telescope, anchored by ALMA, will produce progressively finer imaging of the gas flow close to the black holes in M87 and Sgr A*. The optimism is high that a massive black hole binary background will be observed soon by the IPTA. The first coalescing neutron stars should be detected within a few years by aLIGO and VIRGO. (As we discussed, coalescing neutron stars are strong candidates for short gamma ray bursts and it will be wonderful if we see confident associations with gamma rays, radio bursts or even $\sim$PeV neutrinos.) I lack the imagination to see how gravitational waves could differ from their standard general relativistic description while preserving the measurement (at the $\sim 10^{-3}$ level) of the radiative power of neutron star binaries, but this is surely important to check. Far more likely is the prospect of our being seriously surprised by the detection rate and new, or under-appreciated, source classes such as accretion induced collapse of white dwarfs or three body massive black hole systems, being identified. Another possibility is that the equation of state of cold nuclear matter, which appears to be relatively stiff, suggests that we will learn more basic nuclear physics from the detailed gravitational radiation waveforms than might once have been expected. If classical mechanics and traditional spectroscopy/photometry can make serious contributions to 21st century fundamental physics, then nuclear physics can surely do so too. Even more interesting to relativists than the neutron star binaries is the possibility of observing a black hole binary merger. Although the rate per unit volume is much smaller for stellar-size black hole mergers, they can be seen to much greater distance than neutron star mergers, which will partially offset this disadvantage. Black hole binaries present the best chance for those who so desire, to put classical, strong field, dynamical relativity to the sword. These are exciting times.

It remains for me to thank all of my colleagues for their spirited contributions and the Solvay family, the Solvay Board of Directors, the Solvay Scientific Committee and the staff of the Hotel Metropole for their gracious hospitality.

References

1. Planck Collaboration; P. Ade et al. arXiv:1502.01582 (2015).
2. P. Ade et. al. *Phys. Rev. Lett.* **114** 101301 (2015).
3. X. Wu, F. Wang & X. Fan *Nature* **518** 512 (2015).
4. E. Nardini et al. *Science* **347** 860 (2015).
5. C. Vlahakis et al. arXiv:1503.02652 (2015).
6. P. Kelly et al. *Science* **347** 1123 (2015).
7. M. Mezcua et al. arXiv:1501.04897 (2015).
8. A. Wolter et al. arXiv:151501.01994 (2015).
9. M. Bachetti et al. *Nature* **514** 202 (2014).
10. J. van Leeuwen et al. *Astrophys.J.* **798** 118 (2015).
11. M. Aartsen et al. arXiv:1503.00598 (2015).
12. E. Petroff et al. *Mon. Not. R. Astr. Soc.* **447** 246 (2015).
13. K. Bechtol et al. arXiv:1503.02583; Drlica-Wagner, A. et al. arXiv:1503.02584.

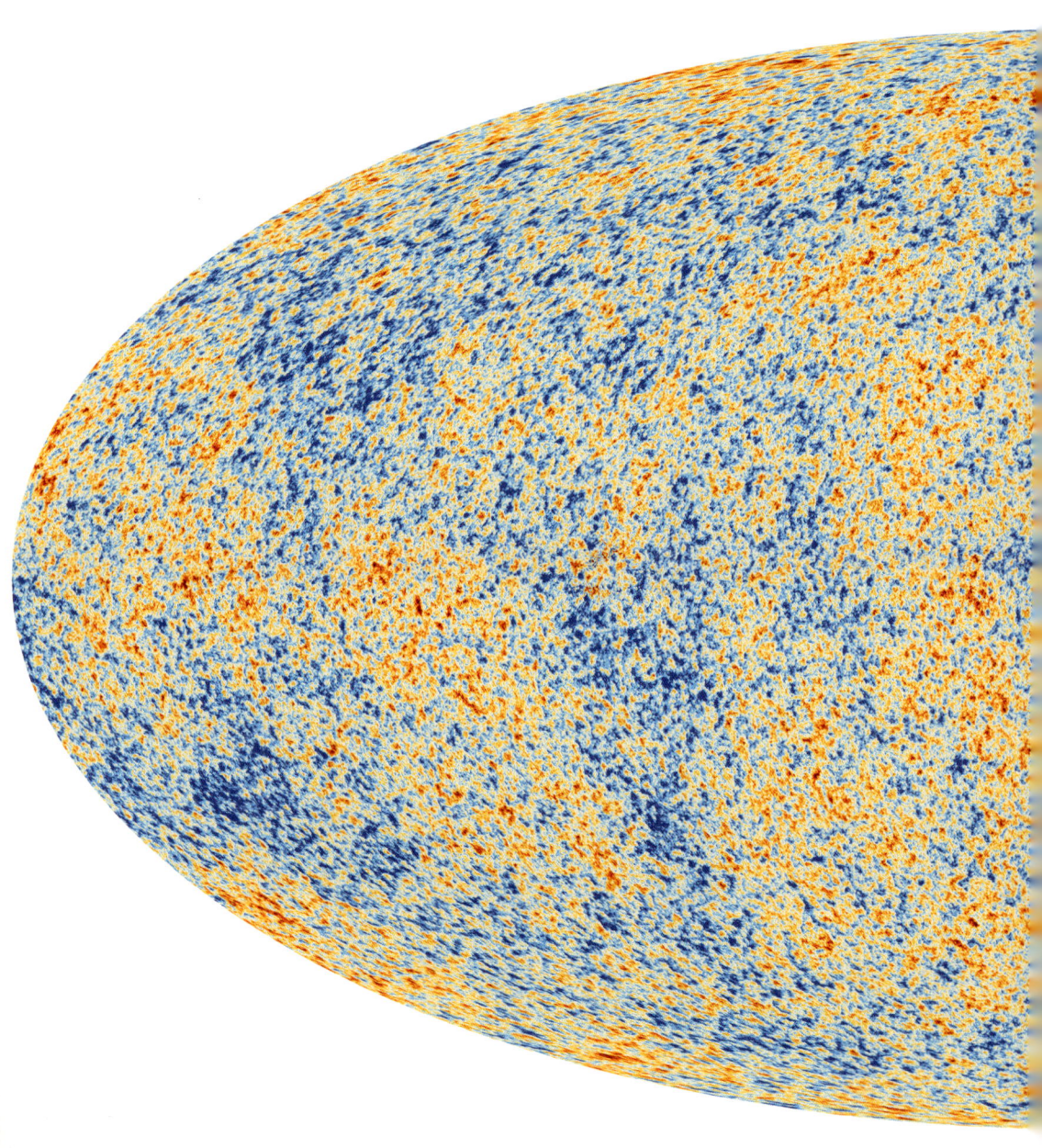

Map of the temperature fluctuations in the Cosmic Microwave Background radiation.

(Courtesy ESA and the Planck collaboration.)

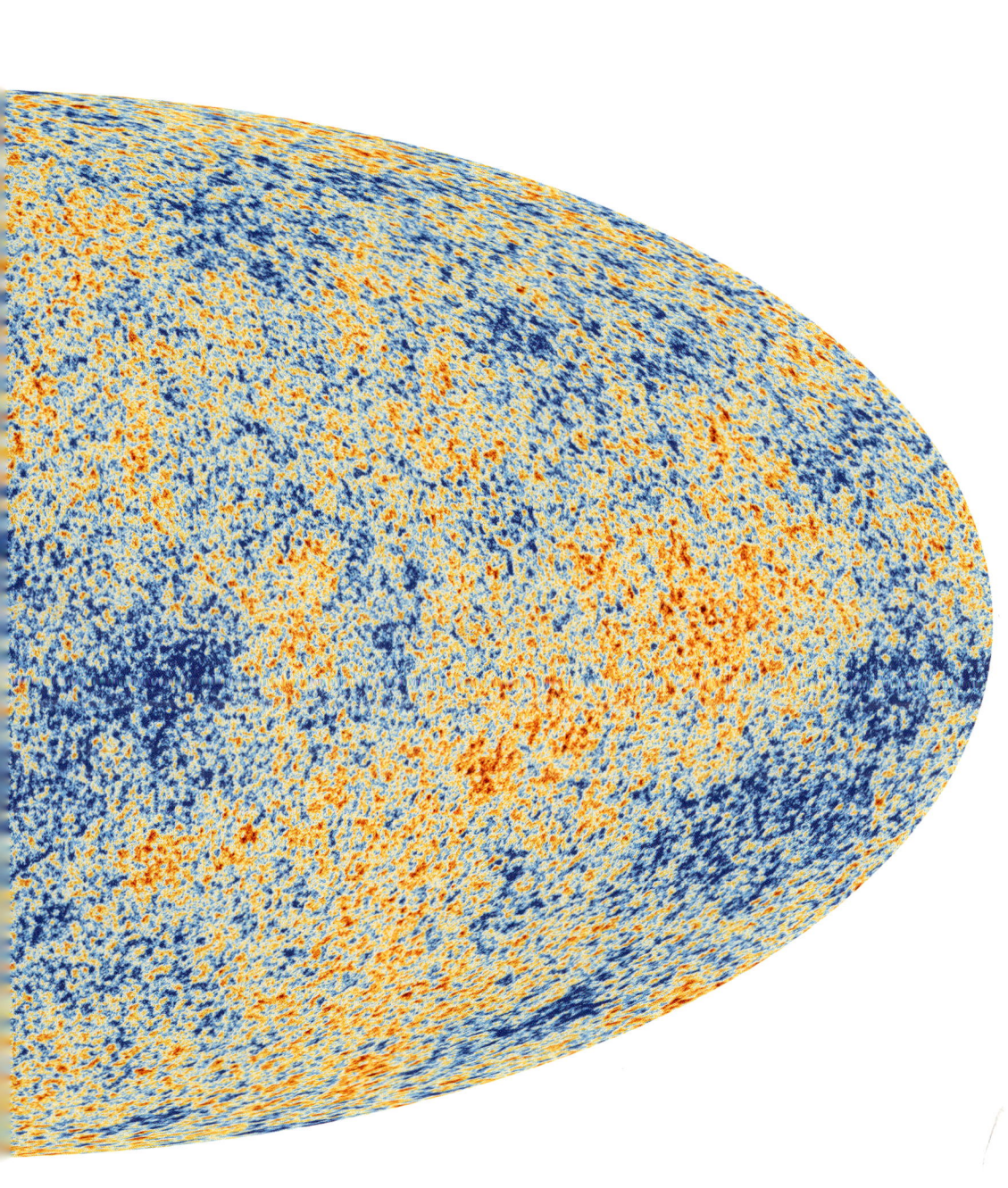

26th International Solvay Conference on Physics
"Astrophysics and Cosmology"

(Brussels) Metropole Hotel, 9 - 11 October 2014